Current Topics in Microbiology and Immunology

Volume 353

Current Topics in Microbiology and Immunology

Previously published volumes

Further volumes can be found at www.springer.com

Vol. 326: **Reddy, Anireddy S. N.; Golovkin, Maxim (Eds.):**
Nuclear pre-mRNA processing in plants. 2008
ISBN 978-3-540-76775-6

Vol. 327: **Manchester, Marianne; Steinmetz Nicole F. (Eds.):**
Viruses and Nanotechnology. 2008.
ISBN 978-3-540-69376-5

Vol. 328: **van Etten, (Ed.):**
Lesser Known Large dsDNA Viruses. 2008.
ISBN 978-3-540-68617-0

Vol. 329: **Griffin, Diane E.; Oldstone, Michael B. A. (Eds.):** Measles. 2009
ISBN 978-3-540-70522-2

Vol. 330: **Griffin, Diane E.; Oldstone, Michael B. A. (Eds.):**
Measles. 2009.
ISBN 978-3-540-70616-8

Vol. 331: **Villiers, E. M. de (Ed.):**
TT Viruses. 2009.
ISBN 978-3-540-70917-8

Vol. 332: **Karasev A. (Ed.):**
Plant produced Microbial Vaccines. 2009.
ISBN 978-3-540-70857-5

Vol. 333: **Compans, Richard W.; Orenstein, Walter A. (Eds.):**
Vaccines for Pandemic Influenza. 2009.
ISBN 978-3-540-92164-6

Vol. 334: **McGavern, Dorian; Dustin, Micheal (Eds.):**
Visualizing Immunity. 2009.
ISBN 978-3-540-93862-0

Vol. 335: **Levine, Beth; Yoshimori, Tamotsu; Deretic, Vojo (Eds.):**
Autophagy in Infection and Immunity. 2009.
ISBN 978-3-642-00301-1

Vol. 336: **Kielian, Tammy (Ed.):**
Toll-like Receptors: Roles in Infection and Neuropathology. 2009.
ISBN 978-3-642-00548-0

Vol. 337: **Sasakawa, Chihiro (Ed.):**
Molecular Mechanisms of Bacterial Infection via the Gut. 2009.
ISBN 978-3-642-01845-9

Vol. 338: **Rothman, Alan L. (Ed.):**
Dengue Virus. 2009.
ISBN 978-3-642-02214-2

Vol. 339: **Spearman, Paul; Freed, Eric O. (Eds.):**
HIV Interactions with Host Cell Proteins. 2009.
ISBN 978-3-642-02174-9

Vol. 340: **Saito, Takashi; Batista, Facundo D. (Eds.):**
Immunological Synapse. 2010.
ISBN 978-3-642-03857-0

Vol. 341: **Bruserud, Øystein (Ed.):**
The Chemokine System in Clinical and Experimental Hematology. 2010.
ISBN 978-3-642-12638-3

Vol. 342: **Arvin, Ann M. (Ed.):**
Varicella-zoster Virus. 2010.
ISBN 978-3-642-12727-4

Vol. 343: **Johnson, John E. (Ed.):**
Cell Entry by Non-Enveloped Viruses. 2010.
ISBN 978-3-642-13331-2

Vol. 344: **Dranoff, Glenn (Ed.):**
Cancer Immunology and Immunotherapy. 2011.
ISBN 978-3-642-14135-5

Vol. 345: **Simon, M. Celeste (Ed.):**
Diverse Effects of Hypoxia on Tumor Progression. 2010.
ISBN 978-3-642-13328-2

Vol. 346: **Christian Rommel; Bart Vanhaesebroeck; Peter K. Vogt (Ed.):**
Phosphoinositide 3-kinase in Health and Disease. 2010.
ISBN 978-3-642-13662-7

Vol. 347: **Christian Rommel; Bart Vanhaesebroeck; Peter K. Vogt (Ed.):**
Phosphoinositide 3-kinase in Health and Disease. 2010.
ISBN 978-3-642-14815-6

Vol. 348: **Lyubomir Vassilev; David Fry (Eds.):**
Small-Molecule Inhibitors of Protein-Protein Interactions. 2011.
ISBN 978-3-642-17082-9

Vol. 349: **Michael Kartin (Ed.):**
NF-kB in Health and Disease. 2011.
ISBN: 978-3-642-16017-2

Vol. 350: **Rafi Ahmed; Tasuku Honjo (Eds.):**
Negative Co-receptors and Ligands. 2011.
ISBN: 978-3-642-19545-7

Vol. 351: **Marcel, B. M. Teunissen (Ed.):**
Intradermal Immunization. 2011.
ISBN: 978-3-642-23689-1

Charles E. Samuel
Editor

Adenosine Deaminases Acting on RNA (ADARs) and A-to-I Editing

Responsible series editor: Michael B. A. Oldstone

Prof. Charles E. Samuel
Department of Molecular, Cellular
and Developmental Biology
University of California
Santa Barbara, CA
USA
e-mail: samuel@lifesci.ucsb.edu

ISSN 0070-217X
ISBN 978-3-642-22800-1 e-ISBN 978-3-642-22801-8
DOI 10.1007/978-3-642-22801-8
Springer Heidelberg Dordrecht London New York

Library of Congress Control Number: 2011937766

Cover design: Deblik, Berlin

Printed on acid-free paper

Springer is part of Springer Science+Business Media (www.springer.com)

Preface

Adenosine deaminases acting on RNA (ADARs) bind double-stranded RNA and catalyze the deamination of adenosine (A), producing inosine (I) in RNA substrates. Because "I" is recognized as "G" instead of "A", nucleotide substitutions are generated that have the potential to amplify genetic diversity and alter gene product function, thereby affecting a broad range of biological processes. A-to-I editing occurs with both cellular and viral RNA substrates, and in both coding and noncoding regions of RNAs. The importance of ADARs for normal development and physiology, both in the absence and presence of pathogen infection, is illustrated by the phenotypes seen in model organisms and cultured cells following genetic disruption of *adar* genes, and either knockdown or over expression of ADAR proteins. This volume of *Current Topics in Microbiology and Immunology* reviews several aspects of ADARs and A-to-I editing. The volume begins with the two chapters that review the biochemical properties of ADAR proteins: their structure and catalytic mechanism, and their nucleic acid binding activities conferred by repeated dsRNA and Z-DNA binding domains. The next four chapters concern A-to-I editing of coding and noncoding RNA transcripts: editing of coding RNAs that affects the open reading frame and subsequently causes changes in ribosome decoding, resulting in protein products with altered function including cellular neurotransmitter receptors and ion channels and viral proteins; and, the editing of noncoding micro RNAs and mRNA 3′-untranslated regions. Bioinformatic strategies to identify new candidate targets of A-to-I editing are next considered. The volume concludes with three chapters that focus on roles that ADARs play that affect virus-host interactions and innate immunity, and mouse development and Drosophila biology.

The objective of this *CTMI* volume is to provide readers with a foundation for understanding what ADARs are and how the act to affect gene expression and product function. It is becoming increasingly apparent that ADARs may function not only as enzymes that deaminate adenosine in RNA substrates with double-stranded character, but also as RNA binding proteins independent of their catalytic property. Future studies of ADARs no doubt will provide us with additional surprises and new insights into the modulation of biological processes by the ADAR family of proteins.

Santa Barbara, April 2011 Charles E. Samuel

Contents

Contributors

Frédéric H.-T. Allain Institute of Molecular Biology and Biophysics, ETH Zurich, 8093 Zürich, Switzerland, e-mail: allain@mol.biol.ethz.ch

Pierre Barraud Institute of Molecular Biology and Biophysics, ETH Zurich, 8093 Zürich, Switzerland

Peter A. Beal Department of Chemistry, University of California, One Shields Ave, Davis, CA 95616, USA, e-mail: beal@chem.ucdavis.edu

Gordon G. Carmichael Department of Genetics and Developmental Biology, University of Connecticut Stem Cell Institute, University of Connecticut Health Center, 400 Farmington Avenue, Farmington, CT 06030-6403, USA, e-mail: carmichael@nso2.uchc.edu

John L. Casey Department of Microbiology and Immunology, Georgetown University Medical Center, Washington, DC, USA, e-mail: caseyj@georgetown.edu

Ling-Ling Chen State Key Laboratory of Molecular Biology, Institute of Biochemistry and Cell Biology, Shanghai Institutes for Biological Sciences, Chinese Academy of Sciences, 320 Yueyang Road, 200031, Shanghai, China, e-mail: linglingchen@sibcb.ac.cn

Eli Eisenberg Raymond and Beverly Sackler School of Physics and Astronomy, Tel-Aviv University, 69978, Tel Aviv, Israel, e-mail: elieis@post.tau.ac.il

Ronald B. Emeson Departments of Pharmacology, Molecular Physiology and Biophysics and Psychiatry, Vanderbilt University School of Medicine, Nashville, TN 37232-8548, USA, e-mail: ron.emeson@vanderbilt.edu

Rena A. Goodman Department of Chemistry, University of California, One Shields Ave, Davis, CA 95616, USA

Jochen C. Hartner TaconicArtemis GmbH, Neurather Ring 1, 51063 Koeln, Germany, e-mail: jochen.hartner@taconicartemis.com

Jennifer L. Hood Training Program in Cellular and Molecular Neuroscience, Vanderbilt University School of Medicine, Nashville, TN 37232-8548, USA

Liam P. Keegan MRC Human Genetics Unit, Institute of Genetics and Molecular Medicine, Western General Hospital, Crewe Road, Edinburgh, EH4 2XU, UK, e-mail: Liam.Keegan@hgu.mrc.ac.uk

Xianghua Li MRC Human Genetics Unit, Institute of Genetics and Molecular Medicine, Western General Hospital, Crewe Road, Edinburgh, EH4 2XU, UK

Brian Liddicoat St. Vincent's Institute of Medical Research and Department of Medicine, St. Vincent's Hospital, University of Melbourne, Melbourne, VIC, 3065, Australia

Mark R. Macbeth Department of Biological Sciences, Carnegie Mellon University, 4400 Fifth Avenue, Pittsburgh, PA 15213 USA, e-mail: macbeth@andrew.cmu.edu

Kazuko Nishikura The Wistar Institute, 3601 Spruce Street, Philadelphia, PA 19104, USA

Mary A. O'Connell MRC Human Genetics Unit, Institute of Genetics and Molecular Medicine, Western General Hospital, Crewe Road, Edinburgh, EH4 2XU, UK

Simona Paro MRC Human Genetics Unit, Institute of Genetics and Molecular Medicine, Western General Hospital, Crewe Road, Edinburgh, EH4 2XU, UK

Charles E. Samuel Department of Molecular, Cellular and Developmental Biology, University of California, Santa Barbara, CA 93106, USA, e-mail: samuel@lifesci.ucsb.edu

Carl R. Walkley St. Vincent's Institute of Medical Research and Department of Medicine, St. Vincent's Hospital, University of Melbourne, Melbourne, VIC, 3065, Australia

Bjorn-Erik Wulff The Wistar Institute, 3601 Spruce Street, Philadelphia, PA 19104, USA, e-mail: wulff@sas.upenn.edu

ADAR Proteins: Structure and Catalytic Mechanism

Rena A. Goodman, Mark R. Macbeth and Peter A. Beal

Abstract Since the discovery of the adenosine deaminase (ADA) acting on RNA (ADAR) family of proteins in 1988 (Bass and Weintraub, Cell 55:1089–1098, 1988) (Wagner et al. Proc Natl Acad Sci U S A 86:2647–2651, 1989), we have learned much about their structure and catalytic mechanism. However, much about these enzymes is still unknown, particularly regarding the selective recognition and processing of specific adenosines within substrate RNAs. While a crystal structure of the catalytic domain of human ADAR2 has been solved, we still lack structural data for an ADAR catalytic domain bound to RNA, and we lack any structural data for other ADARs. However, by analyzing the structural data that is available along with similarities to other deaminases, mutagenesis and other biochemical experiments, we have been able to advance the understanding of how these fascinating enzymes function.

Contents

R. A. Goodman · P. A. Beal (✉)
Department of Chemistry, University of California,
One Shields Ave, Davis, CA 95616, USA
e-mail: beal@chem.ucdavis.edu

M. R. Macbeth (✉)
Department of Biological Sciences, Carnegie Mellon University,
4400 Fifth Avenue, Pittsburgh, PA 15213, USA
e-mail: macbeth@andrew.cmu.edu

Current Topics in Microbiology and Immunology (2012) 353: 1–33
DOI: 10.1007/82_2011_144

Published Online: 17 July 2011

1 The Reaction Catalyzed by ADARs

The overall reaction catalyzed by ADAR enzymes is the conversion of adenosine (A) in RNA to the rare nucleoside inosine (I) (Fig. 1). It is inosine's structural similarity to guanosine (G) (lacking only guanosine's C2 amino group) that is responsible for ADARs' profound effects on the function of RNA substrates. Inosine selectively base pairs with cytidine (C) and therefore functions in translation and replication as guanosine (G). In addition, conversion of adenosine present in A:U pairs in duplex structures leads to I–U mismatches, destabilizing the duplex structure (Bass and Weintraub 1988). Conversely, the ADAR reaction at A–C mismatches leads to more stable I:C pairs. Early work by Bass et al. on the ADAR reaction established key elements of the mechanism. Using ^{18}O labeled water in the reaction and mass spectrometry of the product, they showed that the oxygen atom in the inosine product is from water and that the glycosidic bond of the reacting nucleotide is not broken during the reaction (i.e. ADARs do not use a transglycosylation mechanism) (Polson et al. 1991). This information supported a hydrolytic deamination mechanism for the ADARs similar to that seen previously with the nucleoside modifying enzymes ADA and cytidine deaminase (CDA) (Carter 1995). This type of mechanism implied the existence of a hydrated intermediate along the reaction coordinate and required a means for activation of a water molecule for nucleophilic attack (Fig. 1). Work done with ADARs and other deaminases (discussed in detail below) has helped us to identify the key residues and steps involved in this mechanism.

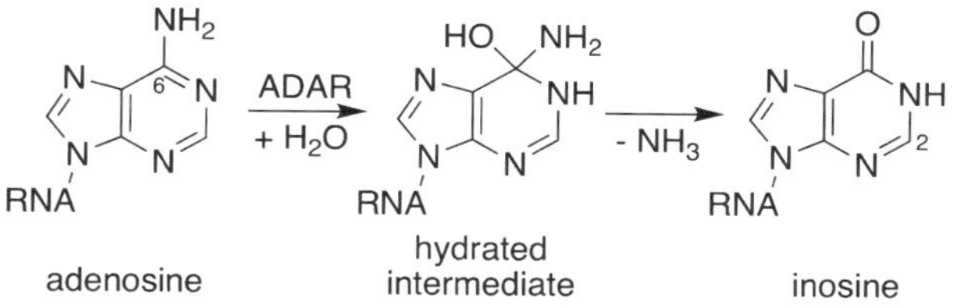

Fig. 1 ADARs catalyze the deamination of adenosine to inosine via addition of water to the 6-position to form a hydrated intermediate. Inosine differs from guanosine only at the 2-position, where inosine lacks guanosine's amino group. Therefore, inosine base-pairs with cytidine and so is recognized as guanosine by the translation machinery

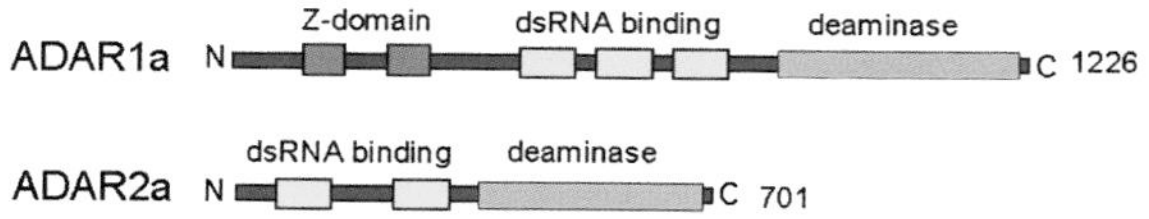

Fig. 2 Domain maps of ADAR1a and ADAR2a. *Yellow boxes* indicate approximate locations of dsRBMs in each protein. *Blue* indicates Z-alpha and Z-beta domains in ADAR1. *Orange boxes* refer to the CDA-like deaminase domains

2 The ADAR Protein Structure is Modular

Cloning and sequencing of ADAR cDNAs allowed for the identification of likely functional domains within the expressed proteins (Kim et al. 1994; Melcher et al. 1996). Indeed, the ADAR proteins are modular in their makeup with multiple independently folded domains that work in concert to achieve efficient and selective RNA editing (Fig. 2). RNA binding is controlled by sequence motifs (dsRBMs) present in multiple copies in both ADAR1 and ADAR2 (Doyle and Jantsch 2002). The C-terminal region of each protein harbors the deaminase domain with the catalytic machinery necessary to convert adenosine to inosine. In addition, ADAR1 has an N-terminal Z-domain similar to other known Z-DNA binding domains (Herbert et al. 1997). Indeed, the Z-alpha subdomain of ADAR1 has Z-DNA and Z-RNA binding activity (Koeris et al. 2005). Below we describe in more detail the structure and function of the ADAR catalytic domain along with a description of our current understanding of the catalytic mechanism. The reader is directed to other reviews in this issue for information about ADARs' dsRBMs and Z-domains.

3 Overview of Structure of ADAR2 Deaminase Domain

The modular nature of ADARs described above has allowed structural biologists to obtain high-resolution structures of both the RNA binding domain and the catalytic domain of ADAR proteins. High resolution structures for two members of the protein family to which ADAR belongs are available: (1) the prokaryotic ADA that acts on tRNA 2 (ADAT2 or TadA) which is a sub-type of the ADAR family that deaminates the wobble position (A34) of $tRNA^{Arg2}$ (Gerber and Keller 1999; Wolf et al. 2002; Kuratani et al. 2005; Kim et al. 2006; Losey et al. 2006; Lee et al. 2007), and (2) the catalytic domain (spanning residues 306–701) of hADAR2 (Macbeth et al. 2005). These structures support phylogenetic data suggesting that ADARs and ADATs belong to the CDA superfamily of enzymes (Kim et al. 1994; Gerber and Keller 1999, 2001). Other deaminases that belong to this family include *E. coli* cytidine nucleoside deaminase and the RNA and DNA editing enzymes apo-lipoprotein B mRNA editing catalytic subunit-1, -2, and -3 (APOBEC1-3) as well as activation induced deaminase (AID) (Navaratnam and Sarwar 2006).

These enzymes catalyze a hydrolytic deamination reaction by employing a zinc coordinated water molecule as the nucleophile. In the proposed mechanism for this reaction, discussed in detail below, a conserved glutamate residue accepts a proton from the nucleophilic water, and the reactive hydroxide ion attacks C6 of adenosine (for ADARs and ADATs) or C4 of cytidine (in CDAs and APOBECs) in their respective substrates. Thus, the enzymes share a common mechanism, a common sequence of residues proximal to the active site zinc, and a striking conservation of structural topology that comprises the 'deaminase motif' (Fig. 3).

The core deaminase motif of CDAs, APOBECs and ADARs consists of a central 5-stranded β-sheet (β1-5) flanked on one face by an α-helix (α1) crossing roughly perpendicular to the β-strands, and two helices (α2–α3) that run parallel to the β-strands on the opposite face of the sheet (Betts et al. 1994; Kuratani et al. 2005; Macbeth et al. 2005; Prochnow et al. 2007; Holden et al. 2008). The two helices contain residues that coordinate the zinc ion as well as the invariant glutamate residue that accepts a proton from the nucleophilic water. In addition, the N termini of α2 and α3 are located at the base of the active site cleft and thus provide positive charge to this region due to the helical dipole. The positive dipole may act to stabilize the developing negative charge on the nucleophile as a proton is transferred, or may facilitate shuttling of a proton between N1 of adenosine, residue E396 and the hydrated intermediate.

It is curious that ADARs bear a resemblance to CDAs but not to the adenosine nucleotide deaminases (ADAs). ADAs, like ADARs, use a zinc-coordinated water as the nucleophile in a hydrolytic deamination reaction at C6 of adenosine nucleotides. However, despite a catalyzing a similar reaction, the three-dimensional (3D) structures are vastly different. The ADA structure is an 8-strand, parallel α/β barrel instead of the helix/β-sheet/2-helix deaminase fold of ADARs and CDAs (Wilson et al. 1991).

The ADAT2 family of deaminases convert the wobble position (A34) of various tRNAs to inosine. In prokaryotes, the reaction is catalyzed by a homodimer of the TadA protein and it is specific for $tRNA^{Arg2}$, while in eukaryotes the modification is catalyzed by a heterodimer of Tad2/Tad3 (ADAT2/3) which deaminates the wobble position of several tRNAs (Gerber and Keller 1999). ADAT2 proteins are 20–25 kDa, lack a dsRBM and have a shortened C

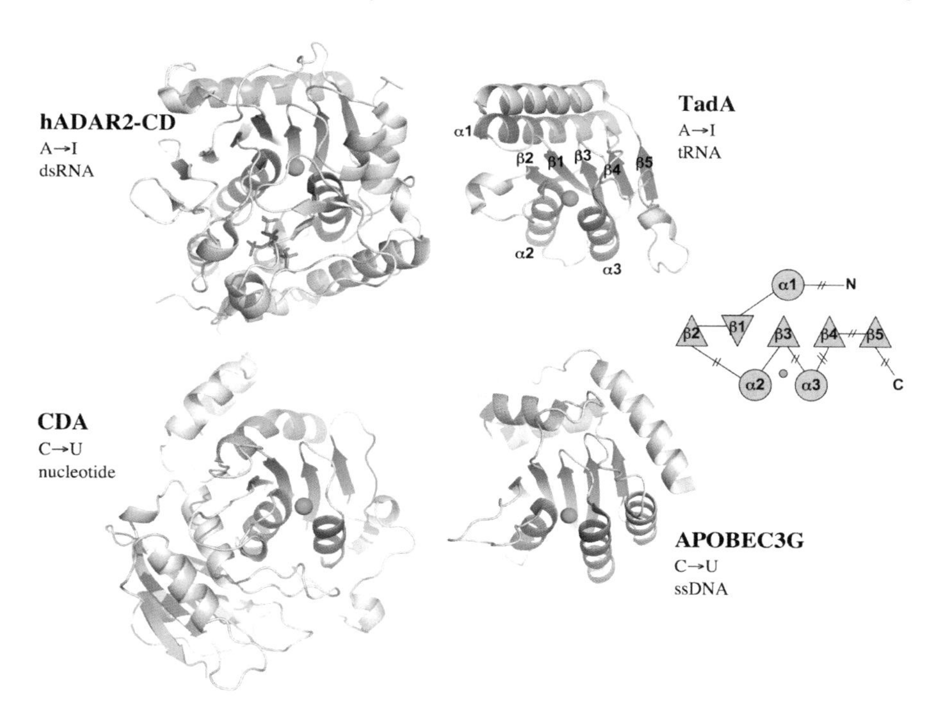

Fig. 3 Comparison of the deaminase motifs of hADAR2-CD, TadA monomer, CDA monomer and APOBEC3G-CD2 (PDB ID: 1ZY7, 1WWR, 1CTU, 3E1U). α-Helices and β-strands that comprise the deaminase motif are *colored orange* and *purple*, respectively, and are labeled in the TadA structure. The catalytic zinc ions are shown as *green spheres* and the IP_6 found in hADAR2 is represented as sticks. To the *right* is a topology diagram of the deaminase motif common to the CDA superfamily with similar color coding and labeling as the structures on the left. *Double hash marks* indicate extended loops between secondary structure elements

terminus compared to the catalytic domain of ADARs. The crystal structure of TadA has been determined both in the apo form and in the presence of the $tRNA^{Arg2}$ anticodon stem-loop (Kuratani et al. 2005; Kim et al. 2006; Losey et al. 2006; Lee et al. 2007). The structures revealed that the enzyme consists primarily of the core deaminase motif with an additional extended helix at the C terminus. The protein exists as a dimer with much of the dimerization interface comprised of the α2 and α3 helices that are part of the deaminase motif and the random coil between β4 and β5. The two subunits are flipped 180° about the dimerization interface, each subunit contributing to the formation of both active sites. One of the monomers provides the zinc and catalytic glutamate residue while the other subunit provides additional residues for recognition of the appropriate substrate (Losey et al. 2006).

The catalytic domain of hADAR2 (hADAR2-CD) is approximately 45 kDa and is comprised of the C-terminal 400 residues. The structure of hADAR2-CD was determined by X-ray crystallography and revealed a globular structure roughly

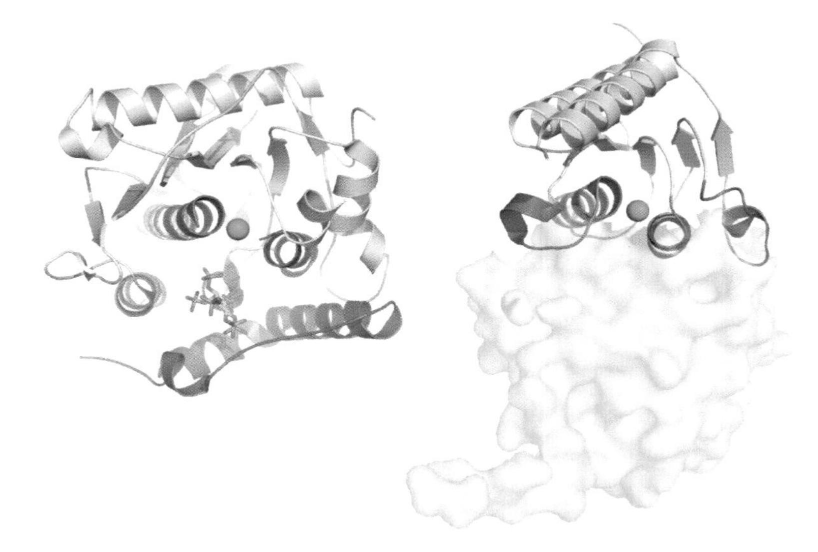

Fig. 4 Comparison of the IP_6 binding pocket of hADAR2-CD to the TadA dimerization interface. *Left* the hADAR2-CD structure highlighting the IP_6 binding pocket in *blue*; the IP_6 molecule is shown as sticks (for clarity, the random coil representing residues 374–393 and 475–515 were cut away). *Right* the TadA homodimer with the elements that contribute to the dimerization interface shown in *blue*. The second subunit of the homodimer is shown as a transparent surface view. The active site zinc ion is in *green*

40 Å in diameter (Macbeth et al. 2005). While the core deaminase motif structure is largely similar to TadA, the intervening loops between secondary structural elements are longer and the central β-sheet is larger, having three more β-strands than TadA. The ADAR2-CD also has an extended C terminus made up of several α-helices that are not found in the TadA protein.

The active site zinc ion of hADAR2-CD is buried in a pocket and coordinated by one histidine and two cysteine residues. The substrate binding surface is surrounded by positive electrostatic potential that likely facilitates binding of dsRNA and it is plausible that residues on the surface may be responsible for the 5′ and 3′ neighbor preferences exhibited by hADAR2 (Macbeth et al. 2005).

One striking feature revealed by the crystal structure of hADAR2-CD was the presence of the metabolite *myo*-inositol-1,2,3,4,5,6-hexa*kis*phosphate (IP_6) buried in the interior of the domain. The IP_6 molecule is 10.5 Å from the active site zinc ion and is linked to it via hydrogen bonding of invariant residues. The binding cavity of IP_6 is on one side of the $\alpha 2$–$\alpha 3$ helices of the deaminase motif, opposite the central β-sheet, and surrounded on the other side by the largely helical C terminus that is missing in TadA. Upon superposition of TadA and hADAR2-CD, the IP_6 binding fold replaces the dimerization interface found in TadA (Fig. 4). The IP_6 pocket is very basic and contains many residues that act as hydrogen bond donors to interact with the phosphate oxygens of IP_6. These residues are conserved

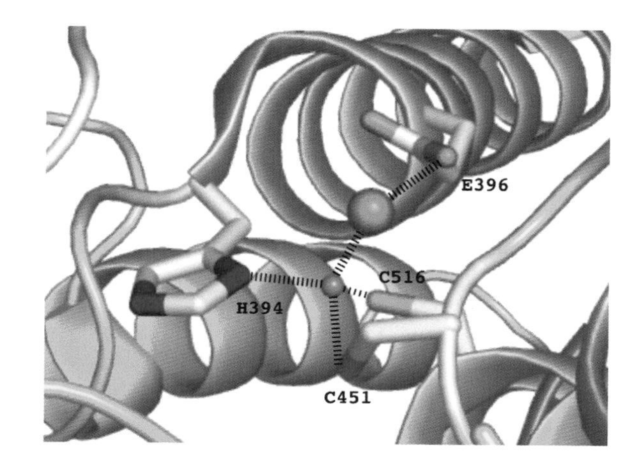

Fig. 5 The Zn(II) in the ADAR2 active site is ligated by Cys516, Cys451, His394 and the nucleophilic water (*red sphere*) that attacks C6 of the edited adenosine. Glu396 is another key active site residue that mediates the proton transfers involved in the reaction

in the ADAR proteins as well as the ADAT1 family of adenosine deaminases that act on tRNA which deaminate A37 of $tRNA^{Ala}$ in eukaryotes. Details of the ADAR2-IP_6 interaction and its role in ADAR function are described below.

4 H394, C451, C516: Zn^{2+} Binding

Other nucleoside/nucleotide deaminases activate water for attack using a zinc ion in their active sites (Wilson et al. 1991; Betts et al. 1994). ADAR2 also uses a zinc-containing active site where the metal ion is ligated by two cysteines (451 and 516) and a histidine (394), along with the water molecule responsible for hydration of the editing site adenosine (Macbeth et al. 2005) (Fig. 5).

Mutation of the equivalent residues in ADAR1 causes a loss of activity (Lai et al. 1995). Zinc ions are commonly found in proteins ligated by some combination of cysteine, histidine, glutamic acid, aspartic acid and water. Zn(II) can be found in both structural and catalytic capacities in proteins. For Zn(II) acting in a catalytic capacity, the preferred ligands are generally His, Glu, Asp and Cys (in that order) (Lee and Lim 2008). Indeed, the active site of the nucleoside deaminase ADA has three histidines, one aspartic acid and the reactive water (Wilson et al. 1991). Carbonic anhydrase also has three histidines ligating the active site zinc (Liljas et al. 1972). CDAs, APOBECs, ADARs and ADATs (including TadA) are somewhat unusual among enzymes with catalytic zinc sites in that the Zn(II) is ligated by two (and sometimes three) cysteines (Betts et al. 1994; Johansson et al. 2002; Chung et al. 2005; Elias and Huang 2005; Macbeth et al. 2005; Teh et al. 2006; Prochnow et al. 2007). Of all the amino acid ligands that Zn(II) could have, cysteine donates the most electron density and therefore decreases the Lewis acidity of the Zn(II) ion to the greatest extent. Lee and Lim analyzed a number of PDB (Protein Data Bank) and CSD (Cambridge Structure

Database) structures containing Zn(II), some of which are known to contain Zn(II) in a structural motif and some of which are known to contain Zn(II) in a catalytic motif. They compared the types of ligands to Zn(II) as a method to determine the function of Zn(II) in an unknown protein, and they found that when there are two or more cysteines ligated to a Zn(II) it usually acts in a structural capacity. However, there are also other effects from the rest of the protein that may contribute to the electronic environment surrounding the Zn(II), and so the distances between Zn(II) and its ligands can also be considered as a method to evaluate the function. They find that the Zn(II)-S and Zn(II)-N distances are longer for catalytic Zn(II) than they are for structural Zn(II) (Lee and Lim 2008). Interestingly, the ADAR2 active site fits into the parameters defined for a structural Zn(II), except for the fact that it has a water bound (Macbeth et al. 2005), which is a characteristic of a catalytic Zn(II). Structural Zn(II) sites, which tend to have shorter bond lengths between the Zn(II) and ligands, are more difficult to target with inhibitors that bind directly to the Zn(II). For catalytic Zn(II) ions, however, the bonds are longer and so it is easier to target the Zn(II) (Lee and Lim 2008). In fact, although the Zn(II) clearly plays a catalytic role in the ADAR enzymes, not only are the ligands and bond lengths more similar to those in a structural Zn(II), but millimolar concentrations of the metal chelator EDTA have no effect on the activity of the protein (Hough and Bass 1994; Saccomanno and Bass 1994). Therefore, it seems that while the Zn(II) in ADARs clearly serves a catalytic function, in some ways acts more like a structural Zn(II), perhaps accounting in part for the slow deamination rates observed for ADARs (see below). This has implications for the design of ADAR inhibitor molecules, since structural zinc sites are more effectively targeted by compounds that interact with the cysteine ligands rather than the zinc itself (Lee and Lim 2008).

Although CDAs have the same 2 Cys, 1 His ligand environment as ADARs and ADATs, their reaction rates are typically faster (Cohen and Wolfenden 1971). However, this may be due to the intrinsic reactivity of adenosine and cytidine to deamination. For instance, the rate constant for uncatalyzed deamination of cytidine in water at 85°C is $8.8 \times 10^{-8}\ s^{-1}$, while the rate constant for deamination of adenosine under the same conditions is an order of magnitude slower at $8.6 \times 10^{-9}\ s^{-1}$ (Frick et al. 1987). Also, the equilibrium constant for hydration of zebularine, a cytosine analog, is about 40-fold more favorable than for hydration of nebularine, an adenosine analog (Wolfenden and Kati 1991; Losey et al. 2006). Hydration of the heterocycle is a key step in deamination reactions catalyzed by these enzymes (see below).

As described above, while the chemical reaction occurring in the ADAR active site is more similar to that of ADA, the ADAR enzymes are more closely related to the family of CDAs (including APOBECs and AID) than to ADA. Indeed, ADARs likely evolved from a CDA ancestor (Kim et al. 1994; Gerber and Keller 2001). Why would modern adenosine deaminases that act on RNA have evolved from a CDA ancestor whose active site appears to be ill equipped for maximum ADA activity? While the active site of ADARs is not apparently capable of high turnover for adenosine deamination (the highest single turnover rate constants we have

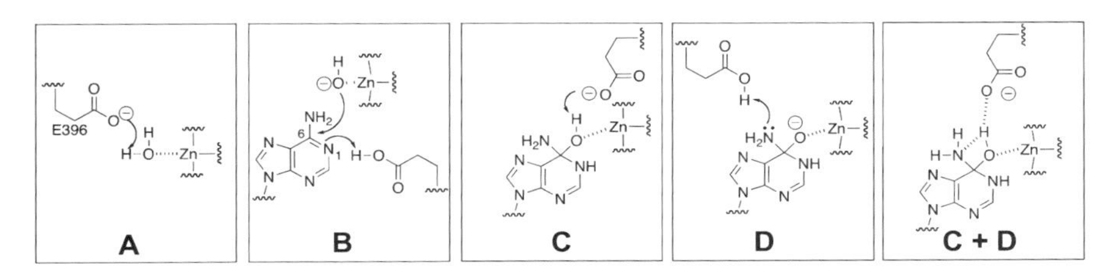

Fig. 6 Proton transfers involving E396 during ADAR2-catalyzed adenosine deamination

measured for model RNA substrates are 1–2 min^{-1}), this may not be necessary for ADAR function. Indeed, the pre-mRNA substrates for ADARs are likely to be present in relatively low concentrations in cells, so tight binding is more important for ADAR function than high turnover. In fact, with their extensive RNA binding surfaces found in multiple domains, the ADARs bind their substrates very tightly (Kd's typically in low nano Molar range). ADA and CDA have turnover numbers $>10\ s^{-1}$, but bind their substrates much less tightly than do ADARs (with K_M's near 10 μM) (Hunt and Hoffee 1982; Ashley and Bartlett 1984). If one estimates ADAR2's catalytic efficiency using a $k_{obs} = 1\ min^{-1}$ and a $Kd = 20$ nM (Stephens et al. 2000), the resulting value of $8.3 \times 10^5\ M^{-1}s^{-1}$ is within an order of magnitude of that calculated for ADA and CDA which are considered highly efficient enzymes (Frick et al. 1987). Thus, the catalytic efficiency of the ADAR enzymes is comparable to that of the nucleoside deaminases.

4.1 E396

A glutamate residue (E396) in the active site of ADAR2 acts as a proton shuttle, similar to the ADAs (Mohamedali et al. 1996), CDAs (Betts et al. 1994) and ADATs (Kuratani et al. 2005) (Fig. 5). It is found within hydrogen bonding distance of the zinc-bound water molecule in the ADAR2 active site (Fig. 5). When this glutamate is mutated to alanine, the enzyme loses all activity (Lai et al. 1995; Haudenschild et al. 2004). Based on studies of both CDA and TadA (see below), E396 of ADAR2 likely deprotonates the zinc-bound water molecule to generate a reactive zinc-hydroxide (Fig. 6a), delivers a proton to N1 (Fig. 6b), serves to remove the proton from the newly formed O6 hydroxyl of the hydrated intermediate (Fig. 6c), and protonates the N6 amino group allowing for departure of ammonia (NH_3) (Fig. 6d). We have treated these latter two proton transfers as distinct steps to illustrate all the transfers required for the reaction to take place. However, Schramm has suggested that E70 of TadA (corresponding to E396 of ADAR2) mediates the transfer of a proton from O6 to N6 via a multi-centered H-bonded array, which combines proton transfers illustrated in C and D into a single step (Fig. 6c + d). This suggestion came from kinetic isotope studies of the TadA reaction that were used to calculate a transition state structure for that enzyme (Luo and Schramm 2008).

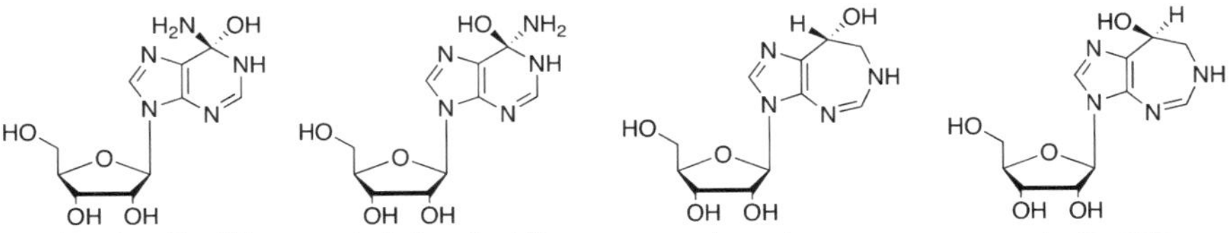

product of pro-R addition product of pro-S addition coformycin epi-coformycin

Fig. 7 ADAR2 and TadA have a mechanism in which the hydroxyl adds to the *pro-S* face of the adenosine, whereas ADA adds to the *pro-R* face of the adenosine. Coformycin is a potent inhibitor of ADA, whereas neither coformycin nor epi-coformycin are inhibitors of ADAR1 even at millimolar concentrations

In that study, the adenosine at the editing site was labeled at several positions: [1′-^{3}H], [5′-^{3}H], [1′-^{14}C], [6-^{13}C], [6-^{15}N, 6-^{13}C] and [1-^{15}N]. Based on the results of those experiments, the authors propose a late nucleophilic aromatic substitution (S_NAr) transition state with the rate-limiting step involving the E70 assisted proton shuttle from O6 to N6 (Fig. 6c + d). The transition state has almost complete N1 protonation and elongation of the N6–C6 bond as it is breaking (Luo and Schramm 2008). The late S_NAr transition state is similar to that observed previously for CDA, underscoring the mechanistic link between these deaminases (Snider et al. 2002; Ireton et al. 2003; Chung et al. 2005). The water attacks from the *pro-S*-face of the adenosine, forming a tetrahedral intermediate that has *S*-stereochemistry. It is likely that this is also true for ADAR2, because it is consistent with modeling nucleotides into the structure of the ADAR2 catalytic domain. This transition state is in contrast to that of adenosine deaminase, which has an early S_NAr transition state and attack from the *pro-R*-face (Sharff et al. 1992; Tyler et al. 2007) (Fig. 7).

Interestingly, ADA is potently inhibited by the natural product coformycin (Ki in pM range), whose structure is a stable mimic of the intermediate for adenosine deamination where the hydroxyl adds to the *pro-R* face of adenosine (Schramm and Baker 1985; Luo et al. 2007) (Fig. 7). Coformycin does not inhibit ADARs, even at millimolar concentrations. This is not surprising since it does not effectively mimic ADARs' *pro-S* attack on adenosine. However, epi-coformycin, which has the opposite configuration at the carbon bearing the hydroxyl and *does* mimic the *pro-S* attack (Fig. 7), is also not an inhibitor of ADARs (Polson et al. 1991). This may be because high affinity requires incorporation into an RNA structure mimicking an ADAR substrate (see below for discussion of 8-azanebularine).

4.2 T375

Roles for the zinc-binding residues (H394, C451, and C516 in ADAR2), and catalytic glutamate (ADAR2 E396) had been suggested prior to solution of the structure of ADAR2's deaminase domain since similar residues are found in other related deaminases (Lai et al. 1995; Melcher et al. 1996). However, when the

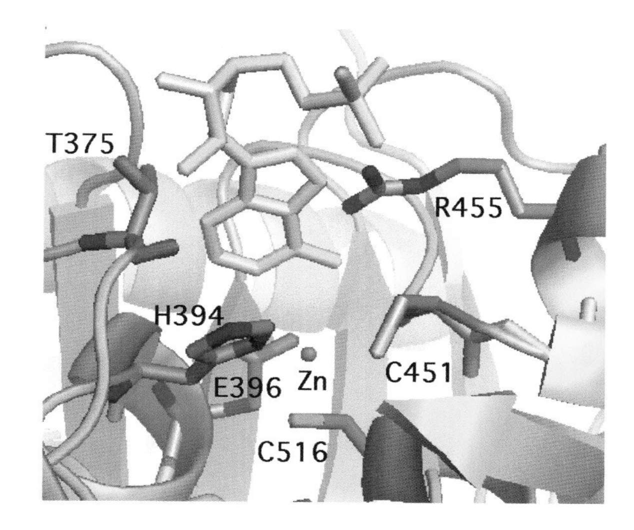

Fig. 8 The potential interactions of T375 and R455 with adenosine at the 2′-hydroxyl and N7-positions, respectively, were observed when the crystal structure of the ADAR2 deaminase domain was solved and AMP was docked into the zinc site

ADAR2 catalytic domain structure was solved, additional residues were identified that could participate in the deamination reaction. For instance, threonine 375 was observed at a position from which it might interact with the edited nucleotide (Macbeth et al. 2005) (Fig. 8). When zebularine (a cytidine analog lacking the C4 amino group) was modeled into the active site, it clashed with this residue, although an adenosine nucleotide did not. The greater size of the purine ring system allows access to the zinc with more shallow penetration of the active site than required for the pyrimidine nucleotide (Macbeth et al. 2005). This explains why ADAR2 does not deaminate cytidine in RNA (Easterwood et al. 2000). The position of this residue also suggested its side chain hydroxyl could form a hydrogen bond with the 2′–OH of the adenosine. Although a threonine is not conserved at this position in ADAR1, the asparagine present in that enzyme (Kim et al. 1994) could still participate in hydrogen bonding.

To explore the role of specific residues in the reaction of ADAR2, our laboratory developed a method to rapidly screen ADAR2 mutant libraries for editing-competent clones by linking editing to the expression of a reporter enzyme in *Saccharomyces cerevisiae* (Pokharel and Beal 2006; Pokharel et al. 2009) (Fig. 9). This was accomplished using a sequence derived from the human GluR B pre-mRNA that forms a duplex recognized by ADAR2. This sequence was mutated such that the editing site falls within an UAG stop codon in frame with sequence encoding the reporter enzyme α-galactosidase. Thus editing converts the UAG stop codon to a UIG (translated as UGG) codon for tryptophan, allowing for expression of the reporter, which is readily detected in yeast colonies grown on X-α-gal plates. When we screened for active mutants at the 375 position of ADAR2, small, hydrophobic residues were selected for, albeit with decreased activity, whereas large residues were selected against. When the residue at position 376 was also varied, only threonine or cysteine was selected at the 375 position (Pokharel and Beal 2006).

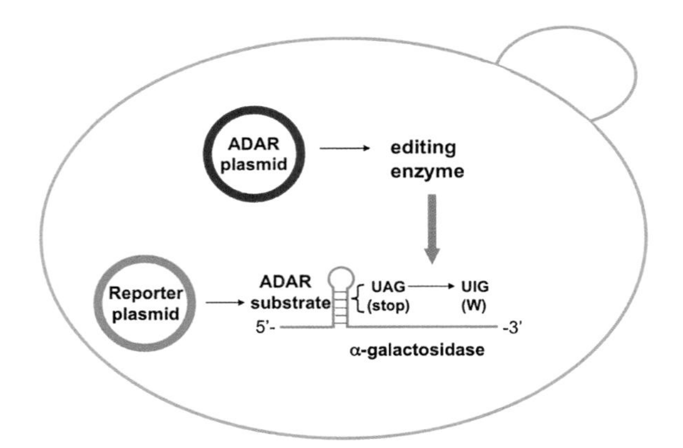

Fig. 9 An α-galactosidase based reporter assay for ADAR2 activity. Upon editing of the substrate RNA, a stop codon is converted to a tryptophan codon. This allows expression of the subsequent in-frame sequence coding for the enzyme α-galactosidase. When yeast expressing α-galactosidase are grown on plates coated with X-α-gal the colonies are colored

In addition, when the 375 position and 455 positions were varied simultaneously, most of the amino acids found to function at position 375 were small hydrogen bond donors, including serine and cysteine in addition to threonine (Pokharel et al. 2009). When the T375C mutant was evaluated in an in vitro assay for deamination kinetics, the rate of deamination was substantially reduced. This activity is still significant, even though reduced, because a similar structural change in the RNA (see below) completely abolishes editing (Jayalath et al. 2009). Both hydroxyl and thiol groups can act as hydrogen bond donors, but only threonine and serine would act as good hydrogen bond acceptors (Zhou et al. 2009). Given these observations, we suggest the residue at position 375 in ADAR2 serves two key roles. First, it prevents cytidine from fully engaging the active site, preventing C to U deamination. Second, the T375 side chain interacts with the 2′-hydroxyl of the edited adenosine, acting as a hydrogen bond donor. We discuss the role of the edited nucleotide 2′-OH in more detail below.

4.3 R455

Another residue in the active site of ADAR2 that appears to be in proximity to the edited adenosine is arginine 455 (Macbeth et al. 2005) (Fig. 8). When AMP is modeled into the active site of the ADAR2 catalytic domain, the R455 side chain appears close to the N7 position of the adenosine. Interestingly, arginine residues are also found in the active sites of CDAs (e.g. R54 of *Bacillus subtilis*, R68 of mouse and human) and are believed to neutralize the excess negative charge in the active site (Carlow et al. 1999; Johansson et al. 2002; Teh et al. 2006). However, when we screened for alternative residues at the 455 position of ADAR2, we recovered active clones with either small residues (A, G, S) or arginine itself (and no lysine) at this location. Indeed, subsequent kinetic studies with the R455A mutant showed it to be nearly as active as wild type ADAR2 with a single turnover rate for deamination within two fold of the wild type rate (Pokharel et al. 2009).

```
ADAR2 RLKENVQFHLYISTSPCGDARIF
ADAR1 QIKKTVSFHLYISTAPCGDGALF
```

Fig. 10 Amino acid sequence alignment of ADAR1 and ADAR2. Conserved residues are marked in *green*, position 455 of ADAR2 and the equivalent position in ADAR1 are marked in *purple*

This indicates that R455, while it may be proximal to the adenosine ring during the reaction, is not required for efficient catalysis. The relatively high reactivity of the ADAR2 R455A mutant allowed us to compare the reactivity of several nucleoside analogues placed at the editing site (discussed in detail below). These results support a model in which adenosine binds into the active site in such a way that R455 approaches the N7 position of the reactive adenosine, but direct interaction is not required. Interestingly, a sequence alignment of the ADAR2 and ADAR1 sequences indicates that ADAR1 has an alanine at the position corresponding to R455 in ADAR2 (position 970 in ADAR1 p150) (Kim et al. 1994) (Fig. 10). This suggests that the pocket we created by mutating R455 to alanine in ADAR2 already exists in ADAR1. Furthermore, one would then expect the wild type form of ADAR1 to tolerate modifications at the 7-position better than the wild type form of ADAR2. This could open up possibilities for designing molecules that would selectively recognize ADAR1 versus ADAR2.

5 Residues Involved in IP_6 Binding

IP_6 is the phosphorylated derivative of *myo*-inositol, and was first characterized as a phosphorous storage mechanism in plant seeds, then later as part of the phospholipase C signal transduction pathway (Berridge and Irvine 1989; Raboy 1997). Phospholipase C cleaves phosphatidylinositol *bis*-phosphate to generate IP_3 and diacylglycerol. IP_3 is then phosphorylated by two kinases in yeast and Drosophila, or three kinases in humans, to make the cellular inositol phosphate derivative IP_6 (York et al. 1999; Seeds et al. 2004; Verbsky et al. 2005).

The crystal structure of hADAR2-CD revealed that IP_6 formed a tight association with the protein (Macbeth et al. 2005). The presence of IP_6 was suggested by a strong electron density generated from 1.7 Å-resolution X-ray diffraction data and was confirmed by positive ion electrospray mass spectrometry of the native protein. ADARs were not suspected of requiring any additional co-factors; when purified to homogeneity, they robustly deaminated dsRNA substrates in vitro in the absence of any additional components. IP_6 co-purifies with hADAR2 upon expression of the protein in the yeast *S. cerevisiae*. It is essential for hADAR2 activity as yeast cells that cannot synthesize IP_6 cannot express active hADAR2 (Macbeth et al. 2005). Additionally, ADARs are notoriously insoluble (and inactive) when expressed in traditional prokaryotic expression systems, likely due to the fact that prokaryotes do not synthesize IP_6.

IP_6 has been implicated in a variety of cellular processes, including two well-studied roles in nucleic acid metabolism. IP_6 associates with the Ku70/80 subunits of DNA dependent protein kinase to stimulate non-homologous end joining of DNA. It is also binds the Gle1 protein to stimulate the DExD/H ATPase Dbp5 and promote mRNA export from the nucleus (York et al. 1999; Hanakahi and West 2002; Weirich et al. 2006). While the precise biochemical role for IP_6 in these interactions is not known, it has been implicated to have a structural role in autocatalysis of clostridial glycosylating toxins and multifunctional autoprocessing (MARTX) toxins from *Clostridium difficile* and *Vibrio cholerae* (Reineke et al. 2007; Prochazkova and Satchell 2008). In these cases, a large bacterial holoprotein translocates through a host cell membrane and then is autoprocessed by a cysteine protease domain to release a functional effector domain (Egerer and Satchell 2010). IP_6 activates the cysteine protease domain by binding to an allosteric site on an opposite surface from the active site (Lupardus et al. 2008; Pruitt et al. 2009). One-dimensional ^{1}H-NMR studies suggest that IP_6 binding to the allosteric site induces a conformational change in the protease domain that confers activity (Pruitt et al. 2009). While there is no structural homology in the IP_6 binding pocket between these cysteine protease domains and hADAR2, it is entirely plausible that IP_6 is playing a structural role to alter the conformation of hADAR2 in a similar manner.

The IP_6 interaction with hADAR2 is unusual compared to other IP_6 requiring proteins in that IP_6 makes a tight association in the core of ADAR2, 8 Å below the protein surface, as opposed to a binding pocket near the surface (Fig. 11a). In this way, the ADAR2-IP_6 interaction is reminiscent of proteins that bind Fe–S complexes or FAD, where the protein envelops the co-factor, although the ADAR2-IP_6 interaction is non-covalent, mediated by H-bonds and completely hydrophilic (Macbeth et al. 2005; Sun et al. 2005; Sazanov and Hinchliffe 2006). IP_6 directly interacts with the side-chains of 12 residues that are conserved in ADARs: R400, R401, K519, R522, S531, K629, Y658, K662, Y668, K672, W687 and K690 (Fig. 11b). In addition, the invariant W523 makes a water mediated H-bond to a phosphate oxygen of P1 of the inositol ring. Twelve residues that are not conserved between ADAR sequences also interact with IP_6 through water mediated H-bonds. In many of these cases, these interactions are via main-chain carbonyl oxygen or secondary amine nitrogen atoms.

At this point the function of the IP_6 co-factor in an ADAR is not known, but it is likely involved in the proper folding of the C terminus of the enzyme. The IP_6 binding pocket in the protein core is exceptionally basic, and the acidic molecule can negate the positive charge of the several lysine and arginine residues. IP_6 may have a more direct role in catalysis by modulating the structure of residues near the active site. IP_6 is 10.5 Å from the active site zinc ion and is linked to it via a hydrogen-bonded relay of invariant residues, including K519, D392, and K483 (Fig. 11c). The latter makes a H-bond to the thiolate of C516 which is a zinc coordinating side-chain. The H-bond between K483 and C516 displaces electron density from the zinc–sulfur bond and allows the zinc to accommodate the developing negative charge of the hydroxide ion during the reaction (Macbeth et al. 2005). Mutation of this residue to

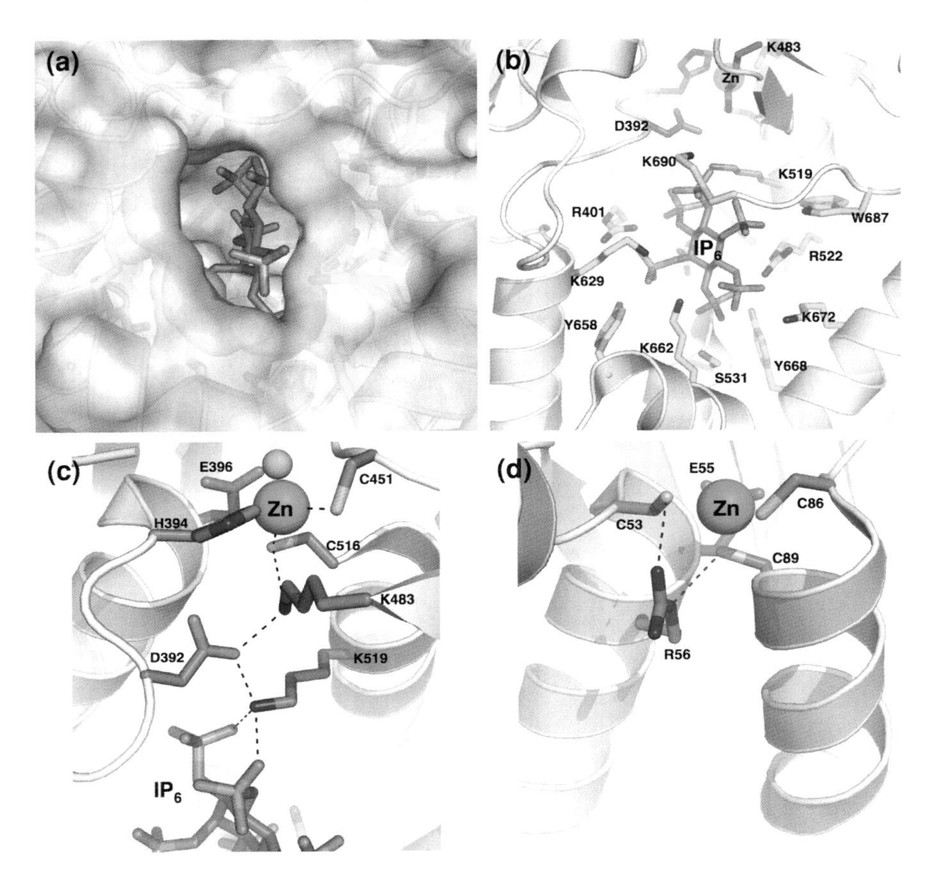

Fig. 11 **a** Surface representation of the IP_6 binding cavity. The hADAR2-CD surface is shown as transparent *gray*, IP_6 is shown as *blue*, *orange* and *red sticks* (for carbon, phosphorous and oxygen, respectively). **b** Same view of hADAR2-CD as in **a**, except as a cartoon structure with the surface removed. *Yellow sticks* represent conserved residues whose side-chains H-bond directly to IP_6. Magenta sticks are invariant residues that participate in a H-bonded chain (*dashes*) between the IP_6 molecule and the active site Zn^{2+} (*green sphere*). **c** Detailed view of the IP_6-Zn^{2+} H-bond relay, with the nucleophilic water shown as an aqua sphere. **d** Structure of one monomer of T-CDA from *B. subtilis*. Shown are the three Zn-coordinating cysteine residues and the conserved R56 that makes H-bonds (*dashes*) to C53 and C89. PDB codes: 1ZY7 and 1JTK

alanine (as well as the mutations D392 N and K519A in the H-bond relay between the active site and IP_6) inactivates hADAR2 in vitro, though the effect of these mutations on the global structure of the protein must be assessed (Macbeth, unpublished observations). The interaction between a basic residue and a zinc-coordinating thiolate also occurs in the tetrameric CDA (T-CDA) of *B. subtilis*. In T-CDAs the zinc ion is coordinated by three cysteine residues; a metal-binding mode that potentially decreases activity of the enzyme due to the extensive electron density contributed by the three thiolates (see above). To overcome this, R56, a conserved residue in T-CDAs, forms two H-bonds with the thiolate sulfurs of C53 and C89, withdrawing the electron density between the zinc–sulfur bond and allowing charge to develop on the nucleophile (Fig. 11d) (Johansson et al. 2002). Mutation of R56 to alanine and glutamine supports this hypothesis as *B. subtilis* CDA harboring these mutations is less active (Johansson et al. 2004).

The yeast protein Tad1p, a member of the ADAT1 family of deaminases that converts A37 to I37 of eukaryotic $tRNA^{Ala}$, also requires IP_6 for optimal activity (Macbeth et al. 2005). This enzyme is approximately 45 kDa and has the extended C terminus present in ADARs but not in ADAT2. Sequence alignments show that many, but not all, of the conserved residues that bind IP_6 in ADARs are conserved in ADAT1. When the activity of yeast ADAT1 was assayed in extracts prepared from mutants that could not synthesize IP_6, activity was decreased to about 5% of normal levels. However, this activity could be recovered to approximately 50% of maximum by the addition of exogenous IP_6. This suggests IP_6 is not essential during folding and that the *apo* form of ADAT1 is in a dynamic state that can bind free IP_6. Alternatively, there are factors present in the extract that could assist folding around the molecule. Curiously, addition of inositol hexa*kis*sulfate (IS_6), the sulfate derivative of inositol, does not recover activity, suggesting that the protein can sense slight differences in size and charge of the co-factor.

Finally one must consider *why* ADARs and ADAT1 evolved to require IP_6, while the evolutionary precursors, TadA/ADAT2 and CDAs, are capable of deaminating their substrates without any additional co-factors. One possibility for the IP_6 requirement is for regulation of activity in a manner similar to the IP_6-dependent proteases described above. When IP_6 is present, ADARs can bind to it and fold into an active conformation, however, when IP_6 levels are low or sequestered away from the ADAR, the enzyme is inactive.

6 Roles of Functional Groups in RNA Substrate at or Near the Deamination Site

One aspect of the ADAR enzymes that has continued to elude the community is the nature of the specificity that these enzymes exhibit. One example of this specificity can be found in the recently discovered editing sites in the pre-mRNA for the DNA repair enzyme NEIL1. This RNA has two editing sites next to each other in an AAA codon for lysine; the central adenosine is preferred by ADAR1 while the other is preferred by ADAR2 (Yeo et al. 2010). Why one enzyme prefers one adenosine over another closely positioned adenosine remains difficult to fully explain. Factors that contribute to specificity include the secondary structure of the RNA substrate and the length of the duplex (Nishikura et al. 1991; Stephens et al. 2000), the positioning of bulges, loops and mismatches (Lehmann and Bass 1999; Ohman et al. 2000), the identity and number of dsRBMs involved in the protein/RNA interaction (Liu et al. 1997, 2000; Stephens et al. 2004; Xu et al. 2006), the identity of the catalytic domain (ADAR1 or ADAR2) (Melcher et al. 1996; Burns et al. 1997; Yang et al. 1997), the RNA sequence flanking the edited nucleotide (Polson and Bass 1994; Lehmann and Bass 2000; Dawson et al. 2004), and the base opposite the edited nucleotide in a duplex substrate (Wong et al. 2001; Källman et al. 2003; Yeo et al. 2010). All of these factors are important, but the relative contribution of each to the overall efficiency of the reaction may not be the same for all editing sites. Indeed, the 5-HT2cR D site is an excellent ADAR2 site even though it does not conform to an ideal site based on flanking sequence. This is also true for the GluR B Q/R site. Factors that contribute to selective ADAR binding of these RNAs and that position the catalytic domain precisely may be more important for these substrates than others with preferred flanking sequence, for instance.

7 The Effect of Helix Defects

One factor that clearly influences specificity of the ADAR enzymes is the presence of defects (bulges, loops, mismatches) in the double helix surrounding the edited adenosine (Lehmann and Bass 1999). One example of this specificity is the editing of the serotonin 2C receptor pre-mRNA. This RNA has six editing sites (A–E in the exon and F in the intron), some edited by ADAR1 and some by ADAR2. These editing sites are in very close proximity (the exonic editing sites are all within 13 nucleotides), yet each enzyme has a clear preference for certain sites (Burns et al. 1997). It is likely that this specificity arises from the mismatches, bulges and loops limiting the number of binding modes available to the dsRBMs of the protein, which require approximately 16 base pairs of duplex to bind (Nishikura et al. 1991). Indeed, the internal loop in the editing complimentary sequence (ECS) of the serotonin 2C receptor pre-mRNA is necessary to maintain this specificity. In the presence of the loop ADAR2 has a strong preference for the D site, whereas without the loop the D site is still preferred, but editing at the other sites increases significantly (Schirle et al. 2010). This type of specificity is typical of ADAR substrates.

8 Preferred Flanking Sequence

The two most clearly defined RNA sequence effects are nearest neighbor effects (in the same strand) (Polson and Bass 1994; Lehmann and Bass 2000) and effects based upon the nucleotide in the opposite strand base paired with the edited adenosine (Wong et al. 2001) (discussed in more detail below). ADAR1 has a 5′ nearest neighbor preference of $A = U > C > G$ (Polson and Bass 1994) but no reported 3′ nearest neighbor preference. On the other hand, ADAR2 has a 5′ nearest neighbor preference of $A \approx U > C = G$ and a 3′ nearest neighbor preference of U = G > C = A (Lehmann and Bass 2000). In a recent report involving high throughput transcriptome sequencing, an analysis was performed to identify the 5′ and 3′ nearest neighbor preferences exhibited by a number of newly discovered editing sites (Li et al. 2009). These were grouped together rather than analyzed individually, so the results contain a combination of ADAR1 and ADAR2 sites. This analysis gave results similar to that discovered previously, except that both U and A were *under*-represented as a 3′ nearest neighbor, in contrast to the earlier finding that U is a *preferred* 3′ nearest neighbor of ADAR2. Guanosine was clearly preferred as a 3′ nearest neighbor, and disfavored as a 5′ nearest neighbor, as predicted for both ADAR1 and ADAR2 according to earlier studies (Polson and Bass 1994; Lehmann and Bass 2000; Li et al. 2009). The differences seen could be a result of the combination of ADAR1 and ADAR2 editing sites, or could be due to the fact that the original studies were carried out on long, synthetic duplex RNAs and not natural editing sites.

One explanation for the ADAR2 3′ nearest neighbor preference of a G is presented in a recent report by Stefl et al. (2010). The authors make the case that this preference is, in fact, a result of a sequence-specific interaction between dsRBM2 and the amino group of the guanosine directly 3′ to the edited adenosine. Unfortunately, as mentioned earlier, there is no structure of the catalytic domain of ADAR2 bound to RNA. However, when one examines the catalytic domain structure with an AMP modeled into the active site, it is difficult to imagine that an interaction between Ser258 and the guanosine 3′ nearest neighbor could be maintained when the adenosine is flipped into the active site. It is possible that the interaction between Ser258 and the guanosine is involved in recognition of the RNA by ADAR2, but that the interaction is lost during catalysis. Further studies are necessary to fully identify the specific protein/RNA interactions that lead to nearest neighbor effects.

9 A–C Mismatch at Editing Site

Both ADAR1 and ADAR2 react efficiently with adenosines in A–C mismatches. Indeed, when mutagenesis has been carried out to vary the base opposite the edited adenosine in duplex substrates, the preferred context is A–C. Adenosines in A:U pairs can also be good substrates, whereas A–A and A–G mismatches react poorly (Fig. 12) (Wong et al. 2001; Källman et al. 2003; Yeo et al. 2010).

While the tRNA adenosine deaminases have a different recognition mechanism than the ADARs, involving the anticodon stem loop, the crystal structure of TadA bound to RNA is the only structure of an RNA deaminase bound to RNA. In this structure, the authors observed that when the adenosine is in the active site, the other nucleobases in the anticodon stem loop are splayed out and unstacked (Losey et al. 2006). In the structure of tRNA guanine transglycosylase, another enzyme that interacts with an anticodon stem loop, the surrounding bases are also splayed out and are inserted into individual recognition pockets on the protein (Xie et al.

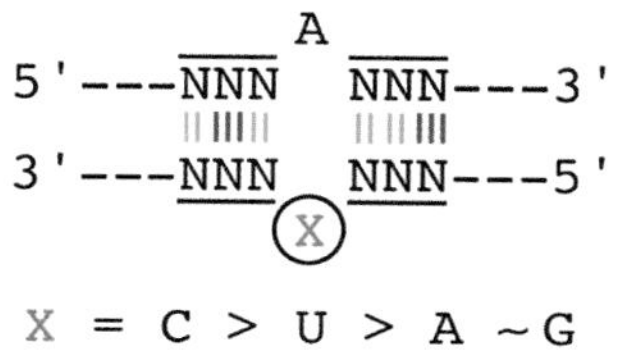

```
          A
5'---NNN   NNN---3'
     |||||  |||||
3'---NNN   NNN---5'
         (X)

X = C > U > A ~ G
```

Fig. 12 For both ADAR1 and ADAR2 an A–C mismatch is the preferred sequence context. In general, an adenosine opposite a pyrimidine is much more efficiently edited than an adenosine opposite a purine

2003). Although there is not yet any evidence that this is the case for ADAR1 or ADAR2, an interaction like this, in which there are specific recognition pockets for surrounding nucleotides, could explain the preference of both of these enzymes for a pyrimidine opposite the edited adenosine and for certain flanking sequences (Fig. 12).

10 2′-Hydroxyl of the Edited Nucleoside

As mentioned above, modeling using the crystal structure of the ADAR2 catalytic domain and activity data for various mutants suggests that T375 may interact with the 2′-position of the edited adenosine. Consistent with this idea, our early work on the reactivity of nucleoside analogs showed that 2′-*O*-methyladenosine at an editing site is deaminated extremely slowly by ADAR2 (>100-fold difference in deamination rate, Fig. 14) (Yi-Brunozzi et al. 1999). More recently we have shown that replacing the 2′-hydroxyl with a 2′-thiol at the edited nucleotide also inhibited the ADAR2 reaction. In fact, we were unable to observe product formation in an ADAR2 reaction where the editing site adenosine was replaced with 2′-deoxy-2′-mercaptoadenosine (Fig. 13) (Jayalath et al. 2009). Thus, while thiol substitution for hydroxyl group was detrimental both in the protein (T375C, discussed above) and in the RNA (2′-hydroxyl to 2′-thiol), a more severe effect was observed with the change in RNA structure (Fig. 13). Since a thiol is a better H-bond donor than acceptor, these results are most consistent with H-bond donation from the threonine side chain to the 2′-hydroxyl of the edited adenosine (Zhou et al. 2009).

Other modifications at the 2′-position of the ribose of the edited adenosine have also been evaluated. 2′-Deoxy-2′-fluoroadenosine was less reactive than adenosine, but only by two fold (Fig. 14) (Yi-Brunozzi et al. 1999). Fluorine has been suggested to be capable of accepting a hydrogen bond, so this result is consistent with the proposed H-bonding interaction (West et al. 1962). However, we also observed only a three fold reduction in rate when adenosine was replaced with 2′-deoxyadenosine (Fig. 14) (Yi-Brunozzi et al. 1999). While the reduction in rate arising from removal of the 2′-hydroxyl indicates a possible interaction, the

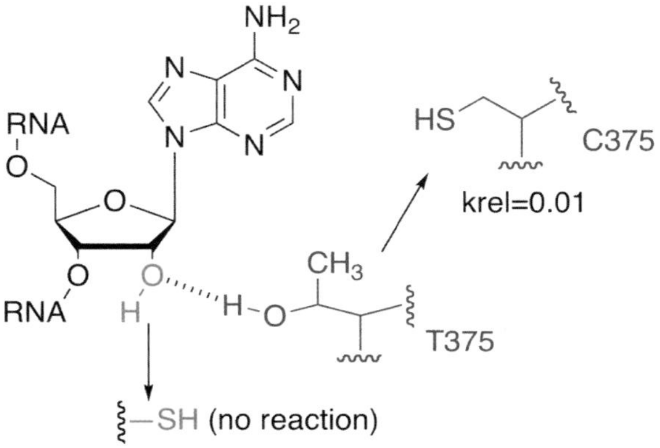

Fig. 13 Mutagenesis and nucleoside analogue data suggest a role for T375 as a hydrogen bond donor and the 2′-hydroxyl as a hydrogen bond acceptor. The hydroxyl on the amino acid could be mutated to a cysteine and maintain activity, albeit highly reduced, but when the hydroxyl on the sugar was mutated to a thiol no product was observed. Since sulfur is a better hydrogen bond donor than acceptor, this supports the model for the role of T375

Fig. 14 Modifications at the 2′-position that change the hydroxyl to either a fluorine or hydrogen are both well tolerated. In contrast, both *O*-methyl and thiol groups inhibit editing

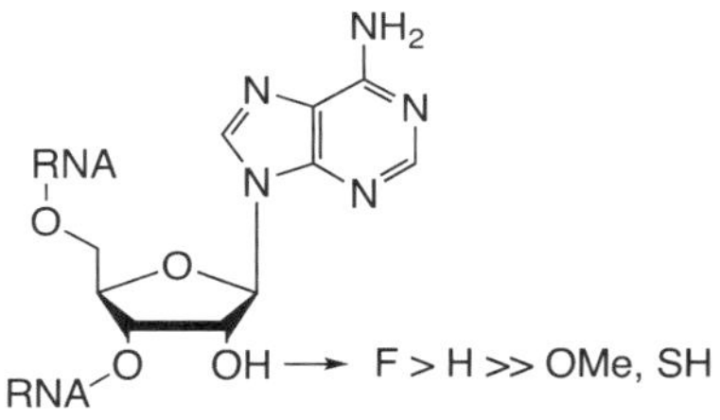

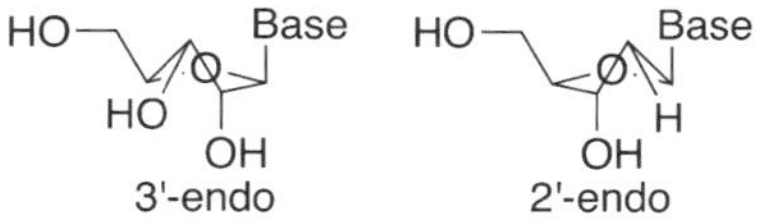

Fig. 15 Sugar puckers adopted by ribose and 2′-deoxyribose. 3′-endo sugar pucker is preferred by ribose due to the presence of an electronegative substituent at the 2′-position

magnitude of the change seems modest considering our proposed role for this group. However, 2′-deoxy nucleotides prefer the 2′-endo ribose conformation unlike the other analogs (Fig. 15). This is the conformation of the nucleoside that is preferred for cytidine deamination catalyzed by CDA. In contrast, ADA prefers substrates in the 3′-endo conformation (Marquez et al. 2009). Since ADAR2 is more closely related to CDAs than ADAs, ADAR2 may prefer substrates that more readily adopt the 2′-endo conformation. Therefore, while adenosine and 2′-fluoroadenosine contain potential hydrogen bond acceptors at the site poised to

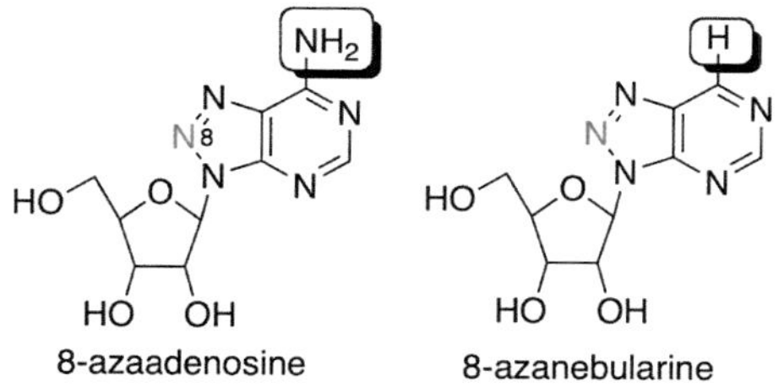

Fig. 16 Structures of 8-azaadenosine and 8-azanebularine. Replacing the carbon at the 8-position with nitrogen increases the reaction rate (8-azaadenosine) and leads to hydration-dependent tight binding (8-azanebularine)

interact with T375, they prefer a 3′-endo sugar pucker because of the electronegative substituents at the 2′-position. However, although 2′-deoxyadenosine does not have a hydrogen bond acceptor, its lack of electronegative substituent allows it to more readily adopt a 2′-endo sugar pucker. These competing affects may account for the similar reaction rates of these substrates.

11 8-Aza Substitution at Edited Adenosine

Substitution of the carbon at position 8 of a purine with nitrogen decreases the barrier to hydration at the 6 position (Erion and Reddy 1998). In addition, the rate of deamination of 8-azaadenosine by ADA is faster than the rate for adenosine (Agarwal et al. 1975) (Fig. 16). This is likely due to the increased electron-withdrawing capability of the nitrogen, which helps to facilitate disruption of the π-electron system. When 8-azaadenosine is placed at the site of the edited adenosine in an ADAR2 substrate, the rate is increased by more than six fold, whereas the binding is comparable. The "aza effect" is greater when the modified adenosine is placed in a poor sequence context for editing. Even though the 8-aza substitution can have a large effect on the deamination rate and is clearly an excellent substrate for ADAR2, 8-azanebularine, which is predicted to form a hydrate mimicking the high energy intermediate of the ADAR reaction, only inhibits the reaction in the millimolar range (Véliz et al. 2003) (Fig. 16). On the other hand, when 8-azanebularine is placed at an editing site in a model ADAR substrate RNA, it binds tightly (Kd in low nano Molar range) to ADAR2 even in the absence of dsRBMI (Haudenschild et al. 2004). These observations underscore the importance of the RNA structure for substrate binding by ADARs. Indeed, ADARs will not deaminate nucleotides or short single stranded oligonucleotides, but require the adenosine to be in the double helical structure of at least ~15 base pairs (Bass and Weintraub 1988; Lehmann and Bass 1999; Doyle and Jantsch 2002).

The high affinity binding observed with 8-azanebularine is dependent upon the catalytic activity of the enzyme. When the critical E396 residue is mutated to alanine, this tight binding is lost (Haudenschild et al. 2004). This suggests that the enzyme carries out the first steps of the reaction to generate a covalent hydrate of 8-azanebularine that cannot proceed down the normal pathway to product because

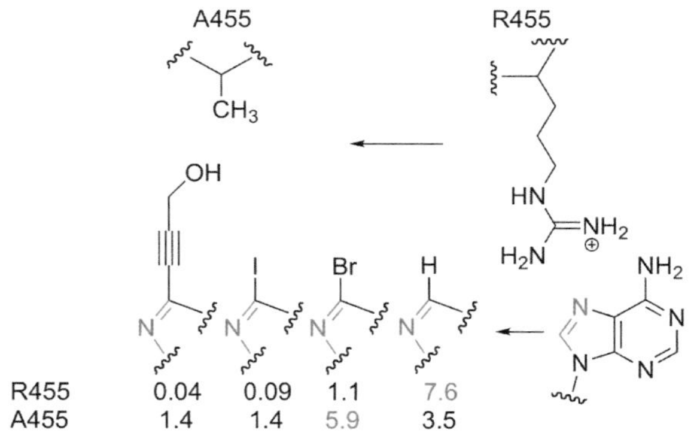

Fig. 17 Deamination rates relative to the rate of adenosine for both R455 and A455 ADAR2 enzymes with modifications at the 7-position [kobs(analogue)/kobs(adenosine)]. There is an increase in reaction rate due to the 8-aza substitution, but a decrease in rate with the more sterically demanding substituents at the 7-position, particularly with the R455 enzyme. Mutation to A455 allows these substituents to be better tolerated

it lacks the C6 amino group. This hydrate has structural similarities to proposed transition states for adenosine deamination and thus binds tightly in the active site (Haudenschild et al. 2004).

12 N7 of Substrate, C7 Substituents on Adenosine Analogs

ADA and TadA both make hydrogen-bonding contacts to all of the purine nitrogens (Wilson et al. 1991; Losey et al. 2006). If ADAR2 makes similar contacts, these interactions are not essential for deamination. When 7-deazaadenosine was incorporated at the position of the edited adenosine, the ADAR2 reaction rate was comparable to that of adenosine (Easterwood et al. 2000), although this molecule is not a substrate for ADA (Frederiksen 1966). Therefore it was possible to incorporate several 7-deaza-8-azaadenosine analogues into the position of the edited adenosine and evaluate the rate of reaction with wild-type ADAR2 compared to the rate of deamination of adenosine (Fig. 17).

When hydration free energies, which are thought to be a reasonable predictor of the ADAR reaction rate, were calculated for these heterocycles, the substituted analogues were calculated to be more stable in the hydrated form. However, for all of these, and particularly for the iodo- and propargyl -alcohol substituted analogues, the change in reaction rate was not as predicted with wild type ADAR2. When we modeled these analogues into the active site of ADAR2, we found that the C7 substituents, particularly those that were more sterically bulky, could clash with the side chain of R455. When the R455 residue was mutated to alanine (described above), this increased the reaction rates of the bulky analogues.

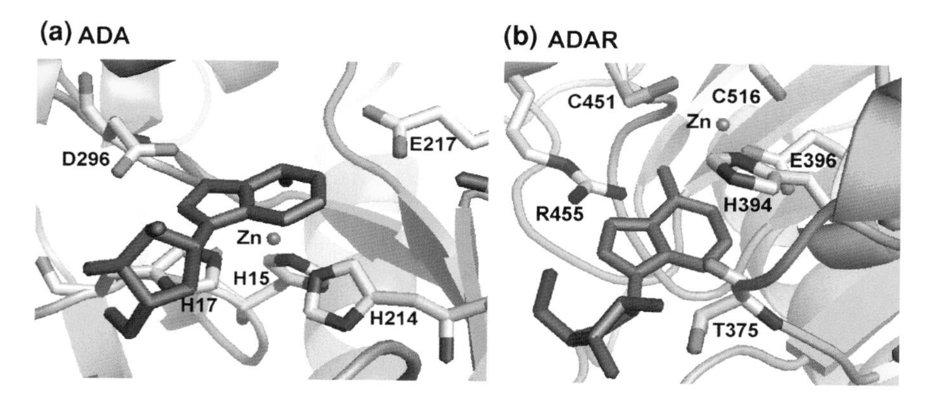

Fig. 18 A comparison of the active site structures of ADA and ADAR2 with nebularine bound (A, ADA) or AMP docked into the structure (B, ADAR2). Hydrogen bonding of D296 of ADA to N7 of adenosine contributes substantially to this reaction, whereas mutation at R455 and use of nucleoside analogues indicate that ADAR2 does not require adenosine N7 interactions

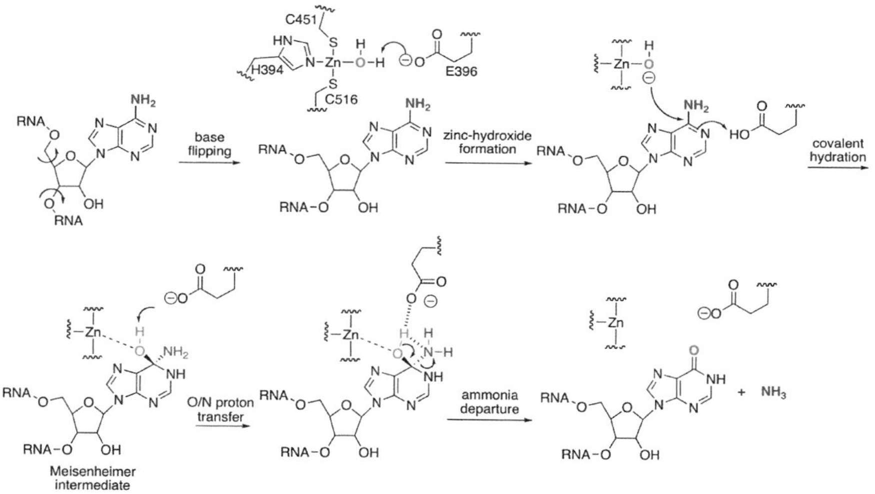

Fig. 19 Active site steps in the proposed ADAR2 mechanism

This suggests that while the R455 residue is close in space to the N7 position of the adenosine in the active site, it does not form an essential interaction with N7, since neither changing N7 to a carbon nor mutating R455 to alanine has a large effect on the rate (Pokharel et al. 2009). This is in contrast to ADA, in which the interaction between an amino acid side chain and N7 of adenosine is critical (Fig. 18). An aspartic acid in the ADA active site (D296 in mouse ADA) hydrogen bonds to

N7 and this residue is necessary for activity (0.001% of wild type k_{cat}/K_m for the D296A mutation) (Sideraki et al. 1996). In addition, removal of N7 from the substrate is highly inhibitory to the ADA reaction and 7-substituted 7-deaza compounds are poor ADA substrates (Frederiksen 1966; Seela and Xu 2007). One can understand the difference in importance of the interaction between enzyme and N7 of the adenine for the different types of adenosine deaminases when one recalls that ADARs bind their substrates tightly via their dsRBMs and are not as dependent on specific interactions with the reacting base to hold the substrate in the active site (as is the case for the nucleoside processing enzyme ADA). Indeed, dissociation constants for RNA binding to the isolated RNA-binding domain of ADAR2 are similar in magnitude to those for the full-length enzyme (Yi-Brunozzi et al. 2001; Stephens et al. 2004). In the case of TadA, a crystal structure of the *S. aureus* enzyme bound to a model substrate RNA did show ordered water molecules interacting with N7 as well as the other purine nitrogens through hydrogen bonding interactions (Losey et al. 2006). However, the effect of removing the N7 interactions for the TadA reaction has not yet been determined.

13 C6- and C2-Positions of the Edited Purine

Analysis of substitutions at the C6- and C2- positions of adenosine allows us to learn more about the placement of the adenosine into the ADAR2 active site. This analysis also shows additional differences between ADAR and ADA. For example, ADA has been shown to deaminate 2,6-diaminopurine only four fold more slowly than adenosine (Chassy and Suhadolnik 1967). However, when 2,6-diaminopurine was placed at the ADAR2 editing site no product was observed (Véliz et al. 2003). 2-Aminopurine (2-AP) containing RNA has also been used to study conformational changes in the substrate that occur during the ADAR2 reaction (see below).

The 6-position of adenosine is somewhat more tolerant of modifications than the 2-position, but still cannot accommodate large increases in steric bulk. ADAR2 is able to process 6-*O*-methylinosine (Easterwood et al. 2000) and N6-methyladenosine at similar rates (Véliz et al. 2003), albeit slower than the rate for adenosine. However, N6-ethyladenosine shows no deamination product when allowed to react with ADAR2, indicating that the ethyl group is too large to be accommodated in the active site (Véliz et al. 2003). Finally, placement of a methyl at the C6-position inhibits formation of the covalent hydrate even in the 8-azanebularine context, as it decreases the binding by nearly 100-fold (Maydanovych and Beal 2006), approaching the affinity of the enzyme for nonspecific binding to RNA (Haudenschild et al. 2004). Methylation at the 6-position of 8-azapurine has been shown to decrease covalent hydration at the same site, and methylation at the site of hydration has actually been shown to inhibit hydration for multiple nitrogen-containing heterocycles (Albert et al. 1976).

13.1 The ADAR Reaction Mechanism

Given the structure/activity relationships obtained for the ADAR2 reaction described above, kinetic isotope effects measured for the TadA reaction and other mechanistic data available, we can propose a catalytic mechanism for the ADAR reaction. This mechanism is summarized below along with discussions of the oligomerization state of the enzyme, substrate recognition, and conformational changes.

13.2 Dimerization

One aspect of the ADAR mechanism in which conflicting data exists in the literature is the question of whether these proteins dimerize, and if so, whether or not this dimerization is RNA binding-dependent and necessary for activity. In the case of *Drosophila melanogaster* ADAR, which has high sequence similarity to human ADAR2, a variant lacking the N-terminal portion of the protein comprising residues 1–46 was shown to retain the ability to bind dsRNA but yet was not able to dimerize and did not retain editing activity. This report also indicated that dADAR dimerization is RNA-dependent (Gallo et al. 2003). However, the N-terminal domain of hADAR2 has been shown to be dispensable for catalytic activity (Macbeth et al. 2004), so perhaps this represents a difference between dADAR and hADAR2. Evidence has also been reported supporting dimerization of hADAR2 both in vivo and in vitro. One report utilizing FRET indicated that ADAR1 and ADAR2 both dimerize in vivo, and in fact can form heterodimers as well. This dimerization was dependent upon the presence of the dsRBMs (Chilibeck et al. 2006). Additional work by Poulsen et al. also gives evidence for in vivo dimerization of ADARs. Using a yeast two-hybrid screen, growth and therefore dimerization was only observed when dsRBM1 was present. When only dsRBM2 or only the catalytic domain was used, no dimerization was observed (Poulsen et al. 2006). On the other hand, Poulsen et al. agree with previous findings that dsRBM2 is most critical for retention of catalytic activity (Macbeth et al. 2004; Poulsen et al. 2006). However, the observations that dsRBM1 is required for binding whereas dsRBM2 is required for catalytic activity are hard to reconcile, considering that RNA binding is necessary for robust catalytic activity (Macbeth et al. 2005). Valente et al. suggest that this may be explained by the choice of mutation made to disrupt the dsRNA binding. They suggest that the A–E mutation made by Poulse et al. could in fact interfere with a hydrophobic residue required for folding, and therefore could entirely disrupt the structure of the dsRBM. Instead, they chose to make a mutation in a conserved KKXXK stretch that is known to be involved in RNA binding in other dsRNA binding proteins. This mutation disrupted RNA binding but in fact still allowed observation of protein dimerization. The authors observed that a dimer in which only one unit was able to bind RNA was not catalytically active, whereas a dimer in which both units were able to bind RNA but only one of the units was catalytically active still exhibited 50% activity. The dimerization of these RNA binding-deficient proteins was confirmed by size-fractionation chromatography (Valente and Nishikura 2007). However, another analysis of the dimerization of ADAR2 by both analytical gel filtration and equilibrium sedimentation indicated that human ADAR2 is a monomer (Macbeth and Bass 2007). Therefore, while ADATs such as ADAT2/3 and TadA have been shown to dimerize (Gerber and Keller 1999; Wolf et al. 2002) and CDA have been shown to form oligomers (Prochnow et al. 2007), the question of dimerization for the ADAR enzymes is still debated. Poulsen and coworkers do suggest (Poulsen et al. 2006), however, that a requirement for dimerization could explain the substrate inhibition phenomenon observed for ADARs (Hough and Bass 1994). Perhaps in the presence of an excess of RNA, only one ADAR binds per RNA molecule, and so dimerization-dependent editing is decreased.

13.3 Substrate Binding and Conformational Changes

ADARs will only bind and edit RNAs containing double helices of sufficient length (Bass and Weintraub 1988; Lehmann and Bass 1999; Doyle and Jantsch 2002). This requirement is due to the presence of the double stranded RNA binding motifs (dsRBMs, also referred to as dsRBDs) in the ADAR structure. The review of Allain and colleagues in this issue addresses the structure of ADAR dsRBMs and their role in editing selectivity. In addition, it appears the ability of a substrate RNA to engage both dsRBMs of ADAR2 contributes to editing specificity. Macbeth and Bass deleted the ADAR2 N-terminal domain including dsRBMI. The resulting enzyme retained the ability to efficiently deaminate at the R/G site of GluR-B. Importantly, this truncation mutant was also able to deaminate short substrates that the full-length enzyme was unable to process. This was rationalized by invoking autoinhibition by ADAR2's N-terminal domain that is relieved upon dsRBMI binding to RNA (Macbeth et al. 2004). Thus, it appears that ADAR2 has a duplex length requirement for its substrates arising from the requirement to bind both dsRBMI and dsRBMII, with dsRBMI binding causing a conformational change in the protein that relieves autoinhibition. This requirement would limit unwanted reaction with other cellular RNA substrates containing short (<15 bp) duplex segments.

One of the consequences of the double helical nature of the ADAR substrate is the requirement for conformational changes in the RNA prior to deamination. It is clear from the structure of the catalytic domain of ADAR2 that the reactive nucleotide must adopt a conformation that removes the edited base from the helical stack before it can access the zinc-containing active site (Macbeth et al. 2005). The issue of conformational changes in the ADAR substrate was addressed using RNAs bearing the fluorescent base 2-AP at different positions, including at a known editing site. Stacking into a duplex quenches the fluorescence of 2-AP

(Ward et al. 1969). Thus, 2-AP can be used as a probe of the stacking environment of a nucleotide under different experimental conditions. These studies demonstrated that ADAR2 causes a conformational change in an RNA substrate consistent with flipping the reactive base from the helix into the enzyme active site. The change in 2-AP fluorescence was significantly faster than deamination, indicating that this step was not rate limiting for the substrate studied (Stephens et al. 2000). Molecular dynamics simulations were also used to study base flipping processes for adenosines in different duplex RNA sequence environments. These efforts demonstrated that an adenosine at a known editing site (R/G of GluR-B) is more prone to move out of the helical stack than other adenosines present in the simulated duplex (Hart et al. 2005). Thus, the local structure of the RNA may facilitate the base-flipping step in the editing reaction. Protein conformational changes have also been studied by monitoring differences in the tryptophan fluorescence of ADAR2 when RNA binds. The results point to a coupling of RNA substrate binding and conformational rearrangements in the ADAR2 catalytic domain (Yi-Brunozzi et al. 2001).

13.4 Deamination in the ADAR Active Site and Product Release

Once adenosine occupies the active site, the adenine base engages the catalytic residues. For ADAR2, E396 (as glutamate) is predicted to deprotonate the water molecule bound to zinc. The resulting hydroxide then attacks the purine at C6 with protonation at N1 via the acid form of E396. These steps form a high energy intermediate with a tetrahedral center at C6. This type of mechanism is referred to as Substitution, Nucleophilic, Aromatic or S_NAr and the intermediate formed after attack of the nucleophile on the aromatic system but before departure of the leaving group is called the Meisenheimer intermediate (Pietra 1969). This intermediate is substantially higher in energy than either the reactant (adenosine) or product (inosine) because it lacks the stabilization of aromaticity. Because this unstable intermediate is likely similar in structure to the highest energy transition state on the deamination pathway, molecules that mimic the intermediate bind tightly to the deaminase active site. Indeed, several different inhibitors of nucleoside deaminases are known that mimic this structure, including the ADA inhibitor coformycin described above (Fig. 7). Since E396 was required for tight binding of ADAR2 to an RNA-containing 8-azanebularine, we believe the enzyme catalyzes the first steps of the reaction to form hydrated 8-azanebularine, a structure simulating the Meisenheimer intermediate.

For deamination to occur from the Meisenheimer intermediate, a proton must be transferred from the newly formed C6 hydroxyl and the leaving amino group. This is undoubtedly mediated again by E396 (Fig. 6c + d). Once protonated, the amino group leaves as ammonia (NH_3) in a step that also forms the product inosine. At that point inosine exits the active site, product RNA is released and the zinc binds a new water molecule to reset the active site for another turnover.

While no kinetic isotope studies have been carried out with ADARs, Schramm's KIE determination and calculation of the transition state structure for TadA suggests the rate determining step for that enzyme is proton transfer from the newly formed C6 hydroxyl and leaving ammonia (Luo and Schramm 2008). Our lab has shown using a model substrate and ADAR2 that RNA binding steps (both association and dissociation) are fast relative to the rate of deamination. In addition, the 2-AP fluorescence changes induced by ADAR2 were also fast relative to the deamination rate indicating base flipping was unlikely to be rate limiting for the substrate studied (Stephens et al. 2000). Our data with ADAR2 substrate analogs is consistent with formation of the Meisenheimer intermediate as rate determining for this enzyme. However, since ADARs deaminate adenosine in a variety of different RNAs, the rate-determining step may be context dependent (Véliz et al. 2003). For instance, in particularly stable duplex structures, base flipping may be slow and rate determining overall. More studies are necessary to determine how the rate-determining step for the ADAR reaction varies among the different known natural substrates.

In this review we are able to explain many aspects of the ADAR structure and mechanism, yet many questions still remain. The field would benefit tremendously from additional high resolution structural data on ADAR–RNA complexes, particularly involving ADAR catalytic domains, and these studies are underway worldwide. More mechanistic work needs to be done with ADAR1 and different ADAR2 substrates to define general features of these reactions and those that are either ADAR- or substrate-specific. Adaptation of the tools and approaches described here will enable us to continue to answer questions about these fascinating enzymes in the years ahead.

Acknowledgments P.A.B would like to acknowledge support from the National Institutes of Health in the form of grant R01GM061115. R.A.G is supported by a Graduate Research Fellowship from the National Science Foundation.

References

Agarwal RP, Sagar SM, Parks RE (1975) Adenosine deaminase from human erythrocytes: purification and effects of adenosine analogs. Biochem Pharmacol 24:693–701

Albert A, Katritzky AR, Boulton AJ (1976) Covalent hydration in nitrogen heterocycles. In: Advances in heterocyclic chemistry, vol 20. Academic Press, NewYork, pp 117–143

Ashley GW, Bartlett PA (1984) Purification and properties of cytidine deaminase from *Escherichia coli*. J Biol Chem 259:13615–13620

Bass BL, Weintraub H (1988) An unwinding activity that covalently modifies its double-stranded RNA substrate. Cell 55:1089–1098

Berridge MJ, Irvine RF (1989) Inositol phosphates and cell signalling. Nature 341:197–205

Betts L, Xiang S, Short SA, Wolfenden R, Carter CW (1994) Cytidine deaminase. The 2.3 A crystal structure of an enzyme: transition-state analog complex. J Mol Biol 235:635–656

Burns CM, Chu H, Rueter SM, Hutchinson LK, Canton H, Sanders-Bush E, Emeson RB (1997) Regulation of serotonin-2C receptor G-protein coupling by RNA editing. Nature 387:303–308

Carlow DC, Carter CW, Mejlhede N, Neuhard J, Wolfenden R (1999) Cytidine deaminases from *B. subtilis* and *E. coli*: compensating effects of changing zinc coordination and quaternary structure. Biochemistry 38:12258–12265

Carter CW (1995) The nucleoside deaminases for cytidine and adenosine: structure, transition state stabilization, mechanism, and evolution. Biochimie 77:92–98

Chassy BM, Suhadolnik RJ (1967) Adenosine aminohydrolase. Binding and hydrolysis of 2- and 6-substituted purine ribonucleosides and 9-substituted adenine nucleosides. J Biol Chem 242:3655–3658

Chilibeck KA, Wu T, Liang C, Schellenberg MJ, Gesner EM, Lynch JM, MacMillan AM (2006) FRET analysis of in vivo dimerization by RNA-editing enzymes. J Biol Chem 281: 16530–16535

Chung SJ, Fromme JC, Verdine GL (2005) Structure of human cytidine deaminase bound to a potent inhibitor. J Med Chem 48:658–660

Cohen RM, Wolfenden R (1971) Cytidine deaminase from *Escherichia coli*. Purification, properties and inhibition by the potential transition state analog 3, 4, 5, 6-tetrahydrouridine. J Biol Chem 246:7561–7565

Dawson TR, Sansam CL, Emeson RB (2004) Structure and sequence determinants required for the RNA editing of ADAR2 substrates. J Biol Chem 279:4941–4951

Doyle M, Jantsch MF (2002) New and old roles of the double-stranded RNA-binding domain. J Struct Biol 140:147–153

Easterwood L, Véliz E, Beal P (2000) Demethylation of 6-O-methylinosine by an RNA-editing adenosine deaminase. J Am Chem Soc 122:11537–11538

Egerer M, Satchell KJF (2010) Inositol hexakisphosphate-induced autoprocessing of large bacterial protein toxins. PLoS Pathog 6:1–8

Elias Y, Huang RH (2005) Biochemical and structural studies of A-to-I editing by tRNA:A34 deaminases at the wobble position of transfer RNA. Biochemistry 44:12057–12065

Erion M, Reddy M (1998) Calculation of relative hydration free energy differences for heteroaromatic compounds: use in the design of adenosine deaminase and cytidine deaminase inhibitors. J Am Chem Soc 120:3295–3304

Frederiksen S (1966) Specificity of adenosine deaminase toward adenosine and 2′-deoxyadenosine analogues. Arch Biochem Biophys 113:383–388

Frick L, MacNeela JP, Wolfenden R (1987) Transition state stabilization by deaminases: rates of nonenzymatic hydrolysis of adenosine and cytidine. Bioorg Chem 15:100–108

Gallo A, Keegan LP, Ring GM, O'Connell MA (2003) An ADAR that edits transcripts encoding ion channel subunits functions as a dimer. EMBO J 22:3421–3430

Gerber AP, Keller W (1999) An adenosine deaminase that generates inosine at the wobble position of tRNAs. Science 286:1146–1149

Gerber AP, Keller W (2001) RNA editing by base deamination: more enzymes, more targets, new mysteries. Trends Biochem Sci 26:376–384

Hanakahi LA, West SC (2002) Specific interaction of IP6 with human Ku70/80, the DNA-binding subunit of DNA-PK. EMBO J 21:2038–2044

Hart K, Nyström B, Ohman M, Nilsson L (2005) Molecular dynamics simulations and free energy calculations of base flipping in dsRNA. RNA 11:609–618

Haudenschild BL, Maydanovych O, Véliz EA, Macbeth MR, Bass BL, Beal PA (2004) A transition state analogue for an RNA-editing reaction. J Am Chem Soc 126:11213–11219

Herbert A, Alfken J, Kim YG, Mian IS, Nishikura K, Rich A (1997) A Z-DNA binding domain present in the human editing enzyme, double-stranded RNA adenosine deaminase. Proc Natl Acad Sci U S A 94:8421–8426

Holden LG, Prochnow C, Chang YP, Bransteitter R, Chelico L, Sen U, Stevens RC, Goodman MF, Chen XS (2008) Crystal structure of the anti-viral APOBEC3G catalytic domain and functional implications. Nature 456:121–124

Hough RF, Bass BL (1994) Purification of the *Xenopus laevis* double-stranded RNA adenosine deaminase. J Biol Chem 269:9933–9939
Hunt SW, Hoffee PA (1982) Adenosine deaminase from deoxycoformycin-sensitive and-resistant rat hepatoma cells. Purification and characterization. J Biol Chem 257:14239–14244
Ireton GC, Black ME, Stoddard BL (2003) The 1.14 A crystal structure of yeast cytosine deaminase: evolution of nucleotide salvage enzymes and implications for genetic chemotherapy. Structure 11:961–972
Jayalath P, Pokharel S, Véliz E, Beal PA (2009) Synthesis and evaluation of an RNA editing substrate bearing 2′-deoxy-2′-mercaptoadenosine. Nucleosides Nucleotides Nucleic Acids 28:78–88
Johansson E, Mejlhede N, Neuhard J, Larsen S (2002) Crystal structure of the tetrameric cytidine deaminase from *Bacillus subtilis* at 2.0 A resolution. Biochemistry 41:2563–2570
Johansson E, Neuhard J, Willemoës M, Larsen S (2004) Structural, kinetic, and mutational studies of the zinc ion environment in tetrameric cytidine deaminase. Biochemistry 43: 6020–6029
Källman AM, Sahlin M, Ohman M (2003) ADAR2 A →I editing: site selectivity and editing efficiency are separate events. Nucleic Acids Res 31:4874–4881
Kim U, Wang Y, Sanford T, Zeng Y, Nishikura K (1994) Molecular cloning of cDNA for double-stranded RNA adenosine deaminase, a candidate enzyme for nuclear RNA editing. Proc Natl Acad Sci U S A 91:11457–11461
Kim J, Malashkevich V, Roday S, Lisbin M, Schramm VL, Almo SC (2006) Structural and kinetic characterization of *Escherichia coli* TadA, the wobble-specific tRNA deaminase. Biochemistry 45:6407–6416
Koeris M, Funke L, Shrestha J, Rich A, Maas S (2005) Modulation of ADAR1 editing activity by Z-RNA in vitro. Nucleic Acids Res 33:5362–5370
Kuratani M, Ishii R, Bessho Y, Fukunaga R, Sengoku T, Shirouzu M, Sekine S-I, Yokoyama S (2005) Crystal structure of tRNA adenosine deaminase (TadA) from *Aquifex aeolicus*. J Biol Chem 280:16002–16008
Lai F, Drakas R, Nishikura K (1995) Mutagenic analysis of double-stranded RNA adenosine deaminase, a candidate enzyme for RNA editing of glutamate-gated ion channel transcripts. J Biol Chem 270:17098–17105
Lee Y-M, Lim C (2008) Physical basis of structural and catalytic Zn-binding sites in proteins. J Mol Biol 379:545–553
Lee W-H, Kim YK, Nam KH, Priyadarshi A, Lee EH, Kim EE, Jeon YH, Cheong C, Hwang KY (2007) Crystal structure of the tRNA-specific adenosine deaminase from *Streptococcus pyogenes*. Proteins 68:1016–1019
Lehmann KA, Bass BL (1999) The importance of internal loops within RNA substrates of ADAR1. J Mol Biol 291:1–13
Lehmann KA, Bass BL (2000) Double-stranded RNA adenosine deaminases ADAR1 and ADAR2 have overlapping specificities. Biochemistry 39:12875–12884
Li JB, Levanon EY, Yoon J-K, Aach J, Xie B, Leproust E, Zhang K, Gao Y, Church GM (2009) Genome-wide identification of human RNA editing sites by parallel DNA capturing and sequencing. Science 324:1210–1213
Liljas A, Kannan KK, Bergstén PC, Waara I, Fridborg K, Strandberg B, Carlbom U, Järup L, Lövgren S, Petef M (1972) Crystal structure of human carbonic anhydrase C. Nature New Biol 235:131–137
Liu Y, George CX, Patterson JB, Samuel CE (1997) Functionally distinct double-stranded RNA-binding domains associated with alternative splice site variants of the interferon-inducible double-stranded RNA-specific adenosine deaminase. J Biol Chem 272:4419–4428
Liu Y, Lei M, Samuel CE (2000) Chimeric double-stranded RNA-specific adenosine deaminase ADAR1 proteins reveal functional selectivity of double-stranded RNA-binding domains from ADAR1 and protein kinase PKR. Proc Natl Acad Sci USA 97:12541–12546
Losey HC, Ruthenburg AJ, Verdine GL (2006) Crystal structure of *Staphylococcus aureus* tRNA adenosine deaminase TadA in complex with RNA. Nat Struct Mol Biol 13:153–159

Luo M, Schramm VL (2008) Transition state structure of *E. coli* tRNA-specific adenosine deaminase. J Am Chem Soc 130:2649–2655
Luo M, Singh V, Taylor EA, Schramm VL (2007) Transition-state variation in human, bovine, and *Plasmodium falciparum* adenosine deaminases. J Am Chem Soc 129:8008–8017
Lupardus PJ, Shen A, Bogyo M, Garcia KC (2008) Small molecule-induced allosteric activation of the Vibrio cholerae RTX cysteine protease domain. Science 322:265–268
Macbeth MR, Bass BL (2007) Large-scale overexpression and purification of ADARs from *Saccharomyces cerevisiae* for biophysical and biochemical studies. Methods Enzymol 424:319–331
Macbeth MR, Lingam AT, Bass BL (2004) Evidence for auto-inhibition by the N terminus of hADAR2 and activation by dsRNA binding. RNA 10:1563–1571
Macbeth MR, Schubert HL, Vandemark AP, Lingam AT, Hill CP, Bass BL (2005) Inositol hexakisphosphate is bound in the ADAR2 core and required for RNA editing. Science 309:1534–1539
Marquez VE, Schroeder GK, Ludek OR, Siddiqui MA, Ezzitouni A, Wolfenden R (2009) Contrasting behavior of conformationally locked carbocyclic nucleosides of adenosine and cytidine as substrates for deaminases. Nucleosides Nucleotides Nucleic Acids 28:614–632
Maydanovych O, Beal PA (2006) C6-substituted analogues of 8-azanebularine: probes of an RNA-editing enzyme active site. Org Lett 8:3753–3756
Melcher T, Maas S, Herb A, Sprengel R, Seeburg PH, Higuchi M (1996) A mammalian RNA editing enzyme. Nature 379:460–464
Mohamedali KA, Kurz LC, Rudolph FB (1996) Site-directed mutagenesis of active site glutamate-217 in mouse adenosine deaminase. Biochemistry 35:1672–1680
Navaratnam N, Sarwar R (2006) An overview of cytidine deaminases. Int J Hematol 83:195–200
Nishikura K, Yoo C, Kim U, Murray JM, Estes PA, Cash FE, Liebhaber SA (1991) Substrate specificity of the dsRNA unwinding/modifying activity. EMBO J 10:3523–3532
Ohman M, Källman AM, Bass BL (2000) In vitro analysis of the binding of ADAR2 to the pre-mRNA encoding the GluR-B R/G site. RNA 6:687–697
Pietra F (1969) Mechanisms for nucleophilic and photonucleophilic aromatic substitution reactions. Q Rev, Chem Soc 23(4):504–521
Pokharel S, Beal PA (2006) High-throughput screening for functional adenosine to inosine RNA editing systems. ACS Chem Biol 1:761–765
Pokharel S, Jayalath P, Maydanovych O, Goodman RA, Wang SC, Tantillo DJ, Beal PA (2009) Matching active site and substrate structures for an RNA editing reaction. J Am Chem Soc 131:11882–11891
Polson AG, Bass BL (1994) Preferential selection of adenosines for modification by double-stranded RNA adenosine deaminase. EMBO J 13:5701–5711
Polson AG, Crain PF, Pomerantz SC, McCloskey JA, Bass BL (1991) The mechanism of adenosine to inosine conversion by the double-stranded RNA unwinding/modifying activity: a high-performance liquid chromatography-mass spectrometry analysis. Biochemistry 30:11507–11514
Poulsen H, Jorgensen R, Heding A, Nielsen FC, Bonven B, Egebjerg J (2006) Dimerization of ADAR2 is mediated by the double-stranded RNA binding domain. RNA 12:1350–1360
Prochazkova K, Satchell KJF (2008) Structure–function analysis of inositol hexakisphosphate-induced autoprocessing of the *Vibrio cholerae* multifunctional autoprocessing RTX toxin. J Biol Chem 283:23656–23664
Prochnow C, Bransteitter R, Klein MG, Goodman MF, Chen XS (2007) The APOBEC-2 crystal structure and functional implications for the deaminase AID. Nature 445:447–451
Pruitt RN, Chagot B, Cover M, Chazin WJ, Spiller B, Lacy DB (2009) Structure–function analysis of inositol hexakisphosphate-induced autoprocessing in *Clostridium difficile* toxin A. J Biol Chem 284:21934–21940
Raboy V (1997) Accumulation and storage of phosphate and minerals. In: Larkins BA, Vasil IK (eds) Cellular and molecular biology of plant seed development. Kluwer, Dordrecht, pp 441–477

Reineke J, Tenzer S, Rupnik M, Koschinski A, Hasselmayer O, Schrattenholz A, Schild H, von Eichel-Streiber C (2007) Autocatalytic cleavage of *Clostridium difficile* toxin B. Nature 446:415–419

Saccomanno L, Bass BL (1994) The cytoplasm of *Xenopus oocytes* contains a factor that protects double-stranded RNA from adenosine-to-inosine modification. Mol Cell Biol 14:5425–5432

Sazanov LA, Hinchliffe P (2006) Structure of the hydrophilic domain of respiratory complex I from *Thermus thermophilus*. Science 311:1430–1436

Schirle NT, Goodman RA, Krishnamurthy M, Beal PA (2010) Selective inhibition of ADAR2-catalyzed editing of the serotonin 2c receptor pre-mRNA by a helix-threading peptide. Org Biomol Chem 8(21):4898–4904

Schramm VL, Baker DC (1985) Spontaneous epimerization of (S)-deoxycoformycin and interaction of (R)-deoxycoformycin (S)-deoxycoformycin, and 8-ketodeoxycoformycin with adenosine deaminase. Biochemistry 24:641–646

Seeds AM, Sandquist JC, Spana EP, York JD (2004) A molecular basis for inositol polyphosphate synthesis in *Drosophila melanogaster*. J Biol Chem 279:47222–47232

Seela F, Xu K (2007) Pyrazolo[3, 4-d]pyrimidine ribonucleosides related to 2-aminoadenosine and isoguanosine: synthesis, deamination and tautomerism. Org Biomol Chem 5:3034–3045

Sharff AJ, Wilson DK, Chang Z, Quiocho FA (1992) Refined 2.5 A structure of murine adenosine deaminase at pH 6.0. J Mol Biol 226:917–921

Sideraki V, Mohamedali KA, Wilson DK, Chang Z, Kellems RE, Quiocho FA, Rudolph FB (1996) Probing the functional role of two conserved active site aspartates in mouse adenosine deaminase. Biochemistry 35:7862–7872

Snider MJ, Reinhardt L, Wolfenden R, Cleland WW (2002) 15 N kinetic isotope effects on uncatalyzed and enzymatic deamination of cytidine. Biochemistry 41:415–421

Stefl R, Oberstrass FC, Hood JL, Jourdan M, Zimmermann M, Skrisovska L, Maris C, Peng L, Hofr C, Emeson RB, Allain FH-T (2010) The solution structure of the ADAR2 dsRBM-RNA complex reveals a sequence-specific readout of the minor groove. Cell 143:225–237

Stephens OM, Yi-Brunozzi HY, Beal PA (2000) Analysis of the RNA-editing reaction of ADAR2 with structural and fluorescent analogues of the GluR-B R/G editing site. Biochemistry 39:12243–12251

Stephens OM, Haudenschild BL, Beal PA (2004) The binding selectivity of ADAR2's dsRBMs contributes to RNA-editing selectivity. Chem Biol 11:1239–1250

Sun F, Huo X, Zhai Y, Wang A, Xu J, Su D, Bartlam M, Rao Z (2005) Crystal structure of mitochondrial respiratory membrane protein complex II. Cell 121:1043–1057

Teh A-H, Kimura M, Yamamoto M, Tanaka N, Yamaguchi I, Kumasaka T (2006) The 1.48 A resolution crystal structure of the homotetrameric cytidine deaminase from mouse. Biochemistry 45:7825–7833

Tyler PC, Taylor EA, Fröhlich RFG, Schramm VL (2007) Synthesis of 5′-methylthio coformycins: specific inhibitors for malarial adenosine deaminase. J Am Chem Soc 129:6872–6879

Valente L, Nishikura K (2007) RNA binding-independent dimerization of adenosine deaminases acting on RNA and dominant negative effects of nonfunctional subunits on dimer functions. J Biol Chem 282:16054–16061

Véliz EA, Easterwood LM, Beal PA (2003) Substrate analogues for an RNA-editing adenosine deaminase: mechanistic investigation and inhibitor design. J Am Chem Soc 125:10867–10876

Verbsky JW, Chang S-C, Wilson MP, Mochizuki Y, Majerus PW (2005) The pathway for the production of inositol hexakisphosphate in human cells. J Biol Chem 280:1911–1920

Wagner RW, Smith JE, Cooperman BS, Nishikura K (1989) A double-stranded RNA unwinding activity introduces structural alterations by means of adenosine to inosine conversions in mammalian cells and *Xenopus* eggs. Proc Natl Acad Sci U S A 86:2647–2651

Ward DC, Reich E, Stryer L (1969) Fluorescence studies of nucleotides and polynucleotides. I. Formycin, 2-aminopurine riboside, 2, 6-diaminopurine riboside, and their derivatives. J Biol Chem 244:1228–1237

Weirich CS, Erzberger JP, Flick JS, Berger JM, Thorner J, Weis K (2006) Activation of the DExD/H-box protein Dbp5 by the nuclear-pore protein Gle1 and its coactivator InsP6 is required for mRNA export. Nat Cell Biol 8:668–676

West R, Powell D, Wheatley LS, Lee MKT, Schleyer PvR (1962) The relative strengths of alkyl halides as proton acceptor groups in hydrogen bonding. J Am Chem Soc 84(16):3221–3222

Wilson DK, Rudolph FB, Quiocho FA (1991) Atomic structure of adenosine deaminase complexed with a transition-state analog: understanding catalysis and immunodeficiency mutations. Science 252:1278–1284

Wolf J, Gerber AP, Keller W (2002) tadA, an essential tRNA-specific adenosine deaminase from *Escherichia coli*. EMBO J 21:3841–3851

Wolfenden R, Kati W (1991) Testing the limits of protein-ligand binding discrimination with transition-state analogue inhibitors. Acc Chem Res 24:209–215

Wong SK, Sato S, Lazinski DW (2001) Substrate recognition by ADAR1 and ADAR2. RNA 7:846–858

Xie W, Liu X, Huang RH (2003) Chemical trapping and crystal structure of a catalytic tRNA guanine transglycosylase covalent intermediate. Nat Struct Biol 10:781–788

Xu M, Wells KS, Emeson RB (2006) Substrate-dependent contribution of double-stranded RNA-binding motifs to ADAR2 function. Mol Biol Cell 17:3211–3220

Yang JH, Sklar P, Axel R, Maniatis T (1997) Purification and characterization of a human RNA adenosine deaminase for glutamate receptor B pre-mRNA editing. Proc Natl Acad Sci U S A 94:4354–4359

Yeo J, Goodman RA, Schirle NT, David SS, Beal PA (2010) RNA editing changes the lesion specificity for the DNA repair enzyme NEIL1. Proc Natl Acad Sci U S A 107:20715–20719

Yi-Brunozzi HY, Easterwood LM, Kamilar GM, Beal PA (1999) Synthetic substrate analogs for the RNA-editing adenosine deaminase ADAR-2. Nucleic Acids Res 27:2912–2917

Yi-Brunozzi HY, Stephens OM, Beal PA (2001) Conformational changes that occur during an RNA-editing adenosine deamination reaction. J Biol Chem 276:37827–37833

York JD, Odom AR, Murphy R, Ives EB, Wente SR (1999) A phospholipase C-dependent inositol polyphosphate kinase pathway required for efficient messenger RNA export. Science 285:96–100

Zhou P, Tian F, Lv F, Shang Z (2009) Geometric characteristics of hydrogen bonds involving sulfur atoms in proteins. Proteins 76:151–163

ADAR Proteins: Double-stranded RNA and Z-DNA Binding Domains

Pierre Barraud and Frédéric H.-T. Allain

Abstract Adenosine deaminases acting on RNA (ADAR) catalyze adenosine to inosine editing within double-stranded RNA (dsRNA) substrates. Inosine is read as a guanine by most cellular processes and therefore these changes create codons for a different amino acid, stop codons or even a new splice-site allowing protein diversity generated from a single gene. We review here the current structural and molecular knowledge on RNA editing by the ADAR family of protein. We focus especially on two types of nucleic acid binding domains present in ADARs, namely the dsRNA and Z-DNA binding domains.

Contents

P. Barraud · F. H.-T. Allain (✉)
Institute of Molecular Biology and Biophysics,
ETH Zurich 8093, Zürich, Switzerland
e-mail: allain@mol.biol.ethz.ch

Current Topics in Microbiology and Immunology (2012) 353: 35–60
DOI: 10.1007/82_2011_145

Published Online: 5 July 2011

1 Introduction

The published sequence of human, mouse, and rat genomes (Venter et al. 2001; Baltimore 2001) revealed a surprisingly small number of genes, estimated to be around 26,000. Such a small number cannot fully account for the expected molecular complexity of these species and it is now well appreciated that such a complexity is likely to come from the multitude of protein variants created by alternative-splicing and editing of pre-mRNA (Graveley 2001; Pullirsch and Jantsch 2010). For example, the sole *paralytic* gene (a *Drosophila* sodium channel) can generate up to one million mRNA isoforms by combining its 13 alternative exons and its 11 known RNA editing sites (Hanrahan et al. 2000). Moreover, alternatively spliced and edited mRNAs are particularly abundant in the neurons. The finely regulated population of the different isoforms of most neurotransmitter receptors, ion channels, neuronal cell-surface receptors, and adhesion molecules ensure proper brain function. Any imbalance of the gene expression can impair neurological functions and lead to severe diseases such as brain cancer, schizophrenia or neuromuscular, and neurodegenerative syndromes (Maas et al. 2006).

RNA editing is a posttranscriptional modification of pre-mRNA (Gott and Emeson 2000). Editing occurs via insertion or deletion of poly-U sequence (seen in Trypanosome mitochondria (Benne et al. 1986)), or via a single base conversion by deamination, cytidine to uridine (C → U) or adenosine to inosine (A → I) (seen from protozoa to man) (Gott and Emeson 2000). These changes can create a codon for a different amino acid, a stop codon, or even a new splice-site allowing protein diversity to be created from a single gene (Gott and Emeson 2000; Keegan et al. 2001; Bass 2002). A → I editing occurs by hydrolytic deamination of the adenine base (Fig. 1a). Because inosine base-pairs with cytidine (Fig. 1b), inosine is read as a guanine by most cellular processes. RNA editing by adenosine deamination is catalyzed by members of an enzyme family known as adenosine deaminases that act on RNA (ADAR) (Bass et al. 1997).

We review here the current structural and molecular knowledge of RNA editing by the ADAR family of proteins. More comprehensive reviews on ADAR functions are available elsewhere (Gott and Emeson 2000; Keegan et al. 2001; Bass 2002; Wulff and Nishikura 2010; Nishikura 2010). We focus here on the structures of RNA substrates and how these structures are recognized by the double-stranded

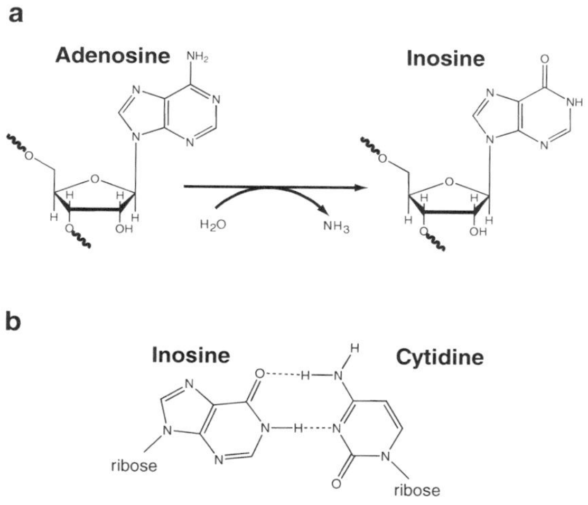

Fig. 1 Deamination of adenosine to inosine by ADAR. **a** Hydrolytic deamination converts adenosine into inosine. **b** Inosine base-pairs with cytidine and is thus read as a guanine by most cellular processes

RNA binding domains (dsRBDs also refer to as dsRBMs for double-stranded RNA binding motifs) present in the ADAR family of protein. We also review the current structural knowledge of another type of nucleic acid binding domain present in ADARs, namely the Z-DNA binding domains.

2 Adenosine Deaminases Acting on RNA Family Members and Their Domain Organization

Adenosine deaminases acting on RNA proteins were first discovered in *Xenopus laevis* (Rebagliati and Melton 1987; Bass and Weintraub 1987, 1988) and have now been characterized in nearly all metazoa from worm to man (Tonkin et al. 2002; Palladino et al. 2000; Slavov et al. 2000; Herbert et al. 1995; Melcher et al. 1996b; O'Connell et al. 1995; Kim et al. 1994; Palavicini et al. 2009), but not in plants, yeast, or fungi. In vertebrates, two functional enzymes (ADAR1 and ADAR2) and one inactive enzyme (ADAR3 (Melcher et al. 1996a; Chen et al. 2000)) have been characterized. ADAR3 most likely originated from ADAR2 to which it is most similar in sequence and domain organization (Fig. 2a). In *Caenorhabditis elegans*, two active ADARs (CeADAR1 and CeADAR2) have been found whereas in *Drosophila melanogaster*, a single ADAR2-like protein (dADAR) was found (Fig. 2a).

ADARs from all organisms have a common modular domain organization that includes from one to three copies of a dsRNA binding domain (dsRBD) in their N-terminal region followed by a C-terminal adenosine deaminase catalytic domain (Fig. 2a). For detailed information regarding the structure and the catalytic activity of the C-terminal domain, please refer to the chapter by Beal and coworkers.

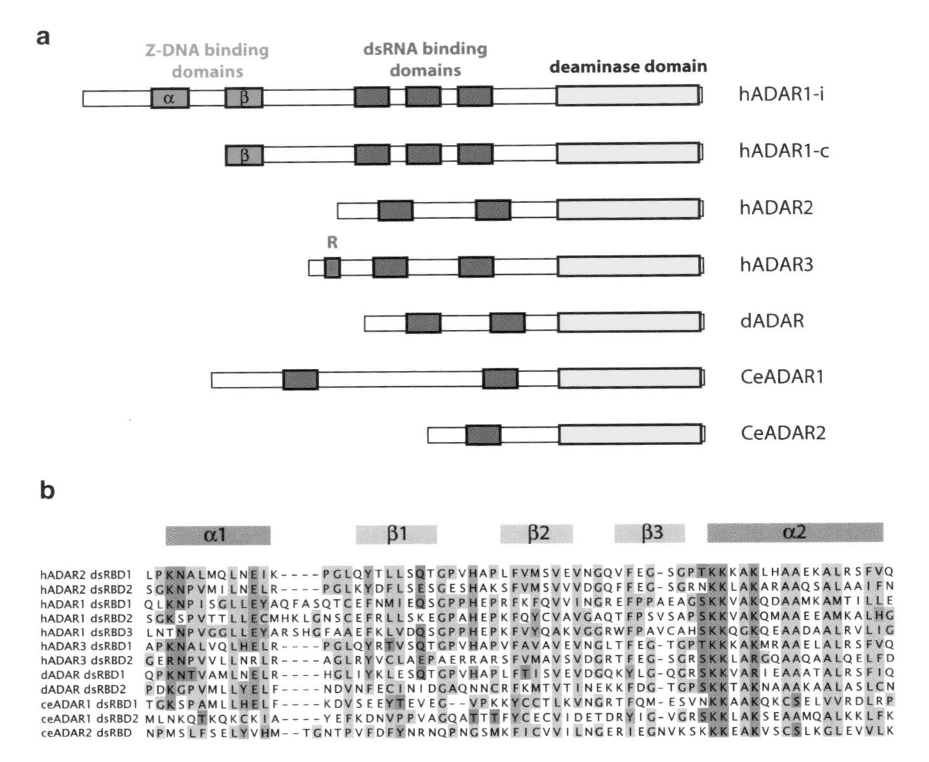

Fig. 2 Domain organization of the ADAR family members. **a** The ADAR family members are represented with their domain structure organization. Three ADARs are found in vertebrates (ADAR1-3). One ADAR is found in *D. melanogaster* (dADAR) and two in *C. elegans* (CeADAR1-2). ADARs have a conserved C-terminal deaminase domain (in *yellow*) and diverse numbers of dsRNA binding domains (in *blue*). In addition ADAR1 has one or two copies of Z-DNA binding domains (in *green*). The long isoform of ADAR1 is interferon-inducible (ADAR1-i), whereas the short isoform is constitutively expressed (ADAR1-c). ADAR3 has an arginine-rich R-domain (in *red*). **b** Sequence alignment of dsRBDs from the ADAR family members. The alignment is coloured by amino acid conservation and properties. hADAR2 dsRBD1 secondary structure elements are shown on top of the alignment

In addition to this common feature, ADAR1 exhibit Z-DNA binding domains in its most N-terminal part, Zα and Zβ (Herbert et al. 1997). This renders it unique among the members of ADAR protein family (Fig. 2a). Actually, ADAR1 is expressed in two isoforms: the interferon-inducible ADAR1-i (inducible; 150 kDa) and the constitutively expressed ADAR1-c (constitutive; 110 kDa) which is initiated from a downstream methionine as the result of alternative-splicing and skipping of the exon containing the upstream methionine (Patterson and Samuel 1995; Patterson et al. 1995; Kawakubo and Samuel 2000). As a consequence, the short version of ADAR1 lacks the N-terminal Z-DNA binding domain (Fig. 2a). It is important to note that only Zα but not Zβ has the ability to bind Z-DNA, the left-handed form of DNA (Athanasiadis et al. 2005).

ADAR1 and ADAR2 are expressed in humans in most tissues and function as homodimers (Cho et al. 2003). In contrast, ADAR3 expresses only in the central nervous system and does not dimerize (Chen et al. 2000) which could explain its inactivity. Moreover, ADAR3 acts as a repressor of ADAR1 and ADAR2 activity, most probably by sequestering their potential substrates without editing them (Chen et al. 2000). ADAR3 contains also an arginine-rich RNA binding domain (R-domain) in its N-terminal region. It has been shown to be responsible for the binding of ADAR3 to single-stranded RNA (Chen et al. 2000). However, there is no structure of this domain in complex with ssRNA that would reveal the molecular basis of RNA recognition. Interestingly, a recent study showed that an R-domain is also present in a minor splicing variant of ADAR2 (Maas and Gommans 2009).

After the presentation of ADARs editing substrates, the structure and function of the Z-DNA binding domains and the dsRNA binding domains of ADAR will be described in the remaining sections.

3 RNA Editing Substrate

3.1 Specificity of Editing

Adenosine deaminases that act on RNA (ADARs) convert adenosine into inosine in cellular and viral RNA transcripts containing either perfect or imperfect regions of double-stranded RNA (dsRNA) (Gott and Emeson 2000; Bass 2002; Nishikura 2010). A → I modification is nonspecific within perfect dsRNA substrates, deaminating up to 50% of the adenosine residues (Polson and Bass 1994; Nishikura et al. 1991). The nonspecific reaction occurs as long as the double-stranded architecture of the RNA substrate is maintained since ADARs unwind dsRNA by changing A–U base-pairs to I–U mismatches (Bass and Weintraub 1988; Wagner et al. 1989). Such modifications can modulate gene silencing triggered by intramolecular structures in mRNA (Tonkin and Bass 2003), nuclear retention of RNA transcripts (Zhang and Carmichael 2001), or antiviral responses by extensive modification of viral transcripts (Wong et al. 1991). The majority of nonselective editing occurs in untranslated regions (UTRs) and introns where large regular duplexes are formed between inverted repeats of Alu and LINE (Long Interspersed Nucleotides Element in primates) or SINE domains (Small Interspersed Nucleotides Elements found in mouse) (Levanon et al. 2004; Athanasiadis et al. 2004; Osenberg et al. 2010). It is estimated that this constitutes about 15,000 editing events in about 2,000 human genes. The biological function of this major A → I editing event is not fully understood yet (Hundley and Bass 2010).

A → I editing can also be highly specific within imperfect dsRNA regions in modifying a single or limited set of adenosine residues (Gott and Emeson 2000; Bass 2002). Selective editing within pre-mRNAs has been shown to affect the

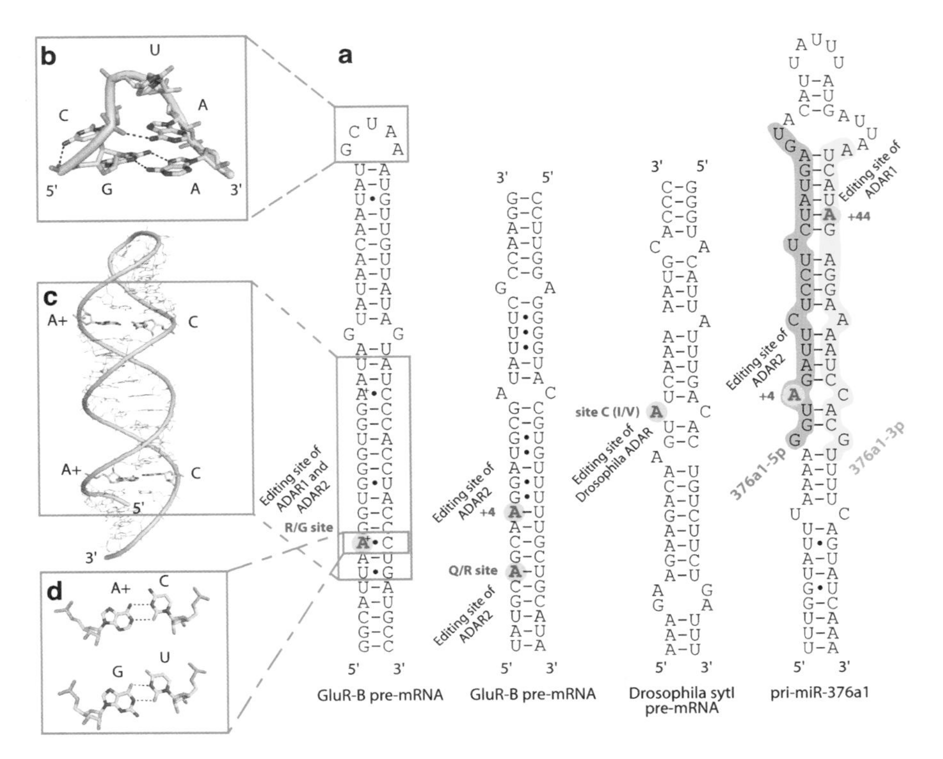

Fig. 3 Structures of various ADAR editing substrates. **a** Secondary structure of ADAR editing substrates: GluR-B R/G and Q/R sites, Drosophila sytI I/V site and pri-miR-376a1. **b** Structure of the GluR-B GCUAA apical pentaloop (PDB code 1YSV). **c** Structure of the RNA helix surrounding the GluR-B R/G site revealing two particular A⁺•C wobble base-pairs (PDB code 2L2 J). **d** Hydrogen bond pattern in an A⁺•C wobble base-pair and comparison with a G•U wobble base-pair

primary amino acid sequence of the resultant protein therefore producing multiple isoforms from a single gene. For example, editing by ADARs produced functionally important isoforms of numerous proteins involved in synaptic neurotransmission, including ligand and voltage-gated ion channels and G-protein coupled receptors. The pre-mRNA encoding the B-subunit of the α-amino-3-hydroxy-5-methyl-4-isoxazole propionic acid (AMPA) subtype of glutamate receptor (GluR-B) is probably the most extensively studied mRNA editing substrate (Seeburg et al. 1998). It is edited at multiple sites and one of these locations is the R/G site, where a genomically encoded AGA is modified to IGA, resulting in an arginine-to-glycine change (the ribosome interprets I as G due to its similar base-pairing properties—Fig. 1b). The R/G site of the GluR-B pre-mRNA is often used as a model system for A → I editing studies as it forms a small and well conserved 70 nucleotide stem-loop containing three mismatches (Aruscavage and Bass 2000), referred to as the R/G stem-loop (Fig. 3).

More recently, specific editing of many pri-miRNAs, pre-miRNA, and miRNAs have been discovered suggesting a cross talk between the RNA editing and RNA interference machineries (Nishikura 2006; Ohman 2007). MicroRNA editing can regulate miRNA expression by affecting pri-microRNA and pre-miRNA processing (Kawahara et al. 2008, 2007a; Heale et al. 2009). MiRNA editing can also affect gene targeting when the seed sequence of the miRNA is edited. This later editing event allows an extension of the number of genes targeted by the miRNAs (Kawahara et al. 2007b). Examples of editing site in miRNA are shown in Fig. 3. For comprehensive information about the modulation of micro-RNA function by ADAR please refer to the chapter by Nishikura and coworkers.

3.2 What Makes a Good Editing Site?

What characterizes a specific A $\rightarrow$ I RNA editing site is a major and long-standing question in the field. It is clear that the targeted adenosine has to be embedded in an RNA stem, and that both the sequence around the adenine and the secondary structure elements present in the RNA stem will have a major impact on the efficiency and the selectivity of editing. The terms *preferences* and *selectivity* are used to describe the properties that enable ADAR proteins to modify a specific adenosine among others (Polson and Bass 1994).

3.2.1 Preferences

Although ADAR dsRBDs are thought to bind unspecifically to any dsRNA, ADARs have small sequence preferences for deaminating particular adenosines among others. Detailed inspection of the editing ability of ADAR1 and ADAR2 on the GluR-B R/G and Q/R sites revealed that these enzymes have overlapping but distinct preferences (Lai et al. 1997; Melcher et al. 1996b; Gerber et al. 1997; Maas et al. 1996). Xenopus and human ADAR1 have a similar preference for $A = U > C > G$ at the 5' of the edited adenosine (Polson and Bass 1994; Lehmann and Bass 2000). Human ADAR2 has also a similar but distinct preference for the 5′ neighbor of the edited adenosine ($U \approx A > C = G$) (Lehmann and Bass 2000). In addition, human ADAR2 has also a 3′ neighbor preference ($G = U > C = A$) (Lehmann and Bass 2000). These initial preference rules were further confirmed and optimized in subsequently discovered targets (Kawahara et al. 2008; Riedmann et al. 2008; Li et al. 2009; Wulff et al. 2011). Chimeric ADARs containing the dsRBDs of ADAR2 and the catalytic domain of ADAR1 and vice versa suggested that the nearest-neighbour preferences come from the deaminase domain (Wong et al. 2001) but recent structures suggest that dsRBDs could also play a role (Stefl et al. 2010). The nucleotide base-pairing with the target adenosine can also drastically influence editing with a preference for a cytidine (forming a AC mismatch which is then converted into a matching I–C pair, like in the GluR-B R/G site, Fig. 3) (Levanon et al. 2004; Athanasiadis et al. 2004; Riedmann et al. 2008; Blow et al. 2004; Wong et al. 2001) over a uridine (like in the GluR-B Q/R site, Fig. 3). Purines are not favoured and a guanosine in some case can severely impair editing (Wong et al. 2001; Kallman et al. 2003; Ohlson et al. 2007). This discrimination between various pairing partners is also determined by the catalytic domain rather than the dsRBDs (Wong et al. 2001).

3.2.2 Selectivity

Obviously, the slight preferences for the identity of neighbouring nucleotides cannot explain the acute specificity observed in some ADAR substrates, like in the GluR-B R/G or Q/R sites, where adenines in good sequence context (as defined by 5′ and 3′ neighbour and pairing partner *preferences*) remain not edited. The property of having adenines in good sequence context that remain not edited defines the concept of *selectivity*. Ultimately, this can result in having a few and even a single edited adenine in an entire dsRNA structure, which one describes as *specificity*. One can easily note that sites of highly specific editing events are never long and perfectly base-paired dsRNA (Fig. 3). The presence of secondary structure elements like terminal loops, internal loops, bulges, and mismatches is very frequent in such substrates. These secondary structured elements are highly conserved during evolution (Aruscavage and Bass 2000; Dawson et al. 2004; Reenan 2005) indicating that the RNA structure is important for the specificity of editing (Aruscavage and Bass 2000; Dawson et al. 2004; Reenan 2005; Ohman et al. 2000; Lehmann and Bass 1999). For example, the presence of internal loops has been shown to increase the selectivity of editing by uncoupling and decreasing the effective length of individual helices which then reduces to a minimum the many ways of binding of ADAR to these substrates (Ohman et al. 2000). However, RNA sequences around highly specific editing sites are also particularly conserved (Aruscavage and Bass 2000; Niswender et al. 1998), and this cannot be explained if only secondary structured elements would define the selectivity of editing. Thus, both the structure and sequence of the RNA editing site determine the selectivity of editing by ADAR. In contrast to their *preferences*, ADARs *selectivity* comes most probably from the binding selectivity of their dsRBDs.

3.3 Structures of Editing Substrates

Structural information about A $\rightarrow$ I RNA editing substrates has been limited so far to the GluR-B R/G site. The GluR-B R/G site is embedded within a 71 nt RNA stem-loop containing three base-pair mismatches and capped with a GCUAA pentaloop. The solution structure of the long human R/G stem-loop has been determined in two fragments by solution NMR.

In the first structure, the apical part of the stem-loop containing the GCU(A/C)A pentaloop has revealed a rigid pentaloop fold, novel for this time (Stefl and Allain 2005). The fold is stabilized by a complex interplay of hydrogen-bonds and stacking interactions (Fig. 3b). The structure of the GCUAA pentaloop explains well the phylogenetic conservation of GCUMA (where M is A/C) (Aruscavage and Bass 2000). The UNCG tetraloops (Cheong et al. 1990; Allain and Varani 1995; Ennifar et al. 2000) and the GCUAA pentaloop are structurally similar. This is particularly interesting considering that the pre-mRNA encoding the R/G site of subunit C of the glutamate receptor that is also specifically edited by ADAR2 has a UCCR tetraloop (Aruscavage and Bass 2000). When the size of the GCUAA pentaloop is changed or the loop is deleted, the level of editing is reduced (Stefl et al. 2006) indicating that this structural element plays an important role in the recognition processes of ADAR2. The role of the loop was subsequently confirmed by using a high throughput method (Pokharel and Beal 2006).

In the second structure, the RNA helix surrounding the editing site that contains two A–C mismatches has revealed an unexpected regular A-form helix (Fig. 3c) (Stefl et al. 2010). Indeed, adenine C2 chemical shifts (a sensitive probe to monitor the protonation of adenine N1) have shown that these two adenines involved in A–C mismatches were protonated at pH below 7.0 and were thus forming a so-called A^{+}•C wobble base-pair similar in its hydrogen bonding pattern to a G•U wobble base-pair (Fig. 3d). Thus A^{+}•C wobble base-pairs generate only little deformation of the helical properties of the stem.

Overall, these structures together with the chemical-shift analysis of the 71 nt R/G site (Stefl et al. 2006, 2010) revealed a rigid RNA stem-loop throughout the sequence. Indeed, the terminal loop is structured (Stefl and Allain 2005) and the three mismatches in the stem (two AC and one GG; Fig. 3) are forming non Watson–Crick pairs leading to rigid and rather regular RNA helix (Stefl et al. 2006, 2010). This is particularly interesting in the context of the selectivity of editing, since it is largely believed that mismatches contribute to the ADAR selectivity. Even if a A^{+}•C wobble base-pair might be more deformable than a regular A–U pair, it seems unlikely that the sole shape recognition of such mismatches could allow the editing of the R/G site to be selective.

4 Adenosine Deaminases Acting on RNA Z-DNA Binding Domains

In the late 1970s, the first atomic resolution structure of DNA, solved by X-ray crystallography was surprisingly different from the expected right-handed B-form helix (Wang et al. 1979). This structure indeed showed a left-handed double helix in which the bases alternate in *anti-* and *syn*-conformations along one strand. As a consequence, there was a zigzag arrangement of the backbone of the molecule. This property gave its name to the Z-DNA conformation observed in this crystal (Wang et al. 1979). For 15 years, many people felt that Z-DNA was a non-functional conformation of DNA, and as a consequence its study rapidly declined (Rich and Zhang 2003). However, the discovery in 1995 of a protein binding specifically and tightly to Z-DNA (Herbert et al. 1995), and two years later the isolation of a small domain responsible for this activity (Herbert et al. 1997), brought back Z-DNA into the limelight. The first Z-DNA binding property was discovered in the vertebrate ADAR1 protein. And so far, even if other proteins have been shown to bind Z-DNA, ADAR1 stays the best-characterized member of the Z-DNA binding protein family. ADAR1 has two related Z-DNA binding domains named Zα and Zβ (Fig. 2), the latter one having no binding capacity for Z-DNA. In this section we review the structural knowledge on ADAR1 Z-DNA binding domains and especially on the binding of Zα to Z-DNA.

4.1 Z-DNA Binding Domain: Structure and Substrate Recognition

The first structural information about Z-DNA binding domains and the molecular basis of the recognition of Z-DNA was revealed together by the crystal structure of the human Zα domain of ADAR1 complexed to DNA (Schwartz et al. 1999b). The Zα domain has a compact α/β fold containing a three-helix bundle (α1 to α3) flanked on one side by a twisted antiparallel β-sheet (β1 to β3) with a $\alpha\beta\alpha\alpha\beta\beta$ topology (Fig. 4). This arrangement of three α-helices and β-strands is known as the helix-turn-helix β-sheet fold (or α+βHTH fold). This fold differs from the related helix-turn-helix (HTH) fold in the fact that it has an additional C-terminal β-sheet packed against the core formed by the α-helices. The interaction of HTH proteins with right-handed B-DNA has been well characterized (Harrison and Aggarwal 1990). The mode of interaction of Zα with Z-DNA has noticeable differences arising from the differences between Z- and B-DNA (Schwartz et al. 1999b). The bound DNA duplex shows a left-handed helix structure with the typical zigzag backbone conformation of the Z-DNA (Fig. 4a). The Zα domain makes contact to only one strand of the DNA molecule in a single continuous recognition surface formed by helix α3 and the C-terminal β-hairpin (Fig. 4a, c). Interestingly, this surface is complementary to the DNA backbone in terms of shape and electrostatic properties. Indeed, all the polar interactions involve direct or water-mediated contacts to the sugar-phosphate backbone of the DNA. This implies that the interaction between Zα and Z-DNA is conformation or shape specific rather than sequence specific. This was further confirmed by recent structures of Zα bound to other DNA molecules of various sequences (Ha et al. 2009). These structures showed almost identical structures regardless of the sequence, confirming that the mode of binding of Zα to Z-DNA is a well conserved shape-specific mode of binding.

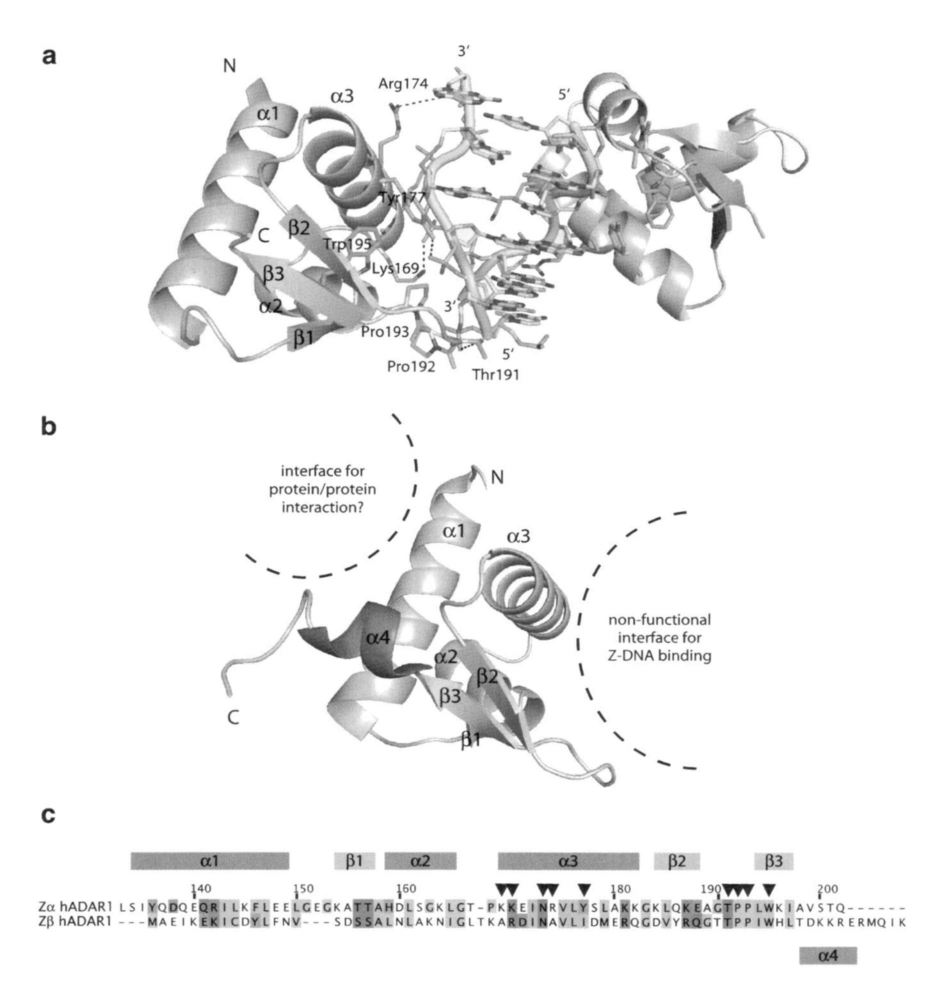

Fig. 4 Structures of ADAR Z-DNA binding domains. **a** Structure of the Zα domain of ADAR1 in complex with Z-DNA $(CG)_3$ showing contacts with the phosphate backbone via helix $\alpha1$ and beta hairpin $\beta2$-$\beta3$. **b** Structure of the Zβ domain of ADAR1 in its free state with a non-functional Z-DNA binding surface and a potential protein/protein interaction surface. The additional helix $\alpha4$ is shown in *red*. **c** Sequence alignment of hADAR1 Zα and Zβ domains. The alignment is coloured by amino acid conservation and properties. Common secondary structure elements are shown on top of the alignment. The position of the additional helix $\alpha4$ is shown below. Residues of Zα involved in direct- or water-mediated contacts with Z-DNA are reported with *black arrows*. Some of these residues are not conserved in Zβ. Residue numbers correspond to the one of Zα

4.2 A Role for ADAR1 Zα Domain?

Although the biological role of Z-DNA binding by Zα has not been clearly defined yet, one possible function might be to direct ADAR1 at actively transcribing genes. Indeed, Z-DNA is stabilized by negative supercoiling (Peck et al. 1982) which is formed transiently upstream of an active RNA polymerase (Liu and Wang 1987). Localizing ADAR1 to the site of transcription would then allow it to efficiently act upon the RNA prior to splicing.

A point particularly interesting in the context of RNA editing is that Zα has also the property to bind to Z-RNA (Brown et al. 2000). These types of domains are thus also referred to as Z-DNA/Z-RNA binding domains. The structure of Zα bound to Z-RNA has revealed a well conserved mode of interaction between Zα and the nucleic acid backbone of Z-DNA and Z-RNA (Placido et al. 2007). This property of Zα might thus help to target ADAR1 to specific sites that are prone to form Z-RNA. For example, the formation of Z-RNA is favoured by alternative purine-pyrimidine sequences, and especially guanosine and cytosine repeats (Herbert and Rich 1999). Interestingly, it was shown that A → I editing by ADAR1-i is substantially increased in a dsRNA substrate containing such Z-forming purine-pyrimidine repeats (Koeris et al. 2005). Moreover, such influence of Z-forming sequences on the level of A → I editing was not observed in the case of ADAR2 which does not contain Z-DNA/Z-RNA binding domain. This suggests a direct role of the Z-DNA/Z-RNA binding domain of ADAR1-i in the enhanced editing activity towards dsRNA with Z-forming sequences. Moreover, editing reactions conducted with ADAR1-i under short incubation time showed a clear positive correlation between the proximity of the edited adenosine to the Z-forming sequence and the number of editing events at those sites (Koeris et al. 2005). So, in contrast to Z-DNA binding, Z-RNA binding by the Zα domain of ADAR1-i has a clear biological role in the context of RNA editing, which is to target ADAR1-i to Z-forming sequences within dsRNA substrates. Such Z-forming RNA structure can be directly encoded in the RNA substrate, but Z-RNA can also be formed in the trail of transcription of RNA viruses, and would allow ADAR1 to more efficiently modify these RNA viruses. Indeed, ADAR1-i has been associated with RNA editing of a wide array of viral genomes (Cattaneo 1994; Horikami and Moyer 1995; Polson et al. 1996; Taylor et al. 2005), and in certain cases depending on virus-host combinations, displays an antiviral action (Samuel 2011).

The structures of Zα bound to Z-DNA and Z-RNA gave crucial understanding of the molecular basis of the zigzag backbone recognition by a specific set of side chains of Zα. However, the mechanism by which Zα converts a right-handed backbone structure (B-form DNA helix or A-form RNA helix) into a left-handed one (B to Z transition or A to Z transition, respectively) has been investigated only recently. The different possible mechanisms as well as the recent proposed model will be discussed in the following section.

4.3 How Does a Z-DNA Binding Domain Bind to Z-DNA?

So far only the B to Z transition of DNA induced by the binding of ADAR1 Zα domain has been experimentally investigated. Two different mechanisms for such a B to Z transition can be imagined (Kim et al. 2000): (1) a *passive* mechanism, in which Zα would bind to the small fraction of Z-DNA present in equilibrium with B-DNA, and because of its high affinity for Z-DNA will then pull the equilibrium towards the formation of Zα/Z-DNA complex; (2) an *active* mechanism, in which Zα would bind to B-DNA and will then actively convert it into Z-DNA.

The structure of the free Zα domain of ADAR1 has been determined in solution by NMR (Schade et al. 1999). The comparison of this structure with that of Zα bound to Z-DNA (Schwartz et al. 1999b) has revealed that most Z-DNA contacting residues are pre-positioned in the free Zα domain to fit Z-DNA. Of the nine Zα side chains contacting Z-DNA in the crystal structure, seven are well-ordered and already pre-positioned in free Zα, which is thus pre-shaped to fit Z-DNA. Moreover, structural comparison of Zα with homologous proteins that bind B-DNA suggested that binding of Zα to B-DNA is disfavoured by steric hindrance (Schade et al. 1999). Altogether, these strongly suggest that binding of Zα would follow a passive mechanism.

However, recent NMR studies monitoring hydrogen exchange rates of imino protons have made possible the indirect observation of Zα bound to B-DNA in the pathway of binding in different DNA sequence contexts (Kang et al. 2009; Seo et al. 2010). These studies would strongly tend to validate an active mechanism, but the authors could nonetheless not clearly exclude a passive one to occur (Kang et al. 2009). The elucidation of the binding mechanism of Zα to Z-DNA would thus probably deserve further study. It would also be interesting to analyze similarly the binding mechanism to Z-RNA.

4.4 ADAR1 Zβ Domain: a Domain for Protein–Protein Interaction?

Zβ, the second Z-DNA binding domain of ADAR1, is present in both the interferon-induced ADAR1-i and the constitutively expressed ADAR1-c (Fig. 2). In contrast to Zα domain, the Zβ domain of ADAR1 does not interact with Z-DNA (Schwartz et al. 1999a). Nevertheless, the Zβ domain is highly conserved among ADAR1 which thus suggests that the two domains Zα and Zβ probably perform different functions. The Zβ structure has been solved by crystallography and consists of four α-helices and a three-stranded β-sheet with a $\alpha\beta\alpha\alpha\beta\beta\alpha$ topology (Fig. 4b, c) (Athanasiadis et al. 2005). This structure has revealed that Zβ is closely related in structure to Zα and belongs to the same α+βHTH family. However, Zβ has an additional helix, helix $\alpha 4$ (Fig. 4b) and is also lacking several crucial residues important for Z-DNA binding (Fig. 4c). This latter point explains why Zβ does not bind to Z-DNA. Interestingly, there is no steric clash that would prevent Zβ to bind to Z-DNA, and the partial restoration of a Zα sequence in Zβ results in weak Z-DNA binding (Kim et al. 2004). The mapping of Zβ amino acid conservation has revealed a distinct conserved surface involved in metal binding and dimerization (Athanasiadis et al. 2005). However, since no biochemical data support either metal binding or dimerization of $Z\beta$, these properties observed in the crystal might have been influenced by packing forces. Nonetheless, the dimerization of $Z\beta$ is an appealing model, considering that ADAR proteins have been shown to be active as dimers (Cho et al. 2003; Gallo et al. 2003). For example, the N-terminal part of dADAR has been shown to be involved in dimerization (Gallo et al. 2003), but the site of dimerization for vertebrate ADARs remains to be established.

5 Adenosine Deaminases Acting on RNA dsRNA Binding Domains

ADAR dsRNA binding domains are essential components for ADAR activity, since they are directly involved in dsRNA substrate recognition and binding. However, the molecular basis explaining how domains largely thought to bind dsRNA non-specifically are actually targeting very specific adenine positions in certain substrate (like the GluR-B R/G site—Fig. 3) have been a puzzling paradox for many years. Recent structures of ADAR2 dsRBDs bound to a natural substrate have given some critical insights into the sequence-specific recognition of ADAR substrates.

5.1 Structural Characteristics of a dsRNA Binding Domain

The dsRBD is a $\sim$65–75 amino acids domain found in eukaryotic, prokaryotic and even viral proteins which have been shown to interact specifically with dsRNA. dsRBDs were first identified in Staufen, a protein responsible for mRNA localization in *Drosophila* and PKR, a dsRNA-dependent protein kinase (St Johnston et al. 1992; McCormack et al. 1992; Green and Mathews 1992). Since the early 1990s, the list of dsRBD containing proteins has been growing and regroups proteins with a large variety of function as development, RNA interference, RNA transport, RNA processing and of course RNA editing (Fierro-Monti and Mathews 2000; Saunders and Barber 2003; Chang and Ramos 2005; Stefl et al. 2005). The structures of various dsRBDs have been determined uncovering a mixed α/β fold with a conserved $\alpha\beta\beta\beta\alpha$ topology in which the two α-helices are packed against the three-stranded anti-parallel β-sheet (Bycroft et al. 1995; Kharrat et al. 1995). In addition, structures of dsRBDs have been determined in complex with dsRNA, most of which with non-natural RNA duplexes (Ryter and Schultz 1998; Ramos et al. 2000; Gan et al. 2006; Wu et al. 2004). These structures have suggested that dsRBDs recognize A-form helix of dsRNA in a sequence-independent manner, since the majority of dsRBD-RNA interaction involve direct contacts with the 2′-hydroxyl groups of the ribose sugar rings and direct- or water-mediated contacts with non-bridging oxygen of the phosphodiester backbone and a subclass of dsRBDs prefer stem-loop over A-form helices (Ramos et al. 2000; Wu et al. 2004).

More recently, the structures of the two dsRBDs of ADAR2 in complex with a natural dsRNA substrate have been determined (Stefl et al. 2010), revealing a sequence-specific readout of the dsRNA minor groove. These structures will be discussed in the following section.

5.2 Sequence Specific Recognition with dsRBD

Recently, the structures of human ADAR2 dsRBD1 and dsRBD2 have been determined in complex with their respective RNA target on the GluR-B R/G site RNA helix (Stefl et al. 2010). These structures have confirmed the conserved mode of recognition of the A-form RNA helix (Ryter and Schultz 1998; Ramos et al. 2000; Gan et al. 2006; Wu et al. 2004) in which helix $\alpha1$ and $\beta1$–$\beta2$ loop interact with the minor groove of the RNA helix at one turn of interval and in which conserved positively charged residues in the N-terminal end of helix $\alpha2$ interact across the major groove with non-bridging oxygen of the phosphodiester backbone. Strikingly, a detailed inspection of the interaction regions revealed unexpected sequence-specific contacts of both dsRBD to the RNA minor grooves.

The RNA major groove is deep and narrow, and as a consequence bases are inaccessible to protein side chains. In contrast, the RNA minor groove is wide and shallow (Saenger 1984) but stereochemical considerations have suggested that discrimination of some base-pairs would be difficult in the minor groove (Seeman et al. 1976; Steitz 1990). Discrimination in the minor groove can mostly arise from an appreciation of the group lying in position 2 of purine rings, i.e. the amino NH_2 group of a guanine which is a polar hydrogen bond donor, and the aromatic H2 proton of an adenine which is non polar and small and can thus accommodate hydrophobic side chains in its close vicinity, whereas the amino group of a guanine would lead to steric clashes (Fig. 5c, d).

Two sequence-specific contacts at two consecutive RNA minor grooves enable ADAR2 dsRBD1 to bind the GluR-B RNA upper stem-loop (USL) at a single register (Stefl et al. 2010). These specific contacts are on one hand a hydrogen bond to the amino group of G22 in the GG mismatch via the main chain carbonyl of Val104 in the $\beta1$–$\beta2$ loop and on the other hand a hydrophobic contact to the adenine H2 of A32 via the side chain of Met84 in helix $\alpha1$ (Fig. 5a). Similarly, ADAR2 dsRBD2 recognizes the GluR-B RNA lower stem-loop via two sequence specific contacts at two consecutive RNA minor grooves: a hydrogen bond to the amino group of G9, located 3′ to the editing site, via the main chain carbonyl of Ser258 in the $\beta1$–$\beta2$ loop and a hydrophobic contact to the adenine H2 of A18 via the side chain of Met238 in helix $\alpha1$ (Fig. 5b) (Stefl et al. 2010).

The importance of these contacts for the binding affinity of the dsRBDs with their respective RNA partner was further quantified in a solution binding assay and in an in vitro editing assay (Stefl et al. 2010). In mutating any of the bases that are recognized in a sequence-specific manner by the dsRBDs, the binding affinity is reduced compared to the wild-type. Furthermore, when replacing the GG

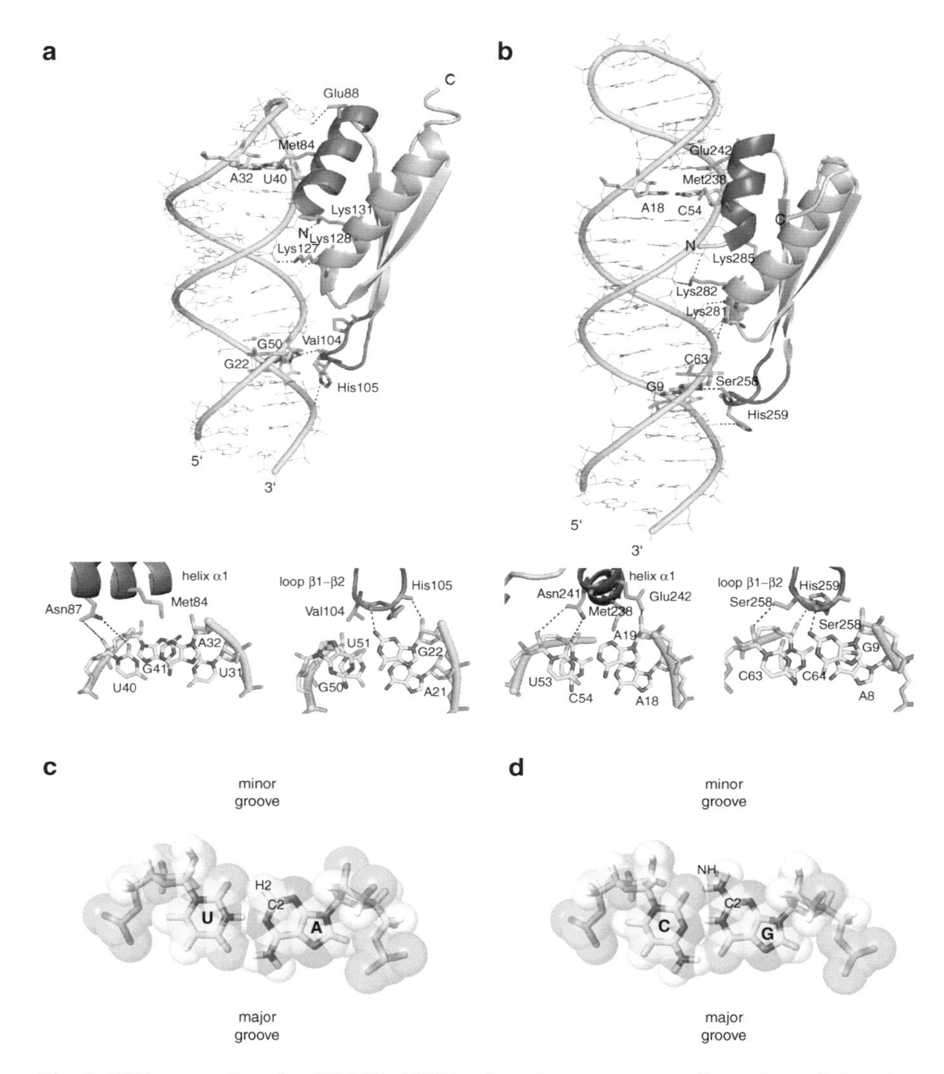

Fig. 5 RNA recognition by ADAR2 dsRBDs through sequence specific readout of the minor groove. **a** Structure of ADAR2 dsRBD1 in complex with the GluR-B R/G upper stem-loop (PDB code 2L3C). Overall structure (*top*) and close-up view of the minor groove sequence-specific recognitions mediated by helix α1 and the β1-β2 loop (*bottom*). **b** Structure of ADAR2 dsRBD2 in complex with the GluR-B R/G lower stem-loop (PDB code 2L2 K). Overall structure (*top*) and close-up view of the minor groove sequence-specific recognitions mediated by helix α1 and the β1-β2 loop (*bottom*). **c** Chemical groups of an A–U pair lying in the major and minor grooves. **d** Chemical groups of a G–C pair lying in the major and minor grooves. Discrimination in the minor groove relies on the appreciation of the group in position 2 of purine rings

mismatch or the two AC mismatches with Watson–Crick pairs that keep intact the specific contacts to the RNA, the binding affinity is less affected than when mutating the specifically recognized bases. In addition, the editing activity of protein mutants affected in the residues involved in sequence-specific contacts (Met in helix $\alpha 1$ and $\beta 1$–$\beta 2$ loop) is reduced to less than 30% of the wild-type protein editing activity. Altogether, this strongly supports the idea that the two dsRBDs of ADAR2 recognize primarily the sequence of the RNA helix rather than its shape.

These structures give the means to reconsider the common beliefs on dsRBDs, and to propose that binding of certain dsRBDs might occur sequence specifically.

5.3 How dsRBDs are Positioned on Substrate?

Structural studies on full-length ADAR proteins in complex with their RNA substrates are essential for understanding the overall editing mechanism process. One interesting point in the field is the requirement of dimerization of ADAR to be active. In vitro studies have shown that editing activity of ADAR1 and ADAR2 (Cho et al. 2003) and of *Drosophila* dADAR (Gallo et al. 2003) needs dimerization. Dimerization of ADAR1 and ADAR2 have been confirmed in vivo by fluorescence and bioluminescence resonance energy transfer studies (Chilibeck et al. 2006; Poulsen et al. 2006). A source of discussion in the field is to know whether this dimerization is dependent on RNA binding (Cho et al. 2003; Gallo et al. 2003; Poulsen et al. 2006; Valente and Nishikura 2007). There are no structural data with a full-length ADAR protein bound to a RNA substrate that would reveal the molecular basis for this dimerization. Nevertheless, the structure of the N-terminal domain of ADAR2 consisting of both dsRBDs have been solved by solution NMR in complex with the GluR-B R/G site RNA (Stefl et al. 2010). In the structure, the two dsRBDs bind one face of the RNA covering approximately 120° of the space around the RNA helix (Fig. 6a, b). This structure is then perfectly in accordance with the dimerization model of ADARs, since another ADAR molecule could bind to the other face of the RNA helix without steric hindrance. Moreover it is clear that in interacting with the guanosine 3' to the edited adenosine (Fig. 6d), ADAR2 dsRBD2 brings the deaminase domain in close proximity to the editing site. When this precise positioning is impaired, specific editing of the GluR-B R/G site is nearly abolished which underlines the functional importance of sequence-specific recognition of RNA by dsRBDs for A $\rightarrow$ I editing (Stefl et al. 2010). Even with these new insights, structural aspects of substrate recognition by ADARs remain a source of questions in the editing field. For example, how the targeted adenosine would be flipped out to reach the catalytic domain, and how after dimerization of ADARs the two catalytic domains would be positioned relative to each other and to the dsRBDs, is of great interest and deserves further structural studies.

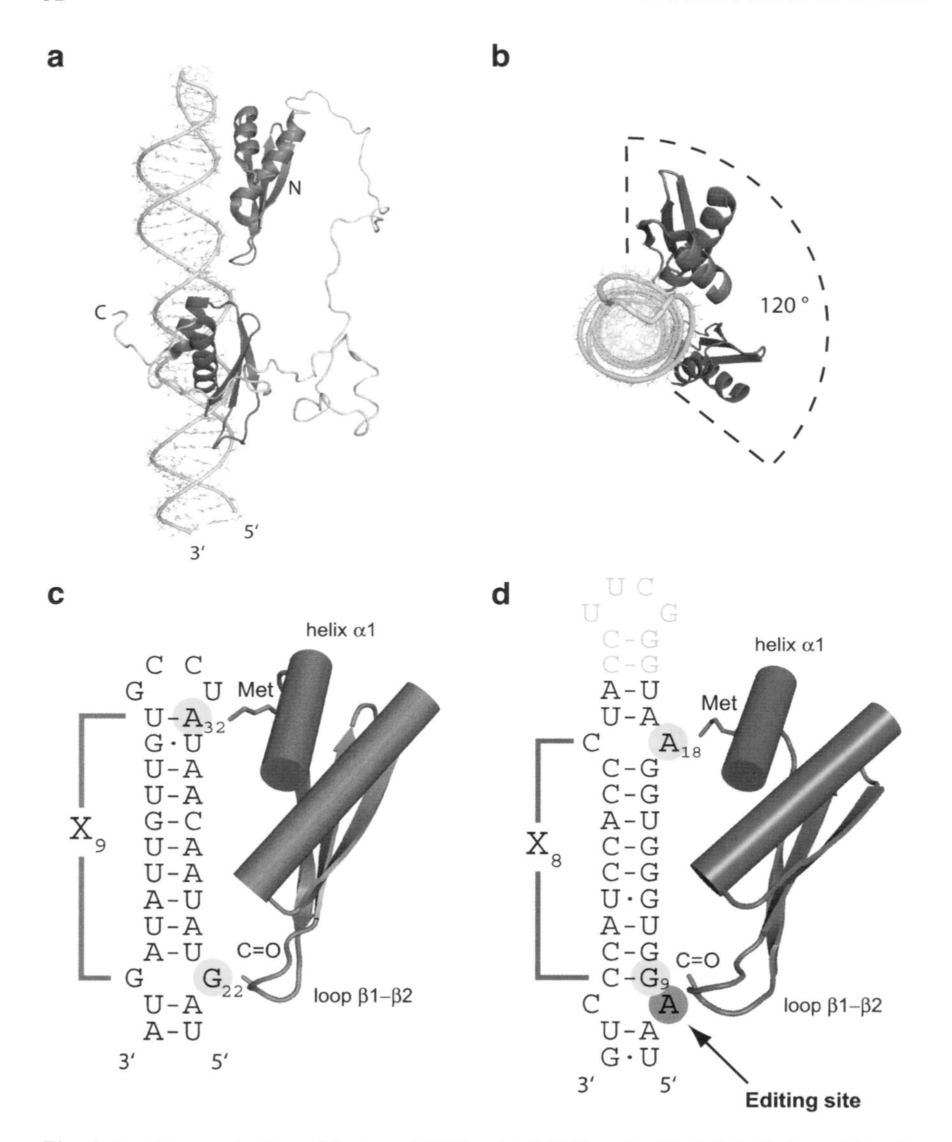

Fig. 6 Spatial organization of the two dsRBDs of ADAR2 on the GluR-B R/G site. **a** *Side view* of the complex (PDB code 2L3 J). **b** *Top view* of the complex showing the portion of the space covered by the two dsRBDs around the RNA helix. **c** Schematic representation of the sequence specific contacts defining the binding register of ADAR2 dsRBD1. **d** Schematic representation of the sequence specific contacts defining the binding register of ADAR2 dsRBD2

5.4 Are the Binding Sites of ADAR dsRBDs Predictable?

An attractive idea would be to transpose the structural knowledge of ADAR2 bound to the GluR-B R/G site to predict the mode of binding of ADAR2 on other substrates. Even more challenging would be the prediction of other ADAR dsRBDs on their respective substrates. The structures of ADAR2 dsRBDs bound to the GluR-B R/G site have shown that the binding is achieved by a direct readout of the RNA sequence in the minor groove of the dsRNA substrate. The two dsRBDs use helix $\alpha 1$ and the $\beta 1$–$\beta 2$ loop as molecular rulers to find their binding register in the RNA minor groove (Fig. 5a, b). While dsRBD1 preferentially recognize G-X_9-A (Fig. 6c), dsRBD2 binds the same sequence but with a different register length: G-X_8-A (Fig. 6d). The length and the relative position of helix $\alpha 1$ relative to the dsRBD fold appear to be the key structural elements that determine the register length of these two dsRBDs (Stefl et al. 2010). Such binding sequence and register for ADAR2 dsRBD2 are present on GluR-B Q/R site (Fig. 3a), but are not always present on ADAR2 substrates (Fig. 3a), and one can thus not exclude that its dsRBDs would adopt a different mode of binding involving different side chains in helix $\alpha 1$ when bound to different substrates. In addition, sequence alignment of diverse ADAR dsRBDs could serve to anticipate similarities and discrepancies between dsRBDs regarding the preferred sequence and register of binding (Fig. 2b). For instance, ADAR1 dsRBDs appear to have a longer helix $\alpha 1$ and lack the ADAR2 equivalent of the methionine involved in the sequence-specific contacts (Figs. 2b and 5). These could explain why ADAR1 and ADAR2 have different substrate specificities (Bass 2002; Lehmann and Bass 2000). Furthermore, dADAR dsRBD2 is very similar to ADAR2 dsRBD2, but whereas their helix $\alpha 1$ are extremely alike, the $\beta 1$–$\beta 2$ loop region is less conserved (Fig. 2b) and a prediction on the sequence and register of binding for this dsRBD remains difficult. Generally, the binding specificities of dsRBDs of other members of the ADAR family are still difficult to predict and would need more structural data involving various dsRBDs.

6 dsRBDs and Z-DNA Binding Domains Act on the Subcellular Localization of ADARs

Whereas ADAR1-i is mostly detected in the cytoplasm (Patterson and Samuel 1995; Poulsen et al. 2001; Desterro et al. 2003), ADAR1-c localizes mainly in the nucleus (Desterro et al. 2003; Sansam et al. 2003). ADAR1-c is mostly located in nucleoli but constantly shuttles between nucleoli and the nucleoplasm, where most ADAR substrates are found (Sansam et al. 2003). A nuclear localization signal (NLS) has been identified in the third dsRBD of human ADAR1 (Eckmann et al. 2001; Strehblow et al. 2002; Poulsen et al. 2001). ADAR1-i harbors also a nuclear export signal (NES) within its most N-terminal Z-DNA binding domain (Zα)

(Poulsen et al. 2001), and has the property to shuttle between the nucleus and the cytoplasm (Eckmann et al. 2001; Poulsen et al. 2001). As most RNA viruses are localizing in the cytoplasm, the unique cytoplasmic localization of ADAR1-i among ADARs, gives additional support for an antiviral function of this protein (George et al. 2011; Samuel 2011). Although lacking the Zα domain, ADAR1-c is also shuttling between the nucleus and the cytoplasm. Transportin-1 is a nuclear import factor for ADAR1, and dsRNA binding of the third dsRBD modulates its interaction with transportin-1 and exportin-5 and thus regulate the nucleocytoplasmic properties of ADAR1 (Fritz et al. 2009).

ADAR2 is localized exclusively in the nucleus, where similar to ADAR1-c it resides mostly in nucleoli, but shuttles constantly to reach the nucleoplasm where ADAR substrates are located (Desterro et al. 2003; Sansam et al. 2003). Although the dsRBDs are not involved in the nuclear localization of ADAR2, they play a crucial role in targeting ADAR2 to the nucleolus, likely through their ability to bind rRNA (Sansam et al. 2003; Xu et al. 2006). However, the biological signification of the nucleolar localization of ADAR2 and ADAR1-c is largely unknown.

7 Concluding Remarks

During the last 10 years, the ADAR family of protein has contributed to a great extent to our general understanding of the molecular basis of nucleic acid recognition for both Z-DNA binding domains and dsRNA binding domains. However, our current understanding of the molecular basis of substrate recognition by ADARs is unfortunately still incomplete. More structures of ADARs substrates alone and in complex with ADARs dsRBDs, would undoubtedly be of great value for the understanding of substrates recognition. In addition, a structure of a full-length ADAR protein revealing how the catalytically active dimer would assemble on an RNA substrate would be fantastic and essential for understanding the overall editing mechanism process. The relatively small amount of structural information obtained to date is a consequence of the demanding biochemical properties of ADAR proteins and also probably of the small number of groups in structural biology working on the editing field. However, piece by piece, we start having a better view of substrate recognition by ADARs. Hopefully, in the near future, more structural information would allow the prediction of ADARs mode of binding on RNA substrates with increased confidence.

Acknowledgments The authors thank Richard Stefl, Florian Oberstrass, Lenka Skrisovska, and Christophe Maris who have done the work performed in the Allain's Lab described in this review. We sincerely apologize to the colleagues whose important work is not cited because of space limitation, or unfortunately because of our negligence. This work was supported by the Swiss National Science Foundation Nr. 31003–133134 and the SNF-NCCR structural biology. PB is supported by the Postdoctoral ETH Fellowship Program.

References

Allain FH, Varani G (1995) Structure of the P1 helix from group I self-splicing introns. J Mol Biol 250:333–353

Aruscavage PJ, Bass BL (2000) A phylogenetic analysis reveals an unusual sequence conservation within introns involved in RNA editing. RNA 6:257–269

Athanasiadis A, Rich A, Maas S (2004) Widespread A-to-I RNA editing of Alu-containing mRNAs in the human transcriptome. PLoS Biol 2:e391

Athanasiadis A, Placido D, Maas S, Brown BA, Lowenhaupt K, Rich A (2005) The crystal structure of the Zbeta domain of the RNA-editing enzyme ADAR1 reveals distinct conserved surfaces among Z-domains. J Mol Biol 351:496–507

Baltimore D (2001) Our genome unveiled. Nature 409:814–816

Bass BL (2002) RNA editing by adenosine deaminases that act on RNA. Annu Rev Biochem 71:817–846

Bass BL, Weintraub H (1987) A developmentally regulated activity that unwinds RNA duplexes. Cell 48:607–613

Bass BL, Weintraub H (1988) An unwinding activity that covalently modifies its double-stranded RNA substrate. Cell 55:1089–1098

Bass BL, Nishikura K, Keller W, Seeburg PH, Emeson RB, O'Connell MA, Samuel CE, Herbert A (1997) A standardized nomenclature for adenosine deaminases that act on RNA. RNA 3:947–949

Benne R, Van den Burg J, Brakenhoff JP, Sloof P, Van Boom JH, Tromp MC (1986) Major transcript of the frameshifted coxII gene from trypanosome mitochondria contains four nucleotides that are not encoded in the DNA. Cell 46:819–826

Blow M, Futreal PA, Wooster R, Stratton MR (2004) A survey of RNA editing in human brain. Genome Res 14:2379–2387

Brown BAn, Lowenhaupt K, Wilbert CM, Hanlon EB, Rich A (2000) The zalpha domain of the editing enzyme dsRNA adenosine deaminase binds left-handed Z-RNA as well as Z-DNA. Proc Natl Acad Sci USA 97:13532–13536

Bycroft M, Grunert S, Murzin AG, Proctor M, St Johnston D (1995) NMR solution structure of a dsRNA binding domain from Drosophila staufen protein reveals homology to the N-terminal domain of ribosomal protein S5. EMBO J 14:3563–3571

Cattaneo R (1994) Biased (A→I) hypermutation of animal RNA virus genomes. Curr Opin Genet Dev 4:895–900

Chang K-Y, Ramos A (2005) The double-stranded RNA-binding motif, a versatile macromolecular docking platform. FEBS J 272:2109–2117

Chen CX, Cho DS, Wang Q, Lai F, Carter KC, Nishikura K (2000) A third member of the RNA-specific adenosine deaminase gene family, ADAR3, contains both single- and double-stranded RNA binding domains. RNA 6:755–767

Cheong C, Varani G, Tinoco I Jr (1990) Solution structure of an unusually stable RNA hairpin, 5'GGAC(UUCG)GUCC. Nature 346:680–682

Chilibeck KA, Wu T, Liang C, Schellenberg MJ, Gesner EM, Lynch JM, MacMillan AM (2006) FRET analysis of in vivo dimerization by RNA-editing enzymes. J Biol Chem 281:16530–16535

Cho D-SC, Yang W, Lee JT, Shiekhattar R, Murray JM, Nishikura K (2003) Requirement of dimerization for RNA editing activity of adenosine deaminases acting on RNA. J Biol Chem 278:17093–17102

Dawson TR, Sansam CL, Emeson RB (2004) Structure and sequence determinants required for the RNA editing of ADAR2 substrates. J Biol Chem 279:4941–4951

Desterro JMP, Keegan LP, Lafarga M, Berciano MT, O'Connell M, Carmo-Fonseca M (2003) Dynamic association of RNA-editing enzymes with the nucleolus. J Cell Sci 116:1805–1818

Eckmann CR, Neunteufl A, Pfaffstetter L, Jantsch MF (2001) The human but not the Xenopus RNA-editing enzyme ADAR1 has an atypical nuclear localization signal and displays the characteristics of a shuttling protein. Mol Biol Cell 12:1911–1924

Ennifar E, Nikulin A, Tishchenko S, Serganov A, Nevskaya N, Garber M, Ehresmann B, Ehresmann C, Nikonov S, Dumas P (2000) The crystal structure of UUCG tetraloop. J Mol Biol 304:35–42
Fierro-Monti I, Mathews MB (2000) Proteins binding to duplexed RNA: one motif, multiple functions. Trends Biochem Sci 25:241–246
Fritz J, Strehblow A, Taschner A, Schopoff S, Pasierbek P, Jantsch MF (2009) RNA-regulated interaction of transportin-1 and exportin-5 with the double-stranded RNA-binding domain regulates nucleocytoplasmic shuttling of ADAR1. Mol Cell Biol 29:1487–1497
Gallo A, Keegan LP, Ring GM, O'Connell MA (2003) An ADAR that edits transcripts encoding ion channel subunits functions as a dimer. EMBO J 22:3421–3430
Gan J, Tropea JE, Austin BP, Court DL, Waugh DS, Ji X (2006) Structural insight into the mechanism of double-stranded RNA processing by ribonuclease III. Cell 124:355–366
George C, Gan Z, Liu Y, Samuel C (2011) Adenosine deaminases acting on RNA, RNA editing, and interferon action. J Interferon Cytokine Res 31:99–117
Gerber A, O'Connell MA, Keller W (1997) Two forms of human double-stranded RNA-specific editase 1 (hRED1) generated by the insertion of an Alu cassette. RNA 3:453–463
Gott JM, Emeson RB (2000) Functions and mechanisms of RNA editing. Annu Rev Genet 34:499–531
Graveley BR (2001) Alternative splicing: increasing diversity in the proteomic world. Trends Genet 17:100–107
Green SR, Mathews MB (1992) Two RNA-binding motifs in the double-stranded RNA-activated protein kinase, DAI. Genes Dev 6:2478–2490
Ha SC, Choi J, Hwang H-Y, Rich A, Kim Y-G, Kim KK (2009) The structures of non-CG-repeat Z-DNAs co-crystallized with the Z-DNA-binding domain, hZ alpha(ADAR1). Nucleic Acids Res 37:629–637
Hanrahan CJ, Palladino MJ, Ganetzky B, Reenan RA (2000) RNA editing of the Drosophila para Na(+) channel transcript. Evolutionary conservation and developmental regulation. Genetics 155:1149–1160
Harrison SC, Aggarwal AK (1990) DNA recognition by proteins with the helix-turn-helix motif. Annu Rev Biochem 59:933–969
Heale BS, Keegan LP, McGurk L, Michlewski G, Brindle J, Stanton CM, Caceres JF, O'Connell MA (2009) Editing independent effects of ADARs on the miRNA/siRNA pathways. EMBO J 28:3145–3156
Herbert A, Rich A (1999) Left-handed Z-DNA: structure and function. Genetica 106:37–47
Herbert A, Lowenhaupt K, Spitzner J, Rich A (1995) Chicken double-stranded RNA adenosine deaminase has apparent specificity for Z-DNA. Proc Natl Acad Sci USA 92:7550–7554
Herbert A, Alfken J, Kim YG, Mian IS, Nishikura K, Rich A (1997) A Z-DNA binding domain present in the human editing enzyme, double-stranded RNA adenosine deaminase. Proc Natl Acad Sci USA 94:8421–8426
Horikami SM, Moyer SA (1995) Double-stranded RNA adenosine deaminase activity during measles virus infection. Virus Res 36:87–96
Hundley HA, Bass BL (2010) ADAR editing in double-stranded UTRs and other noncoding RNA sequences. Trends Biochem Sci 35:377–383
Kallman AM, Sahlin M, Ohman M (2003) ADAR2 A→I editing: site selectivity and editing efficiency are separate events. Nucleic Acids Res 31:4874–4881
Kang Y-M, Bang J, Lee E-H, Ahn H-C, Seo Y-J, Kim KK, Kim Y-G, Choi B-S, Lee J-H (2009) NMR spectroscopic elucidation of the B–Z transition of a DNA double helix induced by the Z alpha domain of human ADAR1. J Am Chem Soc 131:11485–11491
Kawahara Y, Zinshteyn B, Chendrimada TP, Shiekhattar R, Nishikura K (2007a) RNA editing of the microRNA-151 precursor blocks cleavage by the Dicer-TRBP complex. EMBO Rep 8:763–769
Kawahara Y, Zinshteyn B, Sethupathy P, Iizasa H, Hatzigeorgiou AG, Nishikura K (2007b) Redirection of silencing targets by adenosine-to-inosine editing of miRNAs. Science 315: 1137–1140

Kawahara Y, Megraw M, Kreider E, Iizasa H, Valente L, Hatzigeorgiou AG, Nishikura K (2008) Frequency and fate of microRNA editing in human brain. Nucleic Acids Res 36:5270–5280
Kawakubo K, Samuel CE (2000) Human RNA-specific adenosine deaminase (ADAR1) gene specifies transcripts that initiate from a constitutively active alternative promoter. Gene 258:165–172
Keegan LP, Gallo A, O'Connell MA (2001) The many roles of an RNA editor. Nat Rev Genet 2:869–878
Kharrat A, Macias MJ, Gibson TJ, Nilges M, Pastore A (1995) Structure of the dsRNA binding domain of *E. coli RNase* III. EMBO J 14:3572–3584
Kim U, Wang Y, Sanford T, Zeng Y, Nishikura K (1994) Molecular cloning of cDNA for double-stranded RNA adenosine deaminase, a candidate enzyme for nuclear RNA editing. Proc Natl Acad Sci USA 91:11457–11461
Kim YG, Lowenhaupt K, Maas S, Herbert A, Schwartz T, Rich A (2000) The zab domain of the human RNA editing enzyme ADAR1 recognizes Z-DNA when surrounded by B-DNA. J Biol Chem 275:26828–26833
Kim Y-G, Lowenhaupt K, Oh D-B, Kim KK, Rich A (2004) Evidence that vaccinia virulence factor E3L binds to Z-DNA in vivo: implications for development of a therapy for poxvirus infection. Proc Natl Acad Sci USA 101:1514–1518
Koeris M, Funke L, Shrestha J, Rich A, Maas S (2005) Modulation of ADAR1 editing activity by Z-RNA in vitro. Nucleic Acids Res 33:5362–5370
Lai F, Chen CX, Carter KC, Nishikura K (1997) Editing of glutamate receptor B subunit ion channel RNAs by four alternatively spliced DRADA2 double-stranded RNA adenosine deaminases. Mol Cell Biol 17:2413–2424
Lehmann KA, Bass BL (1999) The importance of internal loops within RNA substrates of ADAR1. J Mol Biol 291:1–13
Lehmann KA, Bass BL (2000) Double-stranded RNA adenosine deaminases ADAR1 and ADAR2 have overlapping specificities. Biochemistry 39:12875–12884
Levanon EY, Eisenberg E, Yelin R, Nemzer S, Hallegger M, Shemesh R, Fligelman ZY, Shoshan A, Pollock SR, Sztybel D, Olshansky M et al (2004) Systematic identification of abundant A-to-I editing sites in the human transcriptome. Nat Biotechnol 22:1001–1005
Li JB, Levanon EY, Yoon JK, Aach J, Xie B, LeProust E, Zhang K, Gao Y, Church GM (2009) Genome-wide identification of human rna editing sites by parallel dna capturing and sequencing. Science 324:1210–1213
Liu LF, Wang JC (1987) Supercoiling of the DNA template during transcription. Proc Natl Acad Sci USA 84:7024–7027
Maas S, Gommans WM (2009) Novel exon of mammalian ADAR2 extends open reading frame. PLoS One 4:e4225
Maas S, Melcher T, Herb A, Seeburg PH, Keller W, Krause S, Higuchi M, O'Connell MA (1996) Structural requirements for RNA editing in glutamate receptor pre-mRNAs by recombinant double-stranded RNA adenosine deaminase. J Biol Chem 271:12221–12226
Maas S, Kawahara Y, Tamburro KM, Nishikura K (2006) A-to-I RNA editing and human disease. RNA Biol 3:1–9
McCormack SJ, Thomis DC, Samuel CE (1992) Mechanism of interferon action: identification of a RNA binding domain within the N-terminal region of the human RNA-dependent P1/eIF-2 alpha protein kinase. Virology 188:47–56
Melcher T, Maas S, Herb A, Sprengel R, Higuchi M, Seeburg PH (1996a) RED2, a brain-specific member of the RNA-specific adenosine deaminase family. J Biol Chem 271:31795–31798
Melcher T, Maas S, Herb A, Sprengel R, Seeburg PH, Higuchi M (1996b) A mammalian RNA editing enzyme. Nature 379:460–464
Nishikura K (2006) Editor meets silencer: crosstalk between RNA editing and RNA interference. Nat Rev Mol Cell Biol 7:919–931
Nishikura K (2010) Functions and regulation of RNA editing by ADAR deaminases. Annu Rev Biochem 79:321–349
Nishikura K, Yoo C, Kim U, Murray JM, Estes PA, Cash FE, Liebhaber SA (1991) Substrate specificity of the dsRNA unwinding/modifying activity. EMBO J 10:3523–3532

Niswender CM, Sanders-Bush E, Emeson RB (1998) Identification and characterization of RNA editing events within the 5-HT2C receptor. Ann N Y Acad Sci 861:38–48
O'Connell MA, Krause S, Higuchi M, Hsuan JJ, Totty NF, Jenny A, Keller W (1995) Cloning of cDNAs encoding mammalian double-stranded RNA-specific adenosine deaminase. Mol Cell Biol 15:1389–1397
Ohlson J, Pedersen JS, Haussler D, Ohman M (2007) Editing modifies the GABA(A) receptor subunit alpha3. RNA 13:698–703
Ohman M (2007) A-to-I editing challenger or ally to the microRNA process. Biochimie 89: 1171–1176
Ohman M, Kallman AM, Bass BL (2000) In vitro analysis of the binding of ADAR2 to the pre-mRNA encoding the GluR-B R/G site. RNA 6:687–697
Osenberg S, Paz Yaacov N, Safran M, Moshkovitz S, Shtrichman R, Sherf O, Jacob-Hirsch J, Keshet G, Amariglio N, Itskovitz-Eldor J, Rechavi G (2010) Alu sequences in undifferentiated human embryonic stem cells display high levels of A-to-I RNA editing. PLoS One 5:e11173
Palavicini JP, O'Connell MA, Rosenthal JJ (2009) An extra double-stranded RNA binding domain confers high activity to a squid RNA editing enzyme. RNA 15:1208–1218
Palladino MJ, Keegan LP, O'Connell MA, Reenan RA (2000) dADAR, a Drosophila double-stranded RNA-specific adenosine deaminase is highly developmentally regulated and is itself a target for RNA editing. RNA 6:1004–1018
Patterson JB, Samuel CE (1995) Expression and regulation by interferon of a double-stranded-RNA-specific adenosine deaminase from human cells: evidence for two forms of the deaminase. Mol Cell Biol 15:5376–5388
Patterson JB, Thomis DC, Hans SL, Samuel CE (1995) Mechanism of interferon action: double-stranded RNA-specific adenosine deaminase from human cells is inducible by alpha and gamma interferons. Virology 210:508–511
Peck LJ, Nordheim A, Rich A, Wang JC (1982) Flipping of cloned d(pCpG)n.d(pCpG)n DNA sequences from right- to left-handed helical structure by salt, Co(III), or negative supercoiling. Proc Natl Acad Sci USA 79:4560–4564
Placido D, BA Brown, Lowenhaupt K, Rich A, Athanasiadis A (2007) A left-handed RNA double helix bound by the Z alpha domain of the RNA-editing enzyme ADAR1. Structure 15:395–404
Pokharel S, Beal PA (2006) High-throughput screening for functional adenosine to inosine RNA editing systems. ACS Chem Biol 1:761–765
Polson AG, Bass BL (1994) Preferential selection of adenosines for modification by double-stranded RNA adenosine deaminase. EMBO J 13:5701–5711
Polson AG, Bass BL, Casey JL (1996) RNA editing of hepatitis delta virus antigenome by dsRNA-adenosine deaminase. Nature 380:454–456
Poulsen H, Nilsson J, Damgaard CK, Egebjerg J, Kjems J (2001) CRM1 mediates the export of ADAR1 through a nuclear export signal within the Z-DNA binding domain. Mol Cell Biol 21:7862–7871
Poulsen H, Jorgensen R, Heding A, Nielsen FC, Bonven B, Egebjerg J (2006) Dimerization of ADAR2 is mediated by the double-stranded RNA binding domain. RNA 12:1350–1360
Pullirsch D, Jantsch MF (2010) Proteome diversification by adenosine to inosine RNA editing. RNA Biol 7:205–212
Ramos A, Grunert S, Adams J, Micklem DR, Proctor MR, Freund S, Bycroft M, St Johnston D, Varani G (2000) RNA recognition by a Staufen double-stranded RNA-binding domain. EMBO J 19:997–1009
Rebagliati MR, Melton DA (1987) Antisense RNA injections in fertilized frog eggs reveal an RNA duplex unwinding activity. Cell 48:599–605
Reenan RA (2005) Molecular determinants and guided evolution of species-specific RNA editing. Nature 434:409–413
Rich A, Zhang S (2003) Timeline: Z-DNA: the long road to biological function. Nat Rev Genet 4:566–572
Riedmann EM, Schopoff S, Hartner JC, Jantsch MF (2008) Specificity of ADAR-mediated RNA editing in newly identified targets. RNA 14:1110–1118

Ryter JM, Schultz SC (1998) Molecular basis of double-stranded RNA-protein interactions: structure of a dsRNA-binding domain complexed with dsRNA. EMBO J 17:7505–7513
Saenger W (1984) Principles of nucleic acid structure. Springer, New York
Samuel CE (2011) Adenosine deaminases acting on RNA (ADARs) are both antiviral and proviral. Virology
Sansam CL, Wells KS, Emeson RB (2003) Modulation of RNA editing by functional nucleolar sequestration of ADAR2. Proc Natl Acad Sci USA 100:14018–14023
Saunders LR, Barber GN (2003) The dsRNA binding protein family: critical roles, diverse cellular functions. FASEB J 17:961–983
Schade M, Turner CJ, Kuhne R, Schmieder P, Lowenhaupt K, Herbert A, Rich A, Oschkinat H (1999) The solution structure of the Zalpha domain of the human RNA editing enzyme ADAR1 reveals a prepositioned binding surface for Z-DNA. Proc Natl Acad Sci USA 96:12465–12470
Schwartz T, Lowenhaupt K, Kim YG, Li L, BA Brown, Herbert A, Rich A (1999a) Proteolytic dissection of Zab, the Z-DNA-binding domain of human ADAR1. J Biol Chem 274:2899–2906
Schwartz T, Rould MA, Lowenhaupt K, Herbert A, Rich A (1999b) Crystal structure of the Zalpha domain of the human editing enzyme ADAR1 bound to left-handed Z-DNA. Science 284:1841–1845
Seeburg PH, Higuchi M, Sprengel R (1998) RNA editing of brain glutamate receptor channels: mechanism and physiology. Brain Res Brain Res Rev 26:217–229
Seeman NC, Rosenberg JM, Rich A (1976) Sequence-specific recognition of double helical nucleic acids by proteins. Proc Natl Acad Sci USA 73:804–808
Seo Y-J, Ahn H-C, Lee E-H, Bang J, Kang Y-M, Kim H-E, Lee Y-M, Kim K, Choi B-S, Lee J-H (2010) Sequence discrimination of the Zalpha domain of human ADAR1 during B-Z transition of DNA duplexes. FEBS Lett 584:4344–4350
Slavov D, Clark M, Gardiner K (2000) Comparative analysis of the RED1 and RED2 A-to-I RNA editing genes from mammals, pufferfish and zebrafish. Gene 250:41–51
St Johnston D, Brown NH, Gall JG, Jantsch M (1992) A conserved double-stranded RNA-binding domain. Proc Natl Acad Sci USA 89:10979–10983
Stefl R, Allain FH-T (2005) A novel RNA pentaloop fold involved in targeting ADAR2. RNA 11:592–597
Stefl R, Skrisovska L, Allain FH-T (2005) RNA sequence- and shape-dependent recognition by proteins in the ribonucleoprotein particle. EMBO Rep 6:33–38
Stefl R, Xu M, Skrisovska L, Emeson RB, Allain FH-T (2006) Structure and specific RNA binding of ADAR2 double-stranded RNA binding motifs. Structure 14:345–355
Stefl R, Oberstrass FC, Hood JL, Jourdan M, Zimmermann M, Skrisovska L, Maris C, Peng L, Hofr C, Emeson RB, Allain FH-T (2010) The solution structure of the ADAR2 dsRBM-RNA complex reveals a sequence-specific readout of the minor groove. Cell 143:225–237
Steitz TA (1990) Structural studies of protein-nucleic acid interaction: the sources of sequence-specific binding. Q Rev Biophys 23:205–280
Strehblow A, Hallegger M, Jantsch MF (2002) Nucleocytoplasmic distribution of human RNA-editing enzyme ADAR1 is modulated by double-stranded RNA-binding domains, a leucine-rich export signal, and a putative dimerization domain. Mol Biol Cell 13:3822–3835
Taylor DR, Puig M, Darnell ME, Mihalik K, Feinstone SM (2005) New antiviral pathway that mediates hepatitis C virus replicon interferon sensitivity through ADAR1. J Virol 79:6291–6298
Tonkin LA, Bass BL (2003) Mutations in RNAi rescue aberrant chemotaxis of ADAR mutants. Science 302:1725
Tonkin LA, Saccomanno L, Morse DP, Brodigan T, Krause M, Bass BL (2002) RNA editing by ADARs is important for normal behavior in Caenorhabditis elegans. EMBO J 21:6025–6035
Valente L, Nishikura K (2007) RNA binding-independent dimerization of adenosine deaminases acting on RNA and dominant negative effects of nonfunctional subunits on dimer functions. J Biol Chem 282:16054–16061

Venter JC, Adams MD, Myers EW, Li PW, Mural RJ, Sutton GG, Smith HO, Yandell M, Evans CA, Holt RA, Gocayne JD et al (2001) The sequence of the human genome. Science 291: 1304–1351
Wagner RW, Smith JE, Cooperman BS, Nishikura K (1989) A double-stranded RNA unwinding activity introduces structural alterations by means of adenosine to inosine conversions in mammalian cells and Xenopus eggs. Proc Natl Acad Sci USA 86:2647–2651
Wang AH, Quigley GJ, Kolpak FJ, Crawford JL, van Boom JH, van der Marel G, Rich A (1979) Molecular structure of a left-handed double helical DNA fragment at atomic resolution. Nature 282:680–686
Wong TC, Ayata M, Ueda S, Hirano A (1991) Role of biased hypermutation in evolution of subacute sclerosing panencephalitis virus from progenitor acute measles virus. J Virol 65:2191–2199
Wong SK, Sato S, Lazinski DW (2001) Substrate recognition by ADAR1 and ADAR2. RNA 7:846–858
Wu H, Henras A, Chanfreau G, Feigon J (2004) Structural basis for recognition of the AGNN tetraloop RNA fold by the double-stranded RNA-binding domain of Rnt1p RNase III. Proc Natl Acad Sci USA 101:8307–8312
Wulff B, Nishikura K (2010) Substitutional A-to-I RNA editing. WIREs RNA 1:90–101
Wulff BE, Sakurai M, Nishikura K (2011) Elucidating the inosinome: global approaches to adenosine-to-inosine RNA editing. Nat Rev Genet 12:81–85
Xu M, Wells KS, Emeson RB (2006) Substrate-dependent contribution of double-stranded RNA-binding motifs to ADAR2 function. Mol Biol Cell 17:3211–3220
Zhang Z, Carmichael GG (2001) The fate of dsRNA in the nucleus: a p54(nrb)-containing complex mediates the nuclear retention of promiscuously A-to-I edited RNAs. Cell 106: 465–475

Editing of Neurotransmitter Receptor and Ion Channel RNAs in the Nervous System

Jennifer L. Hood and Ronald B. Emeson

Abstract The central dogma of molecular biology defines the major route for the transfer of genetic information from genomic DNA to messenger RNA to three-dimensional proteins that affect structure and function. Like alternative splicing, the post-transcriptional conversion of adenosine to inosine (A-to-I) by RNA editing can dramatically expand the diversity of the transcriptome to generate multiple, functionally distinct protein isoforms from a single genomic locus. While RNA editing has been identified in virtually all tissues, such post-transcriptional modifications have been best characterized in RNAs encoding both ligand- and voltage-gated ion channels and neurotransmitter receptors. These RNA processing events have been shown to play an important role in the function of the encoded protein products and, in several cases, have been shown to be critical for the normal development and function of the nervous system.

This work was supported by funding from the National Institutes of Health, the Vanderbilt Silvio O. Conte Center for Neuroscience Research and the Vanderbilt Kennedy Center.

J. L. Hood
Training Program in Cellular and Molecular Neuroscience,
Vanderbilt University School of Medicine,
Nashville, TN 37232-8548, USA

R. B. Emeson (✉)
Departments of Pharmacology, Molecular Physiology & Biophysics and Psychiatry,
Vanderbilt University School of Medicine, Nashville, TN 37232-8548, USA
e-mail: ron.emeson@vanderbilt.edu

Current Topics in Microbiology and Immunology (2012) 353: 61–90
DOI: 10.1007/82_2011_157

Published Online: 28 July 2011

Contents

1 Introduction

The conversion of adenosine to inosine (A-to-I) by RNA editing is increasingly identified as a post-transcriptional modification in which genomically-encoded sequences are altered through the site-specific deamination of specific adenosine residue(s) in precursor and mature mRNAs, tRNAs and primary miRNA transcripts (Blow et al. 2006; Gerber et al. 1998; Gott and Emeson 2000). An inosine within the open reading frame (ORF) of an mRNA is read as guanosine during translation, which can lead to specific change(s) in the amino acid coding potential of the mRNA to alter the functional properties of the encoded protein product. Such alterations in the primary nucleotide sequence of mRNA transcripts can affect not only coding potential, but also can alter the structure, stability, translation efficiency and splicing patterns of the modified transcripts, thereby affecting numerous aspects of RNA function in the cell (Gott and Emeson 2000). For specific tRNAs, such editing events are often observed in the wobble position of the anticodon loop (position 34), playing a crucial role in protein synthesis by allowing alternative pairing with U, C, or A in the third position of codons (Crick 1966). Editing also has been shown to play an important role in miRNA expression and function, as A-to-I conversion can modulate the processing of miRNA precursors by Drosha and Dicer, alter the target selectivity of mature miRNAs and decrease miRNA stability (Kawahara et al. 2007; Yang et al. 2006).

2 Identification of A-to-I Editing Targets

A-to-I editing is generally identified as an adenosine to guanosine (A-to-G) discrepancy during comparisons of genomic and cDNA sequences that result from the base-pairing of cytosine to inosine (like guanosine) during reverse transcriptase-mediated first-strand cDNA synthesis. At least eight mRNA transcripts with A-to-I modifications in the ORF were initially noted in mammals, based upon the serendipitous identification of such A-to-G disparities (Table 1). Most notably, subunits of the α-amino-3-hydroxy-5-methyl-isoxazole-4-propionate (AMPA) subtype of ionotropic glutamate receptor (GluR-2, GluR-3 and GluR-4), subunits of the kainate subtype of glutamate-gated ion channel (GluR-5 and GluR-6) and transcripts encoding the 2C-subtype of serotonin receptor ($5HT_{2C}$) were shown to undergo A-to-I modifications that change the amino acid coding potential of the mature mRNAs, producing protein products with altered functional properties (Emeson and Singh 2001; Rueter and Emeson 1998). Recently, numerous laboratories have developed more directed approaches to identify mRNA targets of A-to-I editing, employing both biochemical and computer-based (in silico) strategies (Athanasiadis et al. 2004; Blow et al. 2004; Kikuno et al. 2002; Kim et al. 2004; Levanon et al. 2004; Ohlson et al. 2007). The earliest of these approaches took advantage of proposed sequence conservation surrounding editing sites in two evolutionarily distant species of fruit fly, *Drosophila melanogaster* and *Drosophila pseudoobscura* (Hoopengardner et al. 2003). By comparing sequences for 914 candidate genes that included ion channels, G-protein coupled receptors, proteins involved in fast synaptic neurotransmission and transcription factors, Hoopengardner and colleagues identified 41 genes containing regions within coding sequences that displayed unusually high sequence conservation compared with surrounding sequences. Further characterization of the mRNAs encoded by these genes allowed the identification of 16 additional edited RNA species in *Drosophila*, encoding primarily ligand- and voltage-gated ion channels and components of the synaptic release machinery (Hoopengardner et al. 2003). The identification of numerous editing targets in the *Drosophila* nervous system is consistent with the observation that the prominent phenotype observed in editing-deficient flies involves nervous system dysfunction and neurodegeneration (Hoopengardner et al. 2003; Palladino et al. 2000). Of the newly identified editing events in flies however, only transcripts from the *kcna1* gene, encoding a voltage-gated potassium channel (Kv1.1), have been validated as a target in the mammalian transcriptome (Bhalla et al. 2004; Hoopengardner et al. 2003).

More recently, additional in silico approaches have attempted to identify and validate A-to-G discrepancies between genomic and cDNA sequences on a genome/transcriptome-wide scale, predicting >12,000 editing sites in the human transcriptome of which >94% occur primarily in non-coding regions of RNA transcripts containing short interspersed elements (SINEs) of the Alu and L1 subclass (Athanasiadis et al. 2004; Blow et al. 2004; Kikuno et al. 2002; Kim et al. 2004; Levanon et al. 2004). Although the biologic significance of editing in Alu sequences has not been fully examined, it has been proposed that the editing within this primate-specific class of SINE elements may modulate alternative splicing, chromatin structure and the retention of highly edited RNAs in the nucleus (Chen et al. 2008; Moller-Krull et al. 2008). Despite the success of in silico strategies in the identification and validation of novel editing sites, few codon-altering (recoding) A-to-I modifications were identified in mRNAs.

Table 1 Functional consequences of A-to-L editing

Gene	Protein	AA Change(s)	Function	References
Gria 2	GluR-2 subunit of AMPA subtype of ionotropic glutamate receptor	Q > R R > G	Modulation of Ca^{2+} permeability Modulation of receptor desensitization	Sommer et al. (1991) Lomeli et al. (1994)
Gria 3	GluR-3 subunit of AMPA subtype of ionotropic glutamate receptor	R > G	Modulation of receptor desensitization	Lomeli et al. (1994)
Gria 4	GluR-4 subunit of AMPA subtype of ionotropic glutamate receptor	R > G	Modulation of receptor desensitization	Lomeli et al. (1994)
Grik 1	GluR-5 subunit of AMPA subtype of ionotropic glutamate receptor	Q > R	Modulation of Ca^{2+} permeability	Sailer et al. (1999)
Grik 2	GluR-6 subunit of AMPA subtype of ionotropic glutamate receptor	Q > R I > V Y > C	Modulation of Ca^{2+} permeability	Heinemann et al. (1994) Kohler et al. (1993)
Htr2c	5-hydroxytryptamine (serotonin) receptor 2C	I > V, M N > S, G, D I > V	Modulation of constitutive activity and G-protein coupling efficacy	Burns et al. (1997) Wang et al. (2000)
Kcna1	Voltage-gated K^+ channel, shaker-related subfamily (Kv1.1)	I > M	Modulation of channel inactivation	Bhalla et al. (2004)
Gabra3	γ-aminobutyric acid (GABA) A receptor, subunit $\alpha3$	I > M	Activation/inactivation kinetics Receptor trafficking	Rula et al. (2008) Daniel et al. (2011)
Blcap	Bladder cancer associated protein homolog	Q > R Y > C	Not determined	Levanon et al. (2004)
Flna	Filamin, alpha	Q > R	Not determined	Levanon et al. (2004)
Igfbp7	Insulin-like growth factor binding protein 7	K > R R > G	Not determined	Levanon et al. (2004)
Cyfip2	Cytoplasmic FMR1 interacting protein 2	K > E	Not determined	Levanon et al. (2004)

Among these additional editing targets are transcripts encoding filamin A (FLNA), bladder cancer associated protein (BLCAP), cytoplasmic FMR1 interacting protein 2 (CYFIP2) and insulin-like growth factor binding protein 7 (IGFBP7) (Table 1). Despite their expression in discrete regions of the brain, the role(s) for these proteins in central nervous system function or the consequences of editing have not been determined.

3 Mammalian ADAR Enzymes

A-to-I editing of precursor and mature mRNAs and primary miRNA transcripts is mediated by a family of double-stranded RNA-specific adenosine deaminases (ADARs) that catalyze the hydrolytic deamination of the C-6 position within the purine ring (Polson et al. 1991) and have been the topic of numerous reviews (Bass 2002; Hogg et al. 2011; Nishikura 2010). In mammals, three ADAR proteins (ADAR1, ADAR2 and ADAR3) have been purified and their corresponding genes have been identified (Chen et al. 2000; Hough and Bass 1994; Kim et al. 1994; Liu et al. 1997; O'Connell et al. 1995). ADAR1 and ADAR2 have been shown to be expressed in almost all cell types examined (Melcher et al. 1996b; Wagner et al. 1990) and are able to convert A-to-I in extended regions of duplex RNA within pre-mRNAs, mRNAs, primary miRNA transcripts and viral RNAs (Berg et al. 2001; Luciano et al. 2004; Schaub and Keller 2002; Yang et al. 2006). The expression of ADAR3 has been detected only in post-mitotic neurons in brain regions such as the amygdala and thalamus (Chen et al. 2000), yet ADAR3 has not demonstrated any catalytic activity using synthetic dsRNA or known ADAR substrates (Chen et al. 2000; Maas et al. 2003). ADAR1 and ADAR2 have overlapping yet distinct patterns of editing with some sites edited by only one enzyme and other sites edited equally well by both (Bass 2002; Hogg et al. 2011; Nishikura 2010).

The ADAR1 gene specifies two major protein isoforms, an interferon (IFN) inducible 150 kDa protein (p150) and a ubiquitous, constitutively expressed N-terminally truncated 110 kDa protein (p110), encoded by transcripts with alternative exon 1 structures that initiate from different promoters (George and Samuel 1999). The predicted protein sequence of ADAR1 indicates that it contains three copies of a dsRNA-binding motif (dsRBM), a motif shared among numerous dsRNA-binding proteins (Burd and Dreyfuss 1994; Fierro-Monti and Mathews 2000), a nuclear localization signal in the third dsRBM (Eckmann et al. 2001), and a region homologous to the catalytic domain of other known adenosine and cytidine deaminases. The amino-terminus of the p150 isoform also contains two Z-DNA binding domains, the first of which (Zα) overlaps with a leucine-rich nuclear export signal (Poulsen et al. 2001). The Z-DNA binding domains also have been proposed to tether ADAR1 to sites of transcription (Herbert and Rich 1996) or to mediate interactions between ADARs and other proteins (Poulsen et al. 2001). The locations of the nuclear localization and nuclear export sequences in ADAR1 are consistent with observations that the p150 isoform shuttles between the cytoplasm and the nucleus, whereas the p110 protein is localized predominantly to the nucleus (Eckmann et al. 2001; Fritz et al. 2009; Patterson and Samuel 1995; Strehblow et al. 2002). Alternative splicing within exon 7 of ADAR1 generates two distinct mRNA isoforms that differ by 26 amino acids that encode the linker region between the third double-stranded RNA binding motif and the catalytic domain to affect site-selective editing efficiency (Liu et al. 1997, 1999). Total ablation of ADAR1 expression in mice results in embryonic lethality at day 11.5, manifested by liver disintegration (Hartner et al. 2004) and widespread apoptosis (Wang et al. 2004), suggesting that ADAR1 may promote survival of numerous tissues by editing dsRNAs required for protection against programed cell death. Similar embryonic lethality also results from selective loss of the p150 isoform, indicating a critical role for this ADAR1 isoform in embryonic development (Ward et al. 2011). Humans who are heterozygous for an ADAR1 loss-of-function allele demonstrate dyschromatosis symmetrica hereditaria, a recessive genetic disorder characterized by pea-sized hyperpigmented and hypopigmented macules on the hands and feet (Gao et al. 2005; Miyamura et al. 2003; Suzuki et al. 2005), whereas no such phenotype is observed in mice that are heterozygous for an ADAR1-null allele (Hartner et al. 2004; Wang et al. 2004).

ADAR2 is an 80 kDa protein with structural features similar to those observed for ADAR1 (Melcher et al. 1996b). ADAR2 contains a nuclear localization signal and two dsRNA-binding motifs, sharing approximately 25% amino acid sequence similarity with the dsRBMs of ADAR1. ADAR2 also contains an adenosine deaminase domain sharing 70% amino acid similarity with ADAR1, as well as three zinc-chelating residues conserved in the deaminase domain of both enzymes. As with ADAR1, multiple cDNA isoforms of ADAR2 have been identified in rats, mice and humans including alternative splicing events in mRNA regions encoding the deaminase domain and near the amino-terminus (Gerber et al. 1997; Lai et al. 1997; Rueter et al. 1999). Of particular interest is an alternative splicing event that introduces an additional 47 nucleotides near the 5′-end of the ADAR2 coding region, resulting in a frameshift that is predicted to produce a 9 kDa protein (82 aa) lacking the dsRBMs and catalytic deaminase domain required for protein function (Rueter et al. 1999). This alternative splicing event is dependent upon the ability of ADAR2 to edit its own pre-mRNA, converting an intronic adenosine–adenosine (AA) to an adenosine–inosine (AI) dinucleotide that effectively mimics the highly conserved AG sequence normally found at 3′-splice junctions. These observations indicate that RNA editing may serve as a novel mechanism for the regulation of alternative splicing and provides an autoregulatory strategy by which ADAR2 can modulate its own level of expression (Feng et al. 2006; Rueter et al. 1999). Ablation of ADAR2 expression in mutant mice results in death between post-natal day 0 (P0) and P20, as mutant animals become progressively seizure-prone after P12 (Higuchi et al. 2000).

4 ADAR Substrates in the Central Nervous System

4.1 Glutamate-Gated Ion Channels

4.1.1 AMPA Receptors

Ionotropic glutamate receptors (iGluRs) are involved in fast synaptic neurotransmission and in the establishment and maintenance of synaptic plasticity critical to learning and memory. Three subtypes of iGluRs, named according to selective agonists for each receptor subtype, include *N*-methyl-D-aspartate (NMDA) receptors, α-amino-3-hydroxy-5-methyl-isoxazole-4-propionate (AMPA) receptors and kainate receptors (Ozawa et al. 1998). Ionotropic glutamate receptors are tetrameric and their subunits share a similar core structure: three transmembrane segments (M1, M3 and M4), a pore loop (M2), a large extracellular N-terminal domain, and a highly-regulated, variably-sized C-terminal domain (Fig. 1). The N-terminus and a long hydrophilic region between the M3 and M4 transmembrane segments form a two-domain structure that is responsible for ligand-binding (Wollmuth and Sobolevsky 2004).

The first example of A-to-I editing in mammalian mRNAs was identified in transcripts encoding the GluR-2 subunit of the AMPA receptor in which a genomically-encoded glutamine codon (CAG) was altered to an arginine codon (CIG) (Melcher et al. 1995; Rueter et al. 1995; Sommer et al. 1991; Yang et al. 1995). The substitution of a positively-charged arginine residue for a neutrally-charged glutamine residue at the apex of the membrane reentrant pore loop (M2) changes the conductance properties of channels containing an edited GluR-2(R) subunit (Verdoorn et al. 1991). Heteromeric AMPA channels that contain the edited GluR-2(R) subunit are relatively impermeant to Ca^{2+} ions and show a linear current–voltage (I–V) relationship, whereas channels that lack or contain a non-edited GluR-2(Q) subunit show a double-rectifying I–V relationship and an increased Ca^{2+} conductance (Dingledine et al. 1992; Hollmann et al. 1991; Sommer et al. 1991; Verdoorn et al. 1991). Quantitative PCR analyses of adult rat, mouse and human brain RNA have demonstrated that virtually all GluR-2 transcripts encode this critical arginine residue within M2 while GluR-1, -3 and -4 transcripts encode only a glutamine at the analogous position (Higuchi et al. 2000; Sommer et al. 1991).

RNA editing is also responsible for an A-to-I modification in exon 13 of RNAs encoding the GluR-2, -3 and -4 AMPA receptor subunits to alter a genomically-encoded arginine (AGA) to a glycine (IGA) codon at position 764 (R/G site) to modulate the rate of recovery from receptor desensitization (Fig. 1). Because of faster recovery and a tendency for slower desensitization rates, heteromeric AMPA channels containing edited (glycine-containing) subunits show larger steady-state currents than the non-edited forms (Lomeli et al. 1994). Immediately following this edited codon, an alternative splicing event incorporates one of two mutually-exclusive exons referred to as ‘flip’ and ‘flop’ that encode a portion of the ligand-binding domain

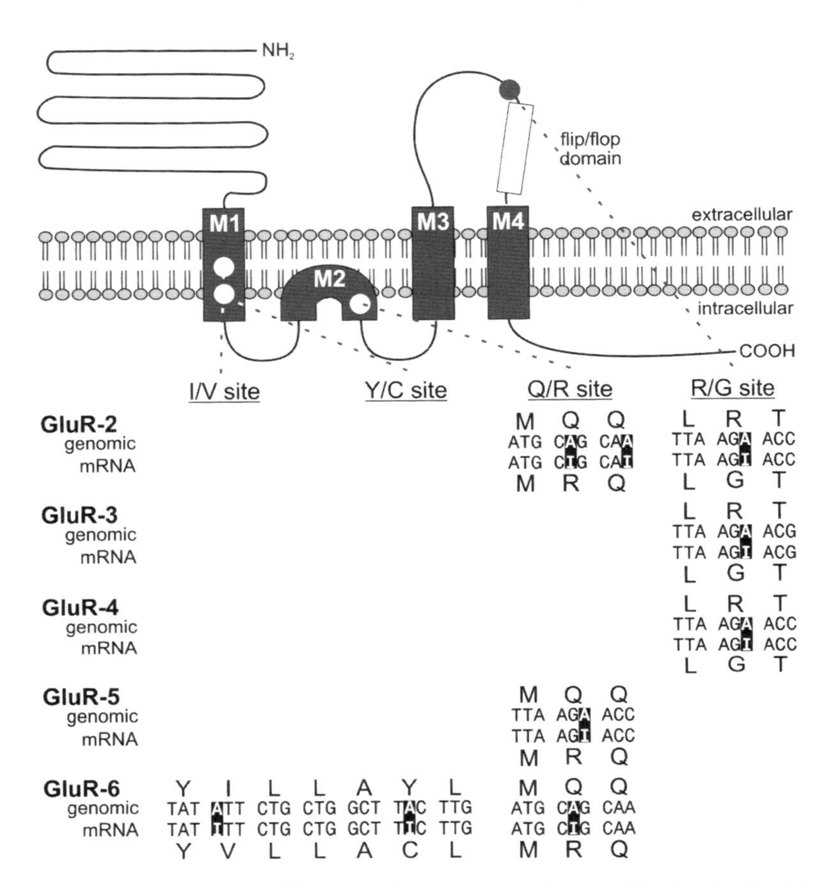

Fig. 1 Summary of the RNA editing events in mouse transcripts encoding ionotropic glutamate receptor subunits. A schematic representation of the proposed topology for GluR subunits is presented, based upon the topology determined for GluR-1 (Hollmann et al. 1991), indicating the relative positions of editing sites. The genomic, mRNA and amino acid sequences surrounding the edited regions are shown and modified nucleosides are presented in inverse lettering

(Fig. 1) (Sommer et al. 1990). Combinations of editing and splicing generate a variety of channels with unique kinetic properties (Koike et al. 2000; Krampfl et al. 2002; Lomeli et al. 1994). Editing at the Q/R and R/G sites together, play a role in receptor trafficking, as editing of the Q/R site attenuates formation of GluR-2 homo-tetramers and leads to retention of the GluR-2(R) subunit in the endoplasmic reticulum (ER) (Greger et al. 2002, 2003, 2007). Indeed, non-edited GluR-2(Q) is released to form homomeric channels on the cell surface while edited subunits remain unassembled in the ER (Greger et al. 2003). Interestingly, several studies have shown that editing at the R/G site (Greger et al. 2006) and flip/flop alternative splicing (Coleman et al. 2006) also play roles in AMPA receptor trafficking.

To determine the biologic significance of Q/R site editing, mutant mice were engineered to solely express the non-edited form of the GluR-2 transcript by

Cre-mediated deletion of the editing complementary sequence (ECS), an intronic region required for formation of the RNA duplex that is essential for A-to-I conversion. A dominant-lethal phenotype was revealed as heterozygous mutant animals appeared healthy until postnatal day 14 (P14), when they begin to develop seizures that lead to death by P20. This phenotype resulted from dramatically increased AMPA receptor permeability to Ca^{2+}, concomitant with neuronal degeneration (Brusa et al. 1995). In subsequent studies, mice homozygous for a null allele of ADAR2 were shown to die of seizures before P20, a phenotype nearly identical to that seen in mice deficient in GluR-2 editing (Q/R site). This early postnatal lethality was rescued by a targeted mutation in which the wild-type GluR-2 allele was modified to express transcripts with a genomically-encoded arginine [GluR-2(R)], thereby circumventing the requirement for editing (Higuchi et al. 2000). Together, these studies highlight the physiologic importance of GluR-2 editing (Q/R site) in normal brain function. However, it also should be noted that Ca^{2+}-permeable AMPA receptors have been identified following mechanical or ischemic brain injury (Rump et al. 1996; Spaethling et al. 2008) or in specific brain regions such as cerebellar Bergmann glia (Burnashev et al. 1992; Iino et al. 2001), yet this Ca^{2+} permeability is thought to result largely from an absence of GluR-2 subunit incorporation into functional AMPA channels rather than the absence of GluR-2 editing.

The editing of GluR-2 mRNA has been implicated recently in the etiology of sporadic amyotrophic lateral sclerosis (ALS), a progressive neurodegenerative disorder involving primarily motor neurons of the cerebral cortex, brain stem and spinal cord, eventually leading to death from respiratory failure (Naganska and Matyja 2011). The editing of GluR-2 transcripts (Q/R site) in spinal motor neurons of ALS patients appears to be inefficient compared to control patients or unaffected neurons (Hideyama et al. 2010), suggesting that the inclusion of the GluR-2(Q) subunit into heteromeric AMPA channels results in a Ca^{2+}-mediated excitotoxicity that contributes to cell death (Kawahara et al. 2003, 2004, 2006; Kwak and Kawahara 2005; Takuma et al. 1999). Support for this hypothesis was recently demonstrated in a mutant mouse line where ADAR2 expression was specifically ablated in $\sim 50\%$ of motor neurons using a conditional ADAR2-null allele in combination with *Cre* recombinase under the control of the vesicular acetylcholine transporter promoter. Motor neurons lacking ADAR2, expressing only non-edited GluR2(Q) subunits, were subject to a slow death, but could be rescued by expression of the Glur-2(R) allele (Hideyama et al. 2010), thus providing the first example of a human neurodegenerative disorder resulting from editing defects and further emphasizing the importance of GluR-2 editing in normal CNS function.

4.1.2 Kainate Receptors

Kainate (KA) receptors, like AMPA receptors, mediate fast excitatory neurotransmission and are widely expressed in a number of brain regions including the neocortex, the caudate/putamen, the CA3 region of the hippocampus, the reticular thalamus and the cerebellar granular layer (Seeburg 1993). There are five KA subunits encoded by distinct genes: GluR-5, GluR-6, GluR-7, KA-1 and KA-2. GluR-5 and GluR-6 form homomeric channels with a high affinity for kainate, but are not activated by AMPA (Bettler et al. 1990; Egebjerg et al. 1991). These two subunits are distinct from the other KA receptor subunits as their RNAs are modified by A-to-I editing (Sommer et al. 1991). Like GluR-2, RNAs for GluR-5 and GluR-6 have a Q/R editing site in a region encoding the hydrophobic pore domain (M2) which can alter the calcium permeability of heteromeric KA receptors containing a GluR-6 subunit (Egebjerg and Heinemann 1993). There are two additional editing sites in the M1 region of GluR-6 and editing leads to the substitution of a valine (ITT) for a genomically-specified isoleucine (ATT) codon (I/V site) and the substitution of a cysteine (TIC) for a tyrosine (TAC) codon (Y/C site). These editing events provide the possibility of eight different edited variants of GluR-6 subunits, all of which are expressed to a varying extent in the CNS, although fully-edited GluR-6 transcripts represent the most abundantly expressed isoform in the adult nervous system (Kohler et al. 1993; Ruano et al. 1995).

Electrophysiologic studies have revealed that channels with editing events in the M1 region of GluR-6 exhibit increased calcium permeability when the M2 pore encodes an arginine at the Q/R site, in direct contrast to the decreased calcium permeability shown for R-containing GluR-2 channels (Kohler et al. 1993). When the M1 region of GluR-6 is not edited, encoding an isoleucine and tyrosine at the I/V and Y/C sites, respectively, the presence of an arginine in M2 does little to alter calcium permeability (Kohler et al. 1993). Recombinant homomeric receptors composed of unedited kainate receptor subunits [GluR-5(Q) and GluR-6(Q)] demonstrate additional functional differences from those containing edited receptor isoforms [GluR-5(R) and GluR-6(R)], including a linear I–V relationship rather than double-rectifying properties, a single low conductance state rather than multiple conductance states and a highly significant increase in the permeability of Cl^- ions (Chittajallu et al. 1999). The physiologic relevance of these functional alterations has yet to be identified however, as studies of mutant mice capable of expressing only the edited GluR-5(R) isoform had no obvious developmental abnormalities or deficits in a number of behavioral paradigms (Sailer et al. 1999). While mutant mice solely expressing the non-edited GluR-6(Q) subunit appeared normal, they exhibited increased NMDA receptor-independent long-term potentiation in hippocampal slices and increased susceptibility to kainate-induced seizures, suggesting a role for GluR-6 (Q/R site) editing the modulation of synaptic plasticity and seizure vulnerability (Vissel et al. 2001).

4.2 The Serotonin 2C Receptor (5HT$_{2C}$)

Serotonin (5-hydroxytryptamine; 5HT) is a monoaminergic neurotransmitter that modulates numerous sensory and motor processes as well as a wide variety of behaviors including sleep, appetite, pain perception, locomotion, thermoregulation, hallucinations and sexual behavior (Werry et al. 2008). The multiple actions of 5HT are mediated by specific interaction with multiple receptor subtypes. Pharmacologic, physiologic and molecular cloning studies have provided evidence for 15 distinct 5HT receptor subtypes which have been subdivided into seven families ($5HT_1$–$5HT_7$) based on relative ligand-binding affinities, genomic structure, amino acid sequence similarities and coupling to specific signal transduction pathways (Barnes and Sharp 1999; Bockaert et al. 2006; Hoyer et al. 1994, 2002). The $5HT_2$ family of receptors includes three receptor subtypes: $5HT_{2A}$, $5HT_{2B}$ and $5HT_{2C}$, which belong to the G-protein-coupled receptor (GPCR) superfamily. The G-protein–$5HT_{2C}$ receptor interactions occur in highly-conserved regions of the second- and third-intracellular loops to potentiate subsequent signal transduction pathways via $G\alpha_{q/11}$, $G\alpha_{12/13}$ and $G\alpha_i$ to modulate effector molecules such as phospholipases C, D and A_2, as well as the extracellular signal-regulated kinases 1 and 2 (Berg et al. 1994, 1998; Werry et al. 2005, 2008). $5HT_{2C}$ mRNA expression has been shown to be widely distributed in neocortical areas, hippocampus, nucleus accumbens, amygdala, choroid plexus, dorsal striatum and substantia nigra (Pasqualetti et al. 1999; Pompeiano et al. 1994), suggestive of physiologic roles in reward behavior, locomotion, energy balance and also when dysregulated, in the development of certain disease states such as obesity, epilepsy, anxiety, sleep disorders and motor dysfunction (Giorgetti and Tecott 2004). Many of these anatomic predictions for $5HT_{2C}$ function have been supported by analyses of $5HT_{2C}$-null mice that exhibit adult-onset obesity, seizures and decreased cocaine-mediated locomotor activity and reward behavior (Abdallah et al. 2009; Brennan et al. 1997; Giorgetti and Tecott 2004; Rocha et al. 2002; Tecott et al. 1995).

RNA transcripts encoding the $5HT_{2C}$ receptor undergo up to five A-to-I editing events that predict alterations in the identity of three amino acids within the second intracellular loop of the receptor to generate as many as 24 receptor isoforms from 32 edited mRNA species (Fig. 2) (Burns et al. 1997; Wang et al. 2000). Sequence analysis of cDNAs isolated from dissected rat, mouse and human brains predicted the region-specific expression of seven major $5HT_{2C}$ isoforms encoded by eleven distinct mRNAs (Abbas et al. 2010; Burns et al. 1997; Morabito et al. 2010; Niswender et al. 1999), suggesting that differentially-edited $5HT_{2C}$ receptors may serve distinct biologic functions in those regions in which they are expressed. Sequencing studies have further revealed that edited mRNAs encoding isoforms with valine, serine and valine (VSV) or valine, asparagine and valine (VNV) at amino acids 157, 159 and 161 are the most highly expressed in a majority of dissected brain regions isolated from human and rat/mouse brains, respectively (Burns et al. 1997; Fitzgerald et al. 1999), whereas the major $5HT_{2C}$ transcripts in the choroid plexus encode the less-edited (INV) and non-edited (INI) receptor isoforms (Burns et al. 1997; Morabito et al. 2010). Functional comparisons in heterologous expression systems, between the non-edited (INI) and the fully-edited (VGV) $5HT_{2C}$ isoforms revealed a 40-fold decrease in serotonergic potency to stimulate phosphoinositide hydrolysis for the VGV isoform due to reduced $G_{q/11}$-protein coupling efficiency and decreased coupling to other

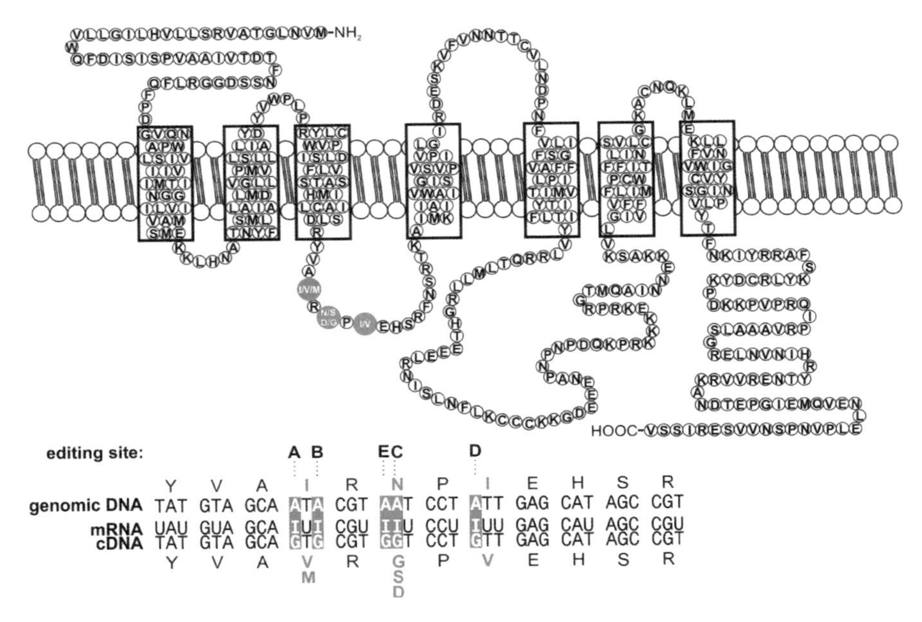

Fig. 2 Summary of RNA recoding events in serotonin 2C receptor RNA and protein isoforms. Schematic representation of the predicted topology and primary amino acid sequence for the mouse $5HT_{2C}$ receptor is presented along with the positions of amino acid alterations within the second intracellular loop resulting from RNA editing events (*colored circles*). Nucleotide and predicted amino acid sequence alignments between $5HT_{2C}$ genomic, mRNA and cDNA sequences are shown; the positions of the five editing sites (A–E) are indicated and nucleotide discrepancies and predicted alterations in amino acid sequence are shown with *colors* corresponding to each codon in which they reside

signaling pathways (Burns et al. 1997; Niswender et al. 1999; Price et al. 2001). In addition, cells expressing more highly edited $5HT_{2C}$ receptors (e.g. VSV and VGV) demonstrate considerably reduced (or absent) constitutive activation in the absence of ligand compared to cells expressing the non-edited isoform (Niswender et al. 1999). This reduction in coupling efficiency and constitutive activity derives from a difference in the ability of edited $5HT_{2C}$ isoforms to spontaneously isomerize to the active R* conformation, a form of the receptor that interacts efficiently with G-proteins in the absence of agonist (Burns et al. 1997; Niswender et al. 1999). As a consequence, the observed potency of agonists with increased affinity for the R* state is disproportionately reduced (Werry et al. 2008) and the 'functional selectivity' of receptor stimulus may be lost (Berg et al. 2001). More recent studies have indicated that alterations in $5HT_{2C}$ editing are observed in suicide victims with a history of major depression (Berg et al. 2001; Gurevich et al. 2002b; Iwamoto and Kato 2003), and in response to anti-depressant and anti-psychotic treatment (Englander et al. 2005; Gurevich et al. 2002b), suggesting that editing of $5HT_{2C}$ transcripts may be involved in psychiatric disorders and also may represent a homeostatic mechanism whereby $5HT_{2C}$ receptor signaling is stabilized in the face of changing synaptic serotonergic input (Englander et al. 2005; Gurevich et al. 2002a).

Chronic administration of IFN-α for the treatment of hepatitis C, hairy-cell leukemia, AIDS-related Kaposi's sarcoma, chronic myelogenous leukemia, and melanoma have been shown to produce depressive symptoms that adversely affect disease outcome because of their negative impact on a patient's quality of life, their interference with treatment adherence and the development of serious complications, including suicide (Ademmer et al. 2001; Valentine et al. 1998; Zdilar et al. 2000). The mechanism by which chronic IFN-α treatment induces depression is yet to be established, although serotonin-mediated effects have been implicated (Cai et al. 2005; Lotrich et al. 2009; Menkes and MacDonald 2000). In vitro studies have demonstrated that IFN-α treatment of glioblastoma cell lines can alter the editing pattern for $5HT_{2C}$ transcripts by increasing the expression of the IFN-inducible isoform of ADAR1 (p150) (Yang et al. 2004), providing a mechanism by which cytokines could induce depression by affecting the editing of numerous ADAR targets in the nervous system to alter subsequent neurotransmitter receptor function.

While the editing of $5HT_{2C}$ transcripts has been shown to modulate multiple aspects of $5HT_{2C}$ receptor signaling and expression in heterologous systems (Berg et al. 2001; Burns et al. 1997; Fitzgerald et al. 1999; Flomen et al. 2004; Marion et al. 2004; Niswender et al. 1999; Price et al. 2001; Wang et al. 2000), until recently, the physiologic importance for the existence of multiple $5HT_{2C}$ isoforms had not been fully explored. Mutant mice solely expressing the fully-edited isoform of the $5HT_{2C}$ receptor display several phenotypic characteristics of Prader–Willi Syndrome (PWS), a maternally imprinted human disorder resulting from a loss of paternal gene expression on chromosome 15q11–13 (Goldstone 2004; Nicholls and Knepper 2001). Mutant mice display a failure to thrive, decreased somatic growth and neonatal muscular hypotonia, followed by post-weaning hyperphagia, in addition to a strain-specific neonatal lethality that is shared with other mouse models of PWS (Chamberlain et al. 2004). These observations are consistent with recent analyses indicating that $5HT_{2C}$ RNA editing is increased in autopsy samples from PWS patients (Kishore and Stamm 2006) and a mouse model (PWS-IC^{del}) that also demonstrates alterations in $5HT_{2C}$-related behaviors (Doe et al. 2009). Previous studies have shown that a maternally-imprinted small nucleolar RNA within the Prader-Willi critical region (*snord115*) can alter both the splicing and editing of $5HT_{2C}$ transcripts (Kishore and Stamm 2006), providing a provoking and straightforward mechanism by which $5HT_{2C}$ RNA processing patterns may be linked with the 15q11–13 locus.

4.3 The α3 Subunit of the $GABA_A$ Receptor (Gabra3)

γ-Aminobutyric acid (GABA) is the main inhibitory neurotransmitter in the vertebrate central nervous system. The $GABA_A$ subtype of GABA receptor is a ligand-gated chloride channel composed of an assembly of five individual subunits, of which there are 19 classes (α1–6, β1–3, γ1–3, δ, ε, ρ1–3, θ, and π) (Olsen and Sieghart 2008, 2009). Different combinations of these subunits generate pharmacologic and functionally distinct isoforms of the $GABA_A$ receptor, which typically contain two α subunits (D'Hulst et al. 2009). The $GABA_A$ receptor is a target for many classes of drugs including barbiturates, benzodiazepenes, ethanol, anti-convulsants and anesthetics (D'Hulst et al. 2009; Fisher 2009; Henschel et al. 2008; Jenkins et al. 2001; Korpi et al. 2007; Low et al. 2000; MacDonald et al. 1989; Meera et al. 2010; Mehta and Ticku 1999; Mohler and Okada 1977; Rudolph et al. 1999; Speth et al. 1980) .

The α3 subunit of $GABA_A$ chloride channels is encoded by the *Gabra3* gene, which is located on the X-chromosome in both humans and mice (Bell et al. 1989; Derry and Barnard 1991). The mouse Gabra3 transcript was first identified as a potential substrate of A-to-I editing using microarray analyses of whole mouse brain RNA that co-immunoprecipitated with ADAR2. This enriched RNA population was further analysed using a bioinformatics paradigm designed to identify extended regions of dsRNA (e.g. stem-loop structures) within the enriched transcripts (Ohlson et al. 2005, 2007). Using this method, an RNA editing event was found in a region encoding sequences immediately adjacent to the extracellular transmembrane 2/3 linker of the α3 subunit (Fig. 3), a region known to be important for channel gating (Ernst et al. 2005). Subsequent sequence analyses verified the presence of an editing site which causes the recoding of a genomically-encoded isoleucine (ATA) to methionine (ATI) codon in cDNA clones isolated from adult mouse brain (Enstero et al. 2010; Rula et al. 2008; Wahlstedt et al. 2009).

To examine functional changes caused by editing, two groups have utilized heterologous expression systems, in which an edited (M) or a non-edited (I) version of the α3 subunit [α3(I) or α3(M)] was expressed along with two additional $GABA_A$ subunits to form a functional $GABA_A$ channel for electrophysiological analyses in human embryonic kidney (HEK) cells. These experiments revealed that $GABA_A$ heteromers containing a non-edited α3(I) subunit have faster activation and deactivation kinetics and slower desensitization than those containing the edited α3(M) isoform. In addition, channels with α3(I) subunits are more outwardly rectifying, assisting in the inhibition of action potentials (Nimmich et al. 2009; Rula et al. 2008). The EC_{50} of GABA for the non-edited channel is approximately 50% of that observed for channels containing the edited α3 subunit, although the effect of various allosteric modulators is not altered (Nimmich et al. 2009). While all six $GABA_A$ α subunits contain a homologous, conserved isoleucine in their third transmembrane domain, only the α3 subunit is edited (Nimmich et al. 2009; Rula et al. 2008). However, mutational analysis revealed that substitution of a methionine for an isoleucine in the analogous position in the α1 subunit resulted in similar functional alterations (Nimmich et al. 2009). In addition to the electrophysiologic changes resulting from the editing of the Gabra3 RNA, more recent studies have revealed that $GABA_A$ receptor trafficking and localization are also affected by A-to-I conversion at the I/M site (Daniel et al. 2011). $GABA_A$ receptors with an α3(M) subunit were expressed to a lesser extent

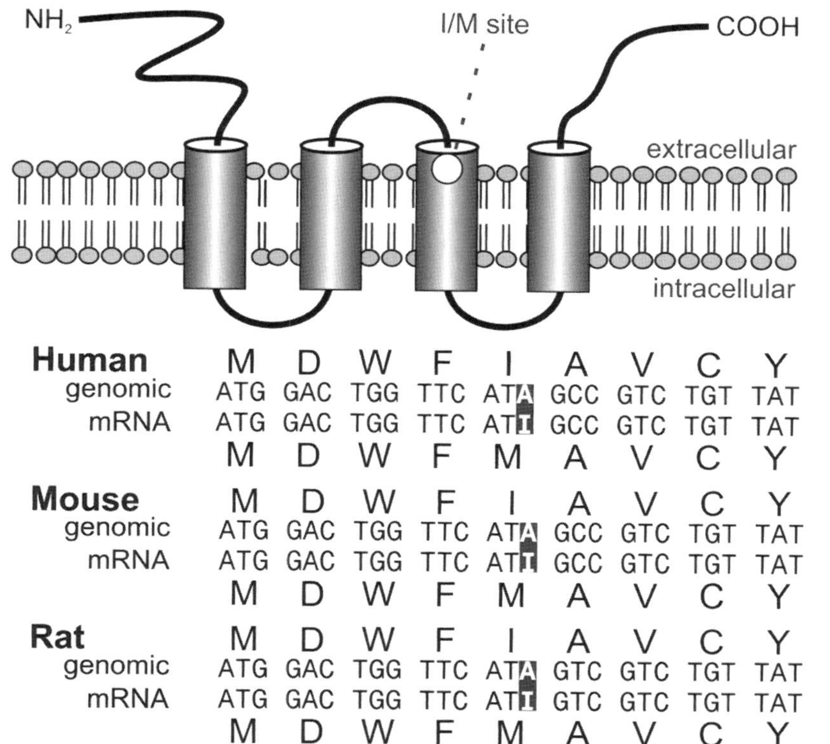

Fig. 3 Conservation of RNA editing in mammalian Gabra3 transcripts. The predicted topology for the α3-subunit of the $GABA_A$ receptor is shown with the position of the editing-dependent amino acid alteration (I/M site) immediately adjacent to the extracellular transmembrane two-third linker. Nucleotide and predicted amino acid sequence alignments between Gabra3 genomic and mRNA from several mammalian species are shown with the positions of the I/M editing site presented with inverse lettering

on the cell surface than heteromeric channels containing a non-edited α3(I) isoform, and α3(M) protein levels were significantly decreased. The substitution of a methionine for an isoleucine in the analogous position of the α1 subunit also modulated receptor cell-surface expression, indicating the importance of this amino acid residue for receptor trafficking (Daniel et al. 2011).

In the mature brain, activation of $GABA_A$ receptors produces a hyperpolarizing influx of chloride ions. While the α1 subunit is the predominant α-subtype in the adult brain, the α2, α3 and α5 subtypes are much more highly expressed in the developing CNS (Laurie et al. 1992). In the developing nervous system however, the chloride gradient in cells is reversed from that of mature neurons, causing chloride ions to flow out from $GABA_A$ channels in response to GABA stimulation. This chloride efflux creates an excitatory response to GABA that can stimulate action potentials as well as relieve the voltage-dependent block of NMDA receptors (Leinekugel et al. 1999). These depolarizing currents occur during a period of robust synaptic development and are crucial for a number of developmental processes including proliferation and synaptogenesis (Ben-Ari 2007; Cancedda et al. 2007; Ge et al. 2006). Concomitant with the age-dependent switch from excitation to inhibition for heteromeric $GABA_A$ receptors is the changing expression of α3 subunit which begins to decline at P7 from the elevated levels observed during embryogenesis (Bosman et al. 2002; Daniel et al. 2011; Hutcheon et al. 2004; Laurie et al. 1992; Rula et al. 2008; Wahlstedt et al. 2009). The editing of Gabra3 transcripts has a reversed developmental pattern however, as editing is low during embryonic development and increases dramatically in adulthood (Ohlson et al. 2007; Rula et al. 2008; Wahlstedt et al. 2009). This developmental pattern of α3 subunit expression provides for the generation of high levels of embryonic $GABA_A$ receptors containing the non-edited a3(I) subunit that are ideally suited to respond to prolonged elevations of GABA, thereby producing robust, long-lived depolarization that may trigger the production of sodium and calcium-mediated action potentials (Rula et al. 2008). Editing of Gabra3 transcripts remains nearly constant after birth in humans (Nicholas et al. 2010), suggesting that the most important role for the non-edited α3(I) isoform is played out during embryogenesis and early development, while the α3(M) subunit, encoded by edited Gabra3 transcripts, is essential throughout the remainder of life for normal inhibitory function.

4.4 Voltage-gated Potassium Channel (Kv1.1)

Voltage-gated K^+ (Kv) channels are key regulators of neuronal membrane excitability, functioning to control resting membrane potentials (Pongs 1999), spontaneous firing rates (Enyedi and Czirjak 2010), the back propagation of action potentials into dendrites (Hoffman et al. 1997) and neurotransmitter release (Ishikawa et al. 2003). Kv channel protein subunits are encoded by at least fourty different genes in humans, and are grouped into 12 subfamilies (Gutman et al. 2005; Jan and Jan 1997). Neurons typically express multiple types of Kv channels with distinct time- and voltage-dependent properties and sub-cellular distributions that differentially contribute to the regulation of firing properties and signal integration. Kv channels are formed by the tetrameric assembly of integral membrane protein subunits (MacKinnon 1991) that each contain six transmembrane segments (S1–6) and intracellular amino- and carboxyl-termini (Fig. 4) (Baldwin et al. 1992; Gutman et al. 2005; MacKinnon 1991). Each functional channel consists of four α-subunits arranged around a central axis that generates an ion conduction pathway formed by the S6 segments, a K^+ ion selectivity filter formed by the connecting loops between the S5 and S6 segments and a transmembrane voltage sensor (S4 segment) responsible for voltage-dependent gating which contains basic amino acid residues at every third position (Doyle et al. 1998; Long et al. 2005).

The first cloned potassium channel gene was encoded by the *Shaker* locus in *Drosophila* (Kamb et al. 1987; Papazian et al. 1987; Tempel et al. 1987).

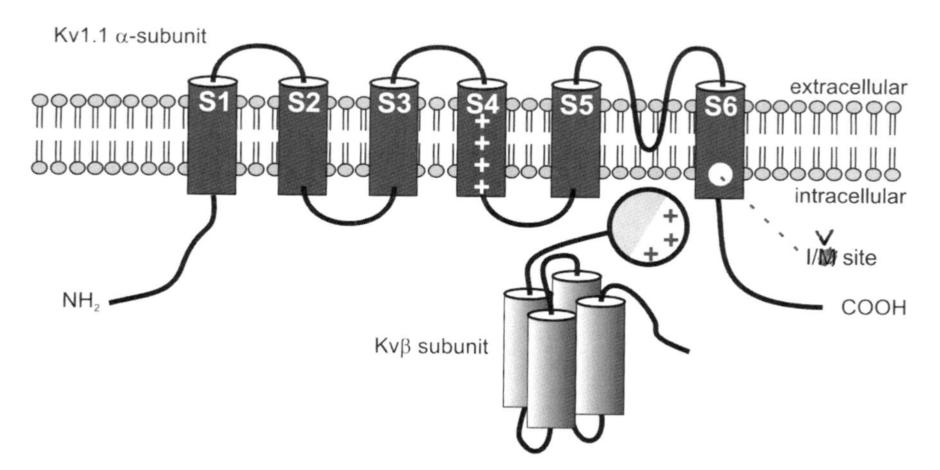

Fig. 4 Schematic diagram of the structures of mammalian Kvα- and β-subunits comprising voltage-gated K^+ channels of the Shaker-related subfamily. Shown is the proposed topology for α-subunits expressing non-inactivating K-channels and a possible topology for Kvβ1 which contains four α-helices and an amino terminal inactivating ball domain with lipophilic (*shaded*) and charged regions. This domain swings upon depolarization into the pore and causes rapid inactivation of the channel. The position of the transmembrane voltage sensor (S4), the K^+ ion selectivity filter (S5–S6 linker) and the I/M editing site in the α-subunit are indicated (Adapted from Heinemann et al. 1994)

Subsequently, a Kv channel gene (*Kcna1*) exhibiting amino acid sequence similar to Shaker was cloned in mice and humans (Curran et al. 1992; Tempel et al. 1988) and shown to encode the Kv1.1 α-subunit, a member of the Kv1 subfamily of voltage-gated potassium channels (Curran et al. 1992; Klocke et al. 1993). The Kv1 subfamily, also known as the *Shaker*-related family, consists of eight different genes (Kv1.1–1.8) (Gutman et al. 2005). Kv1.1 channels are delayed rectifiers that open upon cell depolarization and mediate an outward potassium current, thereby repolarizing the cell and attenuating the action potential (Baranauskas 2007). The Kv1.1-subtype of Kv1 channels is expressed in brain, skeletal muscle, heart, retina and pancreatic islets (Gutman et al. 2005). The Kv1.1 subunit forms both homomeric and heteromeric channels with other members of the Kv1 family, generating extensive functional diversity, as subunit composition greatly affects the kinetic and pharmacologic properties (Al-Sabi et al. 2010; Coleman et al. 1999; Deal et al. 1994; Rasband et al. 2001; Schmidt et al. 1999; Shamotienko et al. 1997; Sokolov et al. 2007). In mammals, the tetramer of α-subunits is joined by four associated β-subunits, giving a 4α–4β stoichiometry (Parcej et al. 1992; Rhodes et al. 1997). An inactivating particle belonging to the β-subunits binds to the inner vestibule of the channel pore region shortly after channel activation to block current flow (Fig. 4), a process known as fast inactivation (Rettig et al. 1994).

In addition to forming a variety of channels types with other Kv1.x subunits, transcripts encoding the Kv1.1 subunit are diversified further by RNA editing.

An isoleucine (ATT) to valine (ITT) codon change occurs as the result of an A-to-I editing event to recode a conserved isoleucine (Ile^{400}) within the Kv family of voltage-gated potassium channels in a region encoding the sixth transmembrane domain (S6), which lines the inner vestibule of the ion-conducting pore (Fig. 4) (Bhalla et al. 2004; Hoopengardner et al. 2003). The extent of editing varies within different regions of the brain, with highest levels in medulla, spinal cord and thalamus (Hoopengardner et al. 2003). Kv1.1 channels containing edited Kv1.1(V) subunits display a 20-fold decrease in the inactivation rate at negative membrane potentials that may result from a reduced affinity for the binding of the inactivating particle of the β-subunits (Bhalla et al. 2004). In addition to reduced inactivation by the gating particle, editing of Kv1.1 RNAs also reduced the blockage of Kv1.1 channels by endogenous, highly-unsaturated signaling lipids such as arachidonic acid, docosahexaenoic acid and anandamide by reducing the affinity of the pore residues for these blocking agents (Decher et al. 2010). Highly-unsaturated fatty acids have been shown previously to convert non-inactivating Kv channels to rapidly-inactivating channels (Honore et al. 1994; Oliver et al. 2004) through occlusion of the permeation pathway, similar to drugs that produce 'open-channel block'. Open-channel block by drugs and lipids was strongly reduced in Kv1.1 channels whose amino acid sequence was altered by RNA editing in the pore cavity and in Kv1.x heteromeric channels containing edited Kv1.1 subunits (Decher et al. 2010). In the *Drosophila* Shaker (Kv1), Shab (Kv2) and squid sqKv2 channels, the position analogous to I^{400} in Kv1.1, are also edited to produce an isoleucine-to-valine substitution that reduces the sensitivity of these channels to highly-unsaturated fatty acids (Bhalla et al. 2004; Decher et al. 2010; Patton et al. 1997; Ryan et al. 2008).

Previous studies have indicated that mutant animals with alterations in the fast inactivation rate of Kv1.1 show behavioral and neurologic deficits associated with hippocampal learning impairment (Giese et al. 1998) and episodic ataxia type-1 (Herson et al. 2003). In a rat model of chronic epilepsy, a fourfold increase in Kv1.1 editing levels was observed in the entorhinal cortex of chronic epileptic animals and a reduced potency for the seizure-inducing Kv open-channel blocker, 4-aminopyridine (4-AP), suggesting that increased editing of Kv1.1 transcripts contributes to the reduced ictogenic potential of 4-AP (Streit et al. 2011). These observations indicate that the combined effects of open-channel block of Kv1 channels by highly-unsaturated lipids together with differential RNA editing can alter the pharmacology of Kv1.x channels and may contribute to the fine-tuning of neuronal signaling in different brain regions.

5 Conclusions

Initially identified as an RNA modification in the anticodon loop of tRNAs from animal, plant and eubacterial origin (Bjork 1995), the deamination of A-to-I has become increasingly recognized as a critical RNA processing event to generate diversity in both the transcriptome and proteome (Blow et al. 2006; Gerber et al. 1998; Gott and Emeson 2000). Early studies found mRNA targets of editing based upon the chance identification of adenosine-to-guanosine (A-to-G) discrepancies between genomic and cDNA sequences that result from the similar base-pairing properties of inosine and guanosine during cDNA synthesis. These mRNA recoding events involved transcripts encoding proteins important for synaptic signaling including ionotropic glutamate receptor subunits (GluR-2, -3, -4, -5 and -6; Fig. 1) and the 2C-subtype of serotonin receptor (Fig. 2). This bias for editing events in the nervous system was thought to reflect either a complexity of neuronal function that requires extensive proteome diversity or simply the small number of neuroscience-focused laboratories that were initially examining this novel RNA processing event. Using primary transcripts encoding the GluR-2 subunit of the AMPA receptor or synthetic duplex RNAs, numerous groups made significant early advances concerning the molecular mechanisms underlying A-to-I conversion including both the *cis*-active regulatory sequences/structures and the enzymes (ADARs) responsible for editing using heterologous expression and in vitro editing systems (Higuchi et al. 1993; Kim and Nishikura 1993; Lomeli et al. 1994; Maas et al. 1996; Melcher et al. 1995, 1996b; O'Connell et al. 1995; Polson and Bass 1994; Rueter et al. 1995; Yang et al. 1995). While the identification of this limited repertoire of substrates proved to be invaluable, the serendipitous nature of such substrate identification raised numerous questions regarding the number of RNAs modified by A-to-I conversion and those tissues in which they were expressed.

While ADAR1 and ADAR2 have shown to be expressed in almost all cell types examined (Melcher et al. 1996b; Wagner et al. 1990), the observation that ADAR expression levels are greatest in the brain (Melcher et al. 1996a, b), accompanied by the development of a candidate-based approach to identify mRNA targets of A-to-I editing in *Drosophila* (Hoopengardner et al. 2003), further supported the idea that editing was enriched in the nervous system. The majority of ADAR targets identified using this strategy were voltage- or ligand-gated ion channels or components of the synaptic vesicular release machinery and independently-identified substrates for editing in *Drosophila* also encoded nervous system signaling components including the paralysis (*para*) voltage-gated Na^+ channel (Hanrahan et al. 2000), the cacophony (*cac*) voltage-gated Ca^{2+} channel (Peixoto et al. 1997; Smith et al. 1998) and the glutamate-gated Cl^- channel (*DrosGluCl-α*) (Semenov and Pak 1999). In addition, flies lacking *Drosophila* ADAR (dADAR) expression exhibited extreme behavioral deficits including temperature-sensitive paralysis, locomotor uncoordination, tremors which increased in severity with age and neurodegeneration (Palladino et al. 2000). Despite this apparent preponderance of editing events in the nervous system, few studies have focused upon editing in peripheral tissues. More recent studies using a transcriptome-wide analysis of neuronal and non-neuronal tissues (cerebellum, frontal lobe, corpus callosum, diencephalon, small intestine, kidney and adrenal gland) identified and validated numerous, novel editing events in multiple transcripts, yet the extent of editing for these RNAs was generally less for tissues outside the central nervous system (Li et al. 2009). The recent development of massively-parallel, high-throughput sequencing strategies (Bentley et al. 2008; Mortazavi et al. 2008) should provide an effective strategy for the identification of A-to-G discrepancies between genomic and cDNA sequences from multiple tissues in a single organism. This experimental paradigm not only provides for a quantitative analysis of novel editing sites, but also reduces the possibility that any observed A-to-G disparity results from a single nucleotide polymorphism.

From alterations in coding potential to changes in structure, stability, translation efficiency and splicing, RNA editing can affect almost all aspects of cellular RNA function. In most cases, RNA editing of protein-coding genes has been shown to generate multiple protein isoforms and to diversify protein function. While the first 10 years of inquiry into the mechanisms of A-to-I conversion provided dramatic advances in our understanding of the enzymatic activities and biochemical mechanisms underlying this RNA processing event, the second decade of study has largely focused upon further identification of ADAR substrates and a determination of how such editing-mediated changes in coding potential can affect protein function. The ultimate biologic relevance of RNA editing resides with the specific substrates that are modified by this process. As we enter a third decade of investigation, many investigators will strive to further identify RNA targets of A-to-I conversion using state-of-the-art sequencing technologies. Other laboratories will examine the functional consequences of identified recoding sites in novel ADAR targets and determine whether dysregulation of editing for specific transcripts is associated with an alteration in phenotype or associations with human disorders.

References

Abbas AI, Urban DJ, Jensen NH, Farrell MS, Kroeze WK, Mieczkowski P, Wang Z, Roth BL (2010) Assessing serotonin receptor mRNA editing frequency by a novel ultra high-throughput sequencing method. Nucleic Acids Res 38:e118

Abdallah L, Bonasera SJ, Hopf FW, O'Dell L, Giorgetti M, Jongsma M, Carra S, Pierucci M, Di Giovanni G, Esposito E, Parsons LH, Bonci A, Tecott LH (2009) Impact of serotonin 2C receptor null mutation on physiology and behavior associated with nigrostriatal dopamine pathway function. J Neurosci Off J Soc Neurosci 29:8156–8165

Ademmer K, Beutel M, Bretzel R, Jaeger C, Reimer C (2001) Suicidal ideation with IFN-alpha and ribavirin in a patient with hepatitis C. Psychosomatics 42:365–367

Al-Sabi A, Shamotienko O, Dhochartaigh SN, Muniyappa N, Le Berre M, Shaban H, Wang J, Sack JT, Dolly JO (2010) Arrangement of Kv1 alpha subunits dictates sensitivity to tetraethylammonium. J Gen Physiol 136:273–282

Athanasiadis A, Rich A, Maas S (2004) Widespread A-to-I RNA editing of Alu-containing mRNAs in the human transcriptome. PLoS Biol 2:e391

Baldwin TJ, Isacoff E, Li M, Lopez GA, Sheng M, Tsaur ML, Yan YN, Jan LY (1992) Elucidation of biophysical and biological properties of voltage-gated potassium channels. Cold Spring Harbor symposia on quantitative biology, vol 57, pp 491–499

Baranauskas G (2007) Ionic channel function in action potential generation: current perspective. Mol Neurobiol 35:129–150

Barnes NM, Sharp T (1999) A review of central 5-HT receptors and their function. Neuropharmacology 38:1083–1152

Bass BL (2002) RNA editing by adenosine deaminases that act on RNA. Annu Rev Biochem 71:817–846

Bell MV, Bloomfield J, McKinley M, Patterson MN, Darlison MG, Barnard EA, Davies KE (1989) Physical linkage of a GABAA receptor subunit gene to the DXS374 locus in human Xq28. Am J Hum Genet 45:883–888

Ben-Ari Y (2007) GABA excites and sculpts immature neurons well before delivery: modulation by GABA of the development of ventricular progenitor cells. Epilepsy Curr/Am Epilepsy Soc 7:167–169

Bentley DR, Balasubramanian S, Swerdlow HP, Smith GP, Milton J, Brown CG, Hall KP, Evers DJ, Barnes CL, Bignell HR, Boutell JM, Bryant J, Carter RJ, Keira Cheetham R, Cox AJ, Ellis DJ, Flatbush MR, Gormley NA, Humphray SJ, Irving LJ, Karbelashvili MS, Kirk SM, Li H, Liu X, Maisinger KS, Murray LJ, Obradovic B, Ost T, Parkinson ML, Pratt MR, Rasolonjatovo IM, Reed MT, Rigatti R, Rodighiero C, Ross MT, Sabot A, Sankar SV, Scally A, Schroth GP, Smith ME, Smith VP, Spiridou A, Torrance PE, Tzonev SS, Vermaas EH, Walter K, Wu X, Zhang L, Alam MD, Anastasi C, Aniebo IC, Bailey DM, Bancarz IR, Banerjee S, Barbour SG, Baybayan PA, Benoit VA, Benson KF, Bevis C, Black PJ, Boodhun A, Brennan JS, Bridgham JA, Brown RC, Brown AA, Buermann DH, Bundu AA, Burrows JC, Carter NP, Castillo N, Chiara ECM, Chang S, Neil Cooley R, Crake NR, Dada OO, Diakoumakos KD, Dominguez-Fernandez B, Earnshaw DJ, Egbujor UC, Elmore DW, Etchin SS, Ewan MR, Fedurco M, Fraser LJ, Fuentes Fajardo KV, Scott Furey W, George D, Gietzen KJ, Goddard CP, Golda GS, Granieri PA, Green DE, Gustafson DL, Hansen NF, Harnish K, Haudenschild CD, Heyer NI, Hims MM, Ho JT, Horgan AM et al (2008) Accurate whole human genome sequencing using reversible terminator chemistry. Nature 456:53–59

Berg KA, Clarke WP, Sailstad C, Saltzman A, Maayani S (1994) Signal transduction differences between 5-hydroxytryptamine type 2A and type 2C receptor systems. Mol Pharmacol 46:477–484

Berg KA, Maayani S, Goldfarb J, Scaramellini C, Leff P, Clarke WP (1998) Effector pathway-dependent relative efficacy at serotonin type 2A and 2C receptors: evidence for agonist-directed trafficking of receptor stimulus. Mol Pharmacol 54:94–104

Berg KA, Cropper JD, Niswender CM, Sanders-Bush E, Emeson RB, Clarke WP (2001) RNA-editing of the 5-HT(2C) receptor alters agonist-receptor-effector coupling specificity. Br J Pharmacol 134:386–392

Bettler B, Boulter J, Hermans-Borgmeyer I, O'Shea-Greenfield A, Deneris ES, Moll C, Borgmeyer U, Hollmann M, Heinemann S (1990) Cloning of a novel glutamate receptor subunit, GluR5: expression in the nervous system during development. Neuron 5:583–595

Bhalla T, Rosenthal JJ, Holmgren M, Reenan R (2004) Control of human potassium channel inactivation by editing of a small mRNA hairpin. Nat Struct Mol Biol 11:950–956

Bjork GR (1995) Genetic dissection of synthesis and function of modified nucleosides in bacterial transfer RNA. Prog Nucleic Acid Res Mol Biol 50:263–338

Blow M, Futreal PA, Wooster R, Stratton MR (2004) A survey of RNA editing in human brain. Genome Res 14:2379–2387

Blow MJ, Grocock RJ, van Dongen S, Enright AJ, Dicks E, Futreal PA, Wooster R, Stratton MR (2006) RNA editing of human microRNAs. Genome Biol 7:R27

Bockaert J, Claeysen S, Becamel C, Dumuis A, Marin P (2006) Neuronal 5-HT metabotropic receptors: fine-tuning of their structure, signaling, and roles in synaptic modulation. Cell Tissue Res 326:553–572

Bosman LW, Rosahl TW, Brussaard AB (2002) Neonatal development of the rat visual cortex: synaptic function of GABAA receptor alpha subunits. J Physiol 545:169–181

Brennan TJ, Seeley WW, Kilgard M, Schreiner CE, Tecott LH (1997) Sound-induced seizures in serotonin 5-HT2c receptor mutant mice. Nat Genet 16:387–390

Brusa R, Zimmermann F, Koh DS, Feldmeyer D, Gass P, Seeburg PH, Sprengel R (1995) Early-onset epilepsy and postnatal lethality associated with an editing-deficient GluR-B allele in mice. Science 270:1677–1680

Burd CG, Dreyfuss G (1994) Conserved structures and diversity of functions of RNA-binding proteins. Science 265:615–621

Burnashev N, Khodorova A, Jonas P, Helm PJ, Wisden W, Monyer H, Seeburg PH, Sakmann B (1992) Calcium-permeable AMPA-kainate receptors in fusiform cerebellar glial cells. Science 256:1566–1570

Burns CM, Chu H, Rueter SM, Hutchinson LK, Canton H, Sanders-Bush E, Emeson RB (1997) Regulation of serotonin-2C receptor G-protein coupling by RNA editing. Nature 387:303–308

Cai W, Khaoustov VI, Xie Q, Pan T, Le W, Yoffe B (2005) Interferon-alpha-induced modulation of glucocorticoid and serotonin receptors as a mechanism of depression. J Hepatol 42:880–887

Cancedda L, Fiumelli H, Chen K, Poo MM (2007) Excitatory GABA action is essential for morphological maturation of cortical neurons in vivo. J Neurosci Off J Soc Neurosci 27:5224–5235

Chamberlain SJ, Johnstone KA, DuBose AJ, Simon TA, Bartolomei MS, Resnick JL, Brannan CI (2004) Evidence for genetic modifiers of postnatal lethality in PWS-IC deletion mice. Hum Mol Genet 13:2971–2977

Chen CX, Cho DS, Wang Q, Lai F, Carter KC, Nishikura K (2000) A third member of the RNA-specific adenosine deaminase gene family, ADAR3, contains both single- and double-stranded RNA binding domains. RNA 6:755–767

Chen LL, DeCerbo JN, Carmichael GG (2008) Alu element-mediated gene silencing. EMBO J 27:1694–1705

Chittajallu R, Braithwaite SP, Clarke VR, Henley JM (1999) Kainate receptors: subunits, synaptic localization and function. Trends Pharmacol Sci 20:26–35

Coleman SK, Newcombe J, Pryke J, Dolly JO (1999) Subunit composition of Kv1 channels in human CNS. J Neurochem 73:849–858

Coleman SK, Moykkynen T, Cai C, von Ossowski L, Kuismanen E, Korpi ER, Keinanen K (2006) Isoform-specific early trafficking of AMPA receptor flip and flop variants. J Neurosci 26:11220–11229

Crick FH (1966) Codon–anticodon pairing: the wobble hypothesis. J Mol Biol 19:548–555

Curran ME, Landes GM, Keating MT (1992) Molecular cloning, characterization, and genomic localization of a human potassium channel gene. Genomics 12:729–737

Daniel C, Wahlstedt H, Ohlson J, Bjork P, Ohman M (2011) Adenosine-to-inosine RNA editing affects trafficking of the gamma-aminobutyric acid type A (GABA(A)) receptor. J Biol Chem 286:2031–2040

Deal KK, Lovinger DM, Tamkun MM (1994) The brain Kv1.1 potassium channel: in vitro and in vivo studies on subunit assembly and posttranslational processing. J Neurosci Off J Soc Neurosci 14:1666–1676

Decher N, Streit AK, Rapedius M, Netter MF, Marzian S, Ehling P, Schlichthorl G, Craan T, Renigunta V, Kohler A, Dodel RC, Navarro-Polanco RA, Preisig-Muller R, Klebe G, Budde T, Baukrowitz T, Daut J (2010) RNA editing modulates the binding of drugs and highly unsaturated fatty acids to the open pore of Kv potassium channels. EMBO J 29:2101–2113

Derry JM, Barnard PJ (1991) Mapping of the glycine receptor alpha 2-subunit gene and the GABAA alpha 3-subunit gene on the mouse X chromosome. Genomics 10:593–597

D'Hulst C, Atack JR, Kooy RF (2009) The complexity of the GABAA receptor shapes unique pharmacological profiles. Drug Discov Today 14:866–875

Dingledine R, Hume RI, Heinemann SF (1992) Structural determinants of barium permeation and rectification in non-NMDA glutamate receptor channels. J Neurosci Off J Soc Neurosci 12:4080–4087

Doe CM, Relkovic D, Garfield AS, Dalley JW, Theobald DE, Humby T, Wilkinson LS, Isles AR (2009) Loss of the imprinted snoRNA mbii-52 leads to increased 5htr2c pre-RNA editing and altered 5HT2CR-mediated behaviour. Hum Mol Genet 18:2140–2148

Doyle DA, Morais Cabral J, Pfuetzner RA, Kuo A, Gulbis JM, Cohen SL, Chait BT, MacKinnon R (1998) The structure of the potassium channel: molecular basis of K^+ conduction and selectivity. Science 280:69–77

Eckmann CR, Neunteufl A, Pfaffstetter L, Jantsch MF (2001) The human but not the *Xenopus* RNA-editing enzyme ADAR1 has an atypical nuclear localization signal and displays the characteristics of a shuttling protein. Mol Biol Cell 12:1911–1924

Egebjerg J, Heinemann SF (1993) Ca_2+ permeability of unedited and edited versions of the kainate selective glutamate receptor GluR6. Proc Natl Acad Sci U S A 90:755–759

Egebjerg J, Bettler B, Hermans-Borgmeyer I, Heinemann S (1991) Cloning of a cDNA for a glutamate receptor subunit activated by kainate but not AMPA. Nature 351:745–748

Emeson R, Singh M (2001) Adenosine-to-inosine RNA editing: substrates and consequences. In: Bass B (ed) RNA editing. Oxford University Press, Oxford, pp 109–138

Englander MT, Dulawa SC, Bhansali P, Schmauss C (2005) How stress and fluoxetine modulate serotonin 2C receptor pre-mRNA editing. J Neurosci Off J Soc Neurosci 25:648–651

Enstero M, Akerborg O, Lundin D, Wang B, Furey TS, Ohman M, Lagergren J (2010) A computational screen for site selective A-to-I editing detects novel sites in neuron specific Hu proteins. BMC Bioinformatics 11:6

Enyedi P, Czirjak G (2010) Molecular background of leak K^+ currents: two-pore domain potassium channels. Physiol Rev 90:559–605

Ernst M, Bruckner S, Boresch S, Sieghart W (2005) Comparative models of GABAA receptor extracellular and transmembrane domains: important insights in pharmacology and function. Mol Pharmacol 68:1291–1300

Feng Y, Sansam CL, Singh M, Emeson RB (2006) Altered RNA editing in mice lacking ADAR2 autoregulation. Mol Cell Biol 26:480–488

Fierro-Monti I, Mathews MB (2000) Proteins binding to duplexed RNA: one motif, multiple functions. Trends Biochem Sci 25:241–246

Fisher JL (2009) The anti-convulsant stiripentol acts directly on the GABA(A) receptor as a positive allosteric modulator. Neuropharmacology 56:190–197

Fitzgerald LW, Iyer G, Conklin DS, Krause CM, Marshall A, Patterson JP, Tran DP, Jonak GJ, Hartig PR (1999) Messenger RNA editing of the human serotonin 5-HT2C receptor. Neuropsychopharmacol Off Publ Am Coll Neuropsychopharmacol 21:82S–90S

Flomen R, Knight J, Sham P, Kerwin R, Makoff A (2004) Evidence that RNA editing modulates splice site selection in the 5-HT2C receptor gene. Nucleic Acids Res 32:2113–2122

Fritz J, Strehblow A, Taschner A, Schopoff S, Pasierbek P, Jantsch MF (2009) RNA-regulated interaction of transportin-1 and exportin-5 with the double-stranded RNA-binding domain regulates nucleocytoplasmic shuttling of ADAR1. Mol Cell Biol 29:1487–1497

Gao M, Wang PG, Yang S, Hu XL, Zhang KY, Zhu YG, Ren YQ, Du WH, Zhang GL, Cui Y, Chen JJ, Yan KL, Xiao FL, Xu SJ, Huang W, Zhang XJ (2005) Two frameshift mutations in the RNA-specific adenosine deaminase gene associated with dyschromatosis symmetrica hereditaria. Arch Dermatol 141:193–196

Ge S, Goh EL, Sailor KA, Kitabatake Y, Ming GL, Song H (2006) GABA regulates synaptic integration of newly generated neurons in the adult brain. Nature 439:589–593

George CX, Samuel CE (1999) Human RNA-specific adenosine deaminase ADAR1 transcripts possess alternative exon 1 structures that initiate from different promoters, one constitutively active and the other interferon inducible. Proc Natl Acad Sci U S A 96:4621–4626

Gerber A, O'Connell MA, Keller W (1997) Two forms of human double-stranded RNA-specific editase 1 (hRED1) generated by the insertion of an Alu cassette. RNA 3:453–463

Gerber A, Grosjean H, Melcher T, Keller W (1998) Tad1p, a yeast tRNA-specific adenosine deaminase, is related to the mammalian pre-mRNA editing enzymes ADAR1 and ADAR2. EMBO J 17:4780–4789

Giese KP, Storm JF, Reuter D, Fedorov NB, Shao LR, Leicher T, Pongs O, Silva AJ (1998) Reduced K^+ channel inactivation, spike broadening, and after-hyperpolarization in Kvbeta1. 1-deficient mice with impaired learning. Learn Mem 5:257–273

Giorgetti M, Tecott LH (2004) Contributions of 5-HT(2C) receptors to multiple actions of central serotonin systems. Eur J Pharmacol 488:1–9

Goldstone AP (2004) Prader–Willi syndrome: advances in genetics, pathophysiology and treatment. Trends Endocrinol Metab TEM 15:12–20

Gott JM, Emeson RB (2000) Functions and mechanisms of RNA editing. Annu Rev Genet 34:499–531
Greger IH, Khatri L, Ziff EB (2002) RNA editing at arg607 controls AMPA receptor exit from the endoplasmic reticulum. Neuron 34:759–772
Greger IH, Khatri L, Kong X, Ziff EB (2003) AMPA receptor tetramerization is mediated by Q/R editing. Neuron 40:763–774
Greger IH, Akamine P, Khatri L, Ziff EB (2006) Developmentally regulated, combinatorial RNA processing modulates AMPA receptor biogenesis. Neuron 51:85–97
Greger IH, Ziff EB, Penn AC (2007) Molecular determinants of AMPA receptor subunit assembly. Trends Neurosci 30:407–416
Gurevich I, Englander MT, Adlersberg M, Siegal NB, Schmauss C (2002a) Modulation of serotonin 2C receptor editing by sustained changes in serotonergic neurotransmission. J Neurosci Off J Soc Neurosci 22:10529–10532
Gurevich I, Tamir H, Arango V, Dwork AJ, Mann JJ, Schmauss C (2002b) Altered editing of serotonin 2C receptor pre-mRNA in the prefrontal cortex of depressed suicide victims. Neuron 34:349–356
Gutman GA, Chandy KG, Grissmer S, Lazdunski M, McKinnon D, Pardo LA, Robertson GA, Rudy B, Sanguinetti MC, Stuhmer W, Wang X (2005) International Union of Pharmacology. LIII. Nomenclature and molecular relationships of voltage-gated potassium channels. Pharmacol Rev 57:473–508
Hanrahan CJ, Palladino MJ, Ganetzky B, Reenan RA (2000) RNA editing of the *Drosophila* para Na+ channel transcript. Evolutionary conservation and developmental regulation. Genetics 155:1149–1160
Hartner JC, Schmittwolf C, Kispert A, Muller AM, Higuchi M, Seeburg PH (2004) Liver disintegration in the mouse embryo caused by deficiency in the RNA-editing enzyme ADAR1. J Biol Chem 279:4894–4902
Heinemann S, Rettig J, Scott V, Parcej DN, Lorra C, Dolly J, Pongs O (1994) The inactivation behaviour of voltage-gated K-channels may be determined by association of alpha- and beta-subunits. J Physiol 88:173–180
Henschel O, Gipson KE, Bordey A (2008) GABAA receptors, anesthetics and anticonvulsants in brain development. CNS Neurol Disord Drug Targets 7:211–224
Herbert A, Rich A (1996) The biology of left-handed Z-DNA. J biol chem 271:11595–11598
Herson PS, Virk M, Rustay NR, Bond CT, Crabbe JC, Adelman JP, Maylie J (2003) A mouse model of episodic ataxia type-1. Nat Neurosci 6:378–383
Hideyama T, Yamashita T, Suzuki T, Tsuji S, Higuchi M, Seeburg PH, Takahashi R, Misawa H, Kwak S (2010) Induced loss of ADAR2 engenders slow death of motor neurons from Q/R site-unedited GluR2. J Neurosci Off J Soc Neurosci 30:11917–11925
Higuchi M, Single FN, Kohler M, Sommer B, Sprengel R, Seeburg PH (1993) RNA editing of AMPA receptor subunit GluR-B: a base-paired intron–exon structure determines position and efficiency. Cell 75:1361–1370
Higuchi M, Maas S, Single FN, Hartner J, Rozov A, Burnashev N, Feldmeyer D, Sprengel R, Seeburg PH (2000) Point mutation in an AMPA receptor gene rescues lethality in mice deficient in the RNA-editing enzyme ADAR2. Nature 406:78–81
Hoffman DA, Magee JC, Colbert CM, Johnston D (1997) K^+ channel regulation of signal propagation in dendrites of hippocampal pyramidal neurons. Nature 387:869–875
Hogg M, Paro S, Keegan LP, O'Connell MA (2011) RNA editing by mammalian ADARs. Adv Genet 73:87–120
Hollmann M, Hartley M, Heinemann S (1991) Ca2+ permeability of KA-AMPA–gated glutamate receptor channels depends on subunit composition. Science 252:851–853
Honore E, Barhanin J, Attali B, Lesage F, Lazdunski M (1994) External blockade of the major cardiac delayed-rectifier K^+ channel (Kv1.5) by polyunsaturated fatty acids. Proc Natl Acad Sci U S A 91:1937–1941
Hoopengardner B, Bhalla T, Staber C, Reenan R (2003) Nervous system targets of RNA editing identified by comparative genomics. Science 301:832–836

Hough RF, Bass BL (1994) Purification of the *Xenopus laevis* double-stranded RNA adenosine deaminase. J biol chem 269:9933–9939
Hoyer D, Clarke DE, Fozard JR, Hartig PR, Martin GR, Mylecharane EJ, Saxena PR, Humphrey PP (1994) International Union of Pharmacology classification of receptors for 5-hydroxytryptamine (Serotonin). Pharmacol Rev 46:157–203
Hoyer D, Hannon JP, Martin GR (2002) Molecular, pharmacological and functional diversity of 5-HT receptors. Pharmacol Biochem Behav 71:533–554
Hutcheon B, Fritschy JM, Poulter MO (2004) Organization of GABA receptor alpha-subunit clustering in the developing rat neocortex and hippocampus. Eur J Neurosci 19:2475–2487
Iino M, Goto K, Kakegawa W, Okado H, Sudo M, Ishiuchi S, Miwa A, Takayasu Y, Saito I, Tsuzuki K, Ozawa S (2001) Glia-synapse interaction through $Ca2^{+}$-permeable AMPA receptors in Bergmann glia. Science 292:926–929
Ishikawa T, Nakamura Y, Saitoh N, Li WB, Iwasaki S, Takahashi T (2003) Distinct roles of Kv1 and Kv3 potassium channels at the calyx of Held presynaptic terminal. J Neurosci Off J Soc Neurosci 23:10445–10453
Iwamoto K, Kato T (2003) RNA editing of serotonin 2C receptor in human postmortem brains of major mental disorders. Neurosci Lett 346:169–172
Jan LY, Jan YN (1997) Cloned potassium channels from eukaryotes and prokaryotes. Annu Rev Neurosci 20:91–123
Jenkins A, Greenblatt EP, Faulkner HJ, Bertaccini E, Light A, Lin A, Andreasen A, Viner A, Trudell JR, Harrison NL (2001) Evidence for a common binding cavity for three general anesthetics within the GABAA receptor. J Neurosci 21:RC136
Kamb A, Iverson LE, Tanouye MA (1987) Molecular characterization of Shaker, a *Drosophila* gene that encodes a potassium channel. Cell 50:405–413
Kawahara Y, Kwak S, Sun H, Ito K, Hashida H, Aizawa H, Jeong SY, Kanazawa I (2003) Human spinal motoneurons express low relative abundance of GluR2 mRNA: an implication for excitotoxicity in ALS. J Neurochem 85:680–699
Kawahara Y, Ito K, Sun H, Aizawa H, Kanazawa I, Kwak S (2004) Glutamate receptors: RNA editing and death of motor neurons. Nature 427:801
Kawahara Y, Sun H, Ito K, Hideyama T, Aoki M, Sobue G, Tsuji S, Kwak S (2006) Underediting of GluR2 mRNA, a neuronal death inducing molecular change in sporadic ALS, does not occur in motor neurons in ALS1 or SBMA. Neurosci Res 54:11–14
Kawahara Y, Zinshteyn B, Sethupathy P, Iizasa H, Hatzigeorgiou AG, Nishikura K (2007) Redirection of silencing targets by adenosine-to-inosine editing of miRNAs. Science 315:1137–1140
Kikuno R, Nagase T, Waki M, Ohara O (2002) HUGE: a database for human large proteins identified in the Kazusa cDNA sequencing project. Nucleic Acids Res 30:166–168
Kim U, Nishikura K (1993) Double-stranded RNA adenosine deaminase as a potential mammalian RNA editing factor. Semin Cell Biol 4:285–293
Kim U, Wang Y, Sanford T, Zeng Y, Nishikura K (1994) Molecular cloning of cDNA for double-stranded RNA adenosine deaminase, a candidate enzyme for nuclear RNA editing. Proc Natl Acad Sci U S A 91:11457–11461
Kim DD, Kim TT, Walsh T, Kobayashi Y, Matise TC, Buyske S, Gabriel A (2004) Widespread RNA editing of embedded alu elements in the human transcriptome. Genome Res 14: 1719–1725
Kishore S, Stamm S (2006) Regulation of alternative splicing by snoRNAs. Cold Spring Harbor symposia on quantitative biology, vol 71, pp 329–334
Klocke R, Roberds SL, Tamkun MM, Gronemeier M, Augustin A, Albrecht B, Pongs O, Jockusch H (1993) Chromosomal mapping in the mouse of eight K(+)-channel genes representing the four Shaker-like subfamilies Shaker, Shab, Shaw, and Shal. Genomics 18:568–574
Kohler M, Burnashev N, Sakmann B, Seeburg PH (1993) Determinants of Ca2+ permeability in both TM1 and TM2 of high affinity kainate receptor channels: diversity by RNA editing. Neuron 10:491–500

Koike M, Tsukada S, Tsuzuki K, Kijima H, Ozawa S (2000) Regulation of kinetic properties of GluR2 AMPA receptor channels by alternative splicing. J Neurosci 20:2166–2174
Korpi ER, Debus F, Linden AM, Malecot C, Leppa E, Vekovischeva O, Rabe H, Bohme I, Aller MI, Wisden W, Luddens H (2007) Does ethanol act preferentially via selected brain GABAA receptor subtypes? the current evidence is ambiguous. Alcohol 41:163–176
Krampfl K, Schlesinger F, Zorner A, Kappler M, Dengler R, Bufler J (2002) Control of kinetic properties of GluR2 flop AMPA-type channels: impact of R/G nuclear editing. Eur J Neurosci 15:51–62
Kwak S, Kawahara Y (2005) Deficient RNA editing of GluR2 and neuronal death in amyotropic lateral sclerosis. J Mol Med 83:110–120
Lai F, Chen CX, Carter KC, Nishikura K (1997) Editing of glutamate receptor B subunit ion channel RNAs by four alternatively spliced DRADA2 double-stranded RNA adenosine deaminases. Mol Cell Biol 17:2413–2424
Laurie DJ, Wisden W, Seeburg PH (1992) The distribution of thirteen GABAA receptor subunit mRNAs in the rat brain III. Embryonic and postnatal development. J Neurosci 12:4151–4172
Leinekugel X, Khalilov I, McLean H, Caillard O, Gaiarsa JL, Ben-Ari Y, Khazipov R (1999) GABA is the principal fast-acting excitatory transmitter in the neonatal brain. Adv Neurol 79:189–201
Levanon EY, Eisenberg E, Yelin R, Nemzer S, Hallegger M, Shemesh R, Fligelman ZY, Shoshan A, Pollock SR, Sztybel D, Olshansky M, Rechavi G, Jantsch MF (2004) Systematic identification of abundant A-to-I editing sites in the human transcriptome. Nat Biotechnol 22:1001–1005
Li JB, Levanon EY, Yoon JK, Aach J, Xie B, Leproust E, Zhang K, Gao Y, Church GM (2009) Genome-wide identification of human RNA editing sites by parallel DNA capturing and sequencing. Science 324:1210–1213
Liu Y, George CX, Patterson JB, Samuel CE (1997) Functionally distinct double-stranded RNA-binding domains associated with alternative splice site variants of the interferon-inducible double-stranded RNA-specific adenosine deaminase. J Biol Chem 272:4419–4428
Liu Y, Emeson RB, Samuel CE (1999) Serotonin-2C receptor pre-mRNA editing in rat brain and in vitro by splice site variants of the interferon-inducible double-stranded RNA-specific adenosine deaminase ADAR1. J Biol Chem 274:18351–18358
Lomeli H, Mosbacher J, Melcher T, Hoger T, Geiger JR, Kuner T, Monyer H, Higuchi M, Bach A, Seeburg PH (1994) Control of kinetic properties of AMPA receptor channels by nuclear RNA editing. Science 266:1709–1713
Long SB, Campbell EB, Mackinnon R (2005) Crystal structure of a mammalian voltage-dependent Shaker family K^+ channel. Science 309:897–903
Lotrich FE, Ferrell RE, Rabinovitz M, Pollock BG (2009) Risk for depression during interferon-alpha treatment is affected by the serotonin transporter polymorphism. Biol Psychiatry 65:344–348
Low K, Crestani F, Keist R, Benke D, Brunig I, Benson JA, Fritschy JM, Rulicke T, Bluethmann H, Mohler H, Rudolph U (2000) Molecular and neuronal substrate for the selective attenuation of anxiety. Science 290:131–134
Luciano DJ, Mirsky H, Vendetti NJ, Maas S (2004) RNA editing of a miRNA precursor. RNA 10:1174–1177
Maas S, Melcher T, Herb A, Seeburg PH, Keller W, Krause S, Higuchi M, O'Connell MA (1996) Structural requirements for RNA editing in glutamate receptor pre-mRNAs by recombinant double-stranded RNA adenosine deaminase. J Biol Chem 271:12221–12226
Maas S, Rich A, Nishikura K (2003) A-to-I RNA editing: recent news and residual mysteries. J Biol Chem 278:1391–1394
MacDonald RL, Rogers CJ, Twyman RE (1989) Barbiturate regulation of kinetic properties of the GABAA receptor channel of mouse spinal neurones in culture. J Physiol 417:483–500
MacKinnon R (1991) Determination of the subunit stoichiometry of a voltage-activated potassium channel. Nature 350:232–235
Marion S, Weiner DM, Caron MG (2004) RNA editing induces variation in desensitization and trafficking of 5-hydroxytryptamine 2C receptor isoforms. J Biol Chem 279:2945–2954

Meera P, Olsen RW, Otis TS, Wallner M (2010) Alcohol- and alcohol antagonist-sensitive human GABAA receptors: tracking delta subunit incorporation into functional receptors. Mol Pharmacol 78:918–924

Mehta AK, Ticku MK (1999) An update on GABAA receptors. Brain Res Brain Res Rev 29: 196–217

Melcher T, Maas S, Higuchi M, Keller W, Seeburg PH (1995) Editing of alpha-amino-3-hydroxy-5-methylisoxazole-4-propionic acid receptor GluR-B pre-mRNA in vitro reveals site-selective adenosine to inosine conversion. J Biol Chem 270:8566–8570

Melcher T, Maas S, Herb A, Sprengel R, Higuchi M, Seeburg PH (1996a) RED2, a brain-specific member of the RNA-specific adenosine deaminase family. J Biol Chem 271:31795–31798

Melcher T, Maas S, Herb A, Sprengel R, Seeburg PH, Higuchi M (1996b) A mammalian RNA editing enzyme. Nature 379:460–464

Menkes DB, MacDonald JA (2000) Interferons, serotonin and neurotoxicity. Psychol Med 30:259–268

Miyamura Y, Suzuki T, Kono M, Inagaki K, Ito S, Suzuki N, Tomita Y (2003) Mutations of the RNA-specific adenosine deaminase gene (DSRAD) are involved in dyschromatosis symmetrica hereditaria. Am J Hum Genet 73:693–699

Mohler H, Okada T (1977) Benzodiazepine receptor: demonstration in the central nervous system. Science 198:849–851

Moller-Krull M, Zemann A, Roos C, Brosius J, Schmitz J (2008) Beyond DNA:RNA editing and steps toward Alu exonization in primates. J Mol Biol 382:601–609

Morabito MV, Abbas AI, Hood JL, Kesterson RA, Jacobs MM, Kump DS, Hachey DL, Roth BL, Emeson RB (2010) Mice with altered serotonin 2C receptor RNA editing display characteristics of Prader–Willi syndrome. Neurobiol Dis 39:169–180

Mortazavi A, Williams BA, McCue K, Schaeffer L, Wold B (2008) Mapping and quantifying mammalian transcriptomes by RNA-Seq. Nat Methods 5:621–628

Naganska E, Matyja E (2011) Amyotrophic lateral sclerosis—looking for pathogenesis and effective therapy. Folia Neuropathol 49:1–13

Nicholas A, de Magalhaes JP, Kraytsberg Y, Richfield EK, Levanon EY, Khrapko K (2010) Age-related gene-specific changes of A-to-I mRNA editing in the human brain. Mech Ageing Dev 131:445–447

Nicholls RD, Knepper JL (2001) Genome organization, function, and imprinting in Prader–Willi and Angelman syndromes. Annu Rev Genomics Hum Genet 2:153–175

Nimmich ML, Heidelberg LS, Fisher JL (2009) RNA editing of the GABA(A) receptor alpha3 subunit alters the functional properties of recombinant receptors. Neurosci Res 63:288–293

Nishikura K (2010) Functions and regulation of RNA editing by ADAR deaminases. Annu Rev Biochem 79:321–349

Niswender CM, Copeland SC, Herrick-Davis K, Emeson RB, Sanders-Bush E (1999) RNA editing of the human serotonin 5-hydroxytryptamine 2C receptor silences constitutive activity. J Biol Chem 274:9472–9478

O'Connell MA, Krause S, Higuchi M, Hsuan JJ, Totty NF, Jenny A, Keller W (1995) Cloning of cDNAs encoding mammalian double-stranded RNA-specific adenosine deaminase. Mol Cell Biol 15:1389–1397

Ohlson J, Enstero M, Sjoberg BM, Ohman M (2005) A method to find tissue-specific novel sites of selective adenosine deamination. Nucleic Acids Res 33:e167

Ohlson J, Pedersen JS, Haussler D, Ohman M (2007) Editing modifies the GABA(A) receptor subunit alpha3. RNA 13:698–703

Oliver D, Lien CC, Soom M, Baukrowitz T, Jonas P, Fakler B (2004) Functional conversion between A-type and delayed rectifier K^+ channels by membrane lipids. Science 304:265–270

Olsen RW, Sieghart W (2008) International Union of Pharmacology. LXX-Subtypes of gamma-aminobutyric acid(A) receptors: classification on the basis of subunit composition, pharmacology, and function update. Pharmacol Rev 60:243–260

Olsen RW, Sieghart W (2009) GABA A receptors: subtypes provide diversity of function and pharmacology. Neuropharmacology 56:141–148

Ozawa S, Kamiya H, Tsuzuki K (1998) Glutamate receptors in the mammalian central nervous system. Prog Neurobiol 54:581–618

Palladino MJ, Keegan LP, O'Connell MA, Reenan RA (2000) dADAR, a *Drosophila* double-stranded RNA-specific adenosine deaminase is highly developmentally regulated and is itself a target for RNA editing. RNA 6:1004–1018

Papazian DM, Schwarz TL, Tempel BL, Jan YN, Jan LY (1987) Cloning of genomic and complementary DNA from Shaker, a putative potassium channel gene from *Drosophila*. Science 237:749–753

Parcej DN, Scott VE, Dolly JO (1992) Oligomeric properties of alpha-dendrotoxin-sensitive potassium ion channels purified from bovine brain. Biochemistry 31:11084–11088

Pasqualetti M, Ori M, Castagna M, Marazziti D, Cassano GB, Nardi I (1999) Distribution and cellular localization of the serotonin type 2C receptor messenger RNA in human brain. Neuroscience 92:601–611

Patterson JB, Samuel CE (1995) Expression and regulation by interferon of a double-stranded-RNA-specific adenosine deaminase from human cells: evidence for two forms of the deaminase. Mol Cell Biol 15:5376–5388

Patton DE, Silva T, Bezanilla F (1997) RNA editing generates a diverse array of transcripts encoding squid Kv2 K^+ channels with altered functional properties. Neuron 19:711–722

Peixoto AA, Smith LA, Hall JC (1997) Genomic organization and evolution of alternative exons in a *Drosophila* calcium channel gene. Genetics 145:1003–1013

Polson AG, Bass BL (1994) Preferential selection of adenosines for modification by double-stranded RNA adenosine deaminase. EMBO J 13:5701–5711

Polson AG, Crain PF, Pomerantz SC, McCloskey JA, Bass BL (1991) The mechanism of adenosine to inosine conversion by the double-stranded RNA unwinding/modifying activity: a high-performance liquid chromatography-mass spectrometry analysis. Biochemistry 30:11507–11514

Pompeiano M, Palacios JM, Mengod G (1994) Distribution of the serotonin 5-HT2 receptor family mRNAs: comparison between 5-HT2A and 5-HT2C receptors. Brain Res Mol Brain Res 23:163–178

Pongs O (1999) Voltage-gated potassium channels: from hyperexcitability to excitement. FEBS Lett 452:31–35

Poulsen H, Nilsson J, Damgaard CK, Egebjerg J, Kjems J (2001) CRM1 mediates the export of ADAR1 through a nuclear export signal within the Z-DNA binding domain. Mol Cell Biol 21:7862–7871

Price RD, Weiner DM, Chang MS, Sanders-Bush E (2001) RNA editing of the human serotonin 5-HT2C receptor alters receptor-mediated activation of G13 protein. J Biol Chem 276: 44663–44668

Rasband MN, Park EW, Vanderah TW, Lai J, Porreca F, Trimmer JS (2001) Distinct potassium channels on pain-sensing neurons. Proc Natl Acad Sci U S A 98:13373–13378

Rettig J, Heinemann SH, Wunder F, Lorra C, Parcej DN, Dolly JO, Pongs O (1994) Inactivation properties of voltage-gated K^+ channels altered by presence of beta-subunit. Nature 369:289–294

Rhodes KJ, Strassle BW, Monaghan MM, Bekele-Arcuri Z, Matos MF, Trimmer JS (1997) Association and colocalization of the Kvbeta1 and Kvbeta2 beta-subunits with Kv1 alpha-subunits in mammalian brain K^+ channel complexes. J Neurosci Off J Soc Neurosci 17: 8246–8258

Rocha BA, Goulding EH, O'Dell LE, Mead AN, Coufal NG, Parsons LH, Tecott LH (2002) Enhanced locomotor, reinforcing, and neurochemical effects of cocaine in serotonin 5-hydroxytryptamine 2C receptor mutant mice. J Neurosci Off J Soc Neurosci 22:10039–10045

Ruano D, Lambolez B, Rossier J, Paternain AV, Lerma J (1995) Kainate receptor subunits expressed in single cultured hippocampal neurons: molecular and functional variants by RNA editing. Neuron 14:1009–1017

Rudolph U, Crestani F, Benke D, Brunig I, Benson JA, Fritschy JM, Martin JR, Bluethmann H, Mohler H (1999) Benzodiazepine actions mediated by specific gamma-aminobutyric acid(A) receptor subtypes. Nature 401:796–800

Rueter S, Emeson R (1998) Adenosine-to-inosine conversion in mRNA. In: Grosjean H, Benne R (eds) Modification and editing of RNA. ASM Press, Washington, D.C, pp 343–361

Rueter SM, Burns CM, Coode SA, Mookherjee P, Emeson RB (1995) Glutamate receptor RNA editing in vitro by enzymatic conversion of adenosine to inosine. Science 267:1491–1494

Rueter SM, Dawson TR, Emeson RB (1999) Regulation of alternative splicing by RNA editing. Nature 399:75–80

Rula EY, Lagrange AH, Jacobs MM, Hu N, Macdonald RL, Emeson RB (2008) Developmental modulation of GABA(A) receptor function by RNA editing. J Neurosci 28:6196–6201

Rump A, Sommer C, Gass P, Bele S, Meissner D, Kiessling M (1996) Editing of GluR2 RNA in the gerbil hippocampus after global cerebral ischemia. J Cereb Blood Flow Metab Off J Int Soc Cereb Blood Flow Metab 16:1362–1365

Ryan MY, Maloney R, Reenan R, Horn R (2008) Characterization of five RNA editing sites in Shab potassium channels. Channels 2:202–209

Sailer A, Swanson GT, Perez-Otano I, O'Leary L, Malkmus SA, Dyck RH, Dickinson-Anson H, Schiffer HH, Maron C, Yaksh TL, Gage FH, O'Gorman S, Heinemann SF (1999) Generation and analysis of GluR5(Q636R) kainate receptor mutant mice. J Neurosci Off J Soc Neurosci 19:8757–8764

Schaub M, Keller W (2002) RNA editing by adenosine deaminases generates RNA and protein diversity. Biochimie 84:791–803

Schmidt K, Eulitz D, Veh RW, Kettenmann H, Kirchhoff F (1999) Heterogeneous expression of voltage-gated potassium channels of the shaker family (Kv1) in oligodendrocyte progenitors. Brain Res 843:145–160

Seeburg PH (1993) The TINS/TiPS Lecture. The molecular biology of mammalian glutamate receptor channels. Trends Neurosci 16:359–365

Semenov EP, Pak WL (1999) Diversification of *Drosophila* chloride channel gene by multiple posttranscriptional mRNA modifications. J Neurochem 72:66–72

Shamotienko OG, Parcej DN, Dolly JO (1997) Subunit combinations defined for K^+ channel Kv1 subtypes in synaptic membranes from bovine brain. Biochemistry 36:8195–8201

Smith LA, Peixoto AA, Hall JC (1998) RNA editing in the *Drosophila* DMCA1A calcium-channel alpha 1 subunit transcript. J Neurogenet 12:227–240

Sokolov MV, Shamotienko O, Dhochartaigh SN, Sack JT, Dolly JO (2007) Concatemers of brain Kv1 channel alpha subunits that give similar K^+ currents yield pharmacologically distinguishable heteromers. Neuropharmacology 53:272–282

Sommer B, Keinanen K, Verdoorn TA, Wisden W, Burnashev N, Herb A, Kohler M, Takagi T, Sakmann B, Seeburg PH (1990) Flip and flop: a cell-specific functional switch in glutamate-operated channels of the CNS. Science 249:1580–1585

Sommer B, Kohler M, Sprengel R, Seeburg PH (1991) RNA editing in brain controls a determinant of ion flow in glutamate-gated channels. Cell 67:11–19

Spaethling JM, Klein DM, Singh P, Meaney DF (2008) Calcium-permeable AMPA receptors appear in cortical neurons after traumatic mechanical injury and contribute to neuronal fate. J Neurotrauma 25:1207–1216

Speth RC, Johnson RW, Regan J, Reisine T, Kobayashi RM, Bresolin N, Roeske WR, Yamamura HI (1980) The benzodiazepine receptor of mammalian brain. Fed Proc 39:3032–3038

Strehblow A, Hallegger M, Jantsch MF (2002) Nucleocytoplasmic distribution of human RNA-editing enzyme ADAR1 is modulated by double-stranded RNA-binding domains, a leucine-rich export signal, and a putative dimerization domain. Mol Biol Cell 13:3822–3835

Streit AK, Derst C, Wegner S, Heinemann U, Zahn RK, Decher N (2011) RNA editing of Kv1.1 channels may account for reduced ictogenic potential of 4-aminopyridine in chronic epileptic rats. Epilepsia 52:645–648

Suzuki N, Suzuki T, Inagaki K, Ito S, Kono M, Fukai K, Takama H, Sato K, Ishikawa O, Abe M, Shimizu H, Kawai M, Horikawa T, Yoshida K, Matsumoto K, Terui T, Tsujioka K, Tomita Y (2005) Mutation analysis of the ADAR1 gene in dyschromatosis symmetrica hereditaria and genetic differentiation from both dyschromatosis universalis hereditaria and acropigmentatio reticularis. J Investig Dermatol 124:1186–1192

Takuma H, Kwak S, Yoshizawa T, Kanazawa I (1999) Reduction of GluR2 RNA editing, a molecular change that increases calcium influx through AMPA receptors, selective in the spinal ventral gray of patients with amyotrophic lateral sclerosis. Ann Neurol 46:806–815

Tecott LH, Sun LM, Akana SF, Strack AM, Lowenstein DH, Dallman MF, Julius D (1995) Eating disorder and epilepsy in mice lacking 5-HT2c serotonin receptors. Nature 374:542–546

Tempel BL, Papazian DM, Schwarz TL, Jan YN, Jan LY (1987) Sequence of a probable potassium channel component encoded at Shaker locus of *Drosophila*. Science 237:770–775

Tempel BL, Jan YN, Jan LY (1988) Cloning of a probable potassium channel gene from mouse brain. Nature 332:837–839

Valentine AD, Meyers CA, Kling MA, Richelson E, Hauser P (1998) Mood and cognitive side effects of interferon-alpha therapy. Semin Oncol 25:39–47

Verdoorn TA, Burnashev N, Monyer H, Seeburg PH, Sakmann B (1991) Structural determinants of ion flow through recombinant glutamate receptor channels. Science 252:1715–1718

Vissel B, Royle GA, Christie BR, Schiffer HH, Ghetti A, Tritto T, Perez-Otano I, Radcliffe RA, Seamans J, Sejnowski T, Wehner JM, Collins AC, O'Gorman S, Heinemann SF (2001) The role of RNA editing of kainate receptors in synaptic plasticity and seizures. Neuron 29:217–227

Wagner RW, Yoo C, Wrabetz L, Kamholz J, Buchhalter J, Hassan NF, Khalili K, Kim SU, Perussia B, McMorris FA et al (1990) Double-stranded RNA unwinding and modifying activity is detected ubiquitously in primary tissues and cell lines. Mol Cell Biol 10:5586–5590

Wahlstedt H, Daniel C, Enstero M, Ohman M (2009) Large-scale mRNA sequencing determines global regulation of RNA editing during brain development. Genome Res 19:978–986

Wang Q, O'Brien PJ, Chen CX, Cho DS, Murray JM, Nishikura K (2000) Altered G protein-coupling functions of RNA editing isoform and splicing variant serotonin2C receptors. J Neurochem 74:1290–1300

Wang Q, Miyakoda M, Yang W, Khillan J, Stachura DL, Weiss MJ, Nishikura K (2004) Stress-induced apoptosis associated with null mutation of ADAR1 RNA editing deaminase gene. J Biol Chem 279:4952–4961

Ward SV, George CX, Welch MJ, Liou LY, Hahm B, Lewicki H, de la Torre JC, Samuel CE, Oldstone MB (2011) RNA editing enzyme adenosine deaminase is a restriction factor for controlling measles virus replication that also is required for embryogenesis. Proc Natl Acad Sci U S A 108:331–336

Werry TD, Gregory KJ, Sexton PM, Christopoulos A (2005) Characterization of serotonin 5-HT2C receptor signaling to extracellular signal-regulated kinases 1 and 2. J Neurochem 93:1603–1615

Werry TD, Loiacono R, Sexton PM, Christopoulos A (2008) RNA editing of the serotonin 5HT2C receptor and its effects on cell signalling, pharmacology and brain function. Pharmacol Ther 119:7–23

Wollmuth LP, Sobolevsky AI (2004) Structure and gating of the glutamate receptor ion channel. Trends Neurosci 27:321–328

Yang JH, Sklar P, Axel R, Maniatis T (1995) Editing of glutamate receptor subunit B pre-mRNA in vitro by site-specific deamination of adenosine. Nature 374:77–81

Yang W, Wang Q, Kanes SJ, Murray JM, Nishikura K (2004) Altered RNA editing of serotonin 5-HT2C receptor induced by interferon: implications for depression associated with cytokine therapy. Brain Res Mol Brain Res 124:70–78

Yang W, Chendrimada TP, Wang Q, Higuchi M, Seeburg PH, Shiekhattar R, Nishikura K (2006) Modulation of microRNA processing and expression through RNA editing by ADAR deaminases. Nat Struct Mol Biol 13:13–21

Zdilar D, Franco-Bronson K, Buchler N, Locala JA, Younossi ZM (2000) Hepatitis C, interferon alfa, and depression. Hepatology 31:1207–1211

Modulation of MicroRNA Expression and Function by ADARs

Bjorn-Erik Wulff and Kazuko Nishikura

Abstract MicroRNAs (miRNAs) are small non-coding RNAs that regulate gene expression by preventing the translation of specific messenger RNAs. Adenosine deaminases acting on RNAs (ADARs) catalyze adenosine-to-inosine (A-to-I) RNA editing, the conversion of adenosines into inosines, in double-stranded RNAs. Because inosine preferentially base pairs with cytidine, this conversion is equivalent to an adenosine to guanosine change. Over the past seven years, an increasing number of edited adenosines have been identified in miRNAs. Editing of miRNAs affects their biogenesis, causes their degradation or alters the set of messenger RNAs that they regulate. Recently, ADARs have been shown to also affect the miRNA phenomenon by sequestering miRNAs or by editing the messenger RNAs they regulate. This article reviews the recent attempts to identify miRNA editing sites and elucidate the effects of ADARs on miRNA expression and function.

Abbreviations

A	Adenosine
I	Inosine
G	Guanosine
U	Uridine
C	Cytidine
UTR	Untranslated region
dsRNA	Double-stranded RNA
ADAR	Adenosine deaminase acting on RNA
miRNA	MicroRNA
pri-miRNA	Primary miRNA

B.-E. Wulff (✉) · K. Nishikura
The Wistar Institute, 3601 Spruce Street, Philadelphia, PA 19104, USA
e-mail: wulff@sas.upenn.edu

Current Topics in Microbiology and Immunology (2012) 353: 91–109
DOI: 10.1007/82_2011_151

Published Online: 15 July 2011

pre-miRNA	Precursor miRNA
mRNA	Messenger RNA
pre-mRNA	Precursor messenger RNA
siRNA	Small interfering RNA
cDNA	Complementary DNA
Tudor-SN	Tudor staphylococcal nuclease
EBV	Epstein-Barr virus
DFFA	DNA fragmentation factor alpha
PRPS1	Phosphoribosyl pyrophosphate synthetase 1

Contents

1 Introduction

Adenosine-to-inosine (A-to-I) RNA editing is a posttranscriptional process catalyzed by enzymes of the adenosine deaminase acting on RNA (ADAR) family (Bass 2002; Nishikura 2010). ADARs are conserved from man to sea anemones (Jin et al. 2009) and have been extensively studied in mammals, *Xenopus*, *Drosophila*, and *C. elegans*. Mammals have three ADAR genes, ADAR1, ADAR2, and ADAR3, though any enzymatic activity of ADAR3 remains to be demonstrated. Transcription from separate promoters generates two different ADAR1 isoforms: the nuclear, constitutively expressed ADAR1p110 and the cytoplasmic, interferon-inducible ADAR1p150. ADARs catalyze the conversion of specific adenosines in double-stranded RNA (dsRNA) structures to inosines, which base pair with cytidine and are therefore usually equivalent to guanosines (Bass 2002; Nishikura 2010).

Encoded in the genomes of animals, plants, and some of their DNA viruses, microRNAs (miRNAs) provide a mechanism for post-transcriptional gene silencing (Kim et al. 2009; Krol et al. 2010; Siomi and Siomi 2010). Following transcription, they are processed into mature miRNAs of $\sim$21 nucleotides (nt) in length. These are loaded onto the RNA-induced silencing complex (RISC) which sequesters or degrades transcripts complementary to its loaded miRNA. This gives miRNAs tremendous influence on the state of the cell.

Because both A-to-I editing and miRNA processing make use of dsRNAs, they have many substrates in common. This allows RNA editing to influence the miRNA phenomenon. For example, editing of primary miRNAs (pri-miRNAs) can affect their processing into mature miRNAs or lead to expression of edited miRNAs, which silence a different set of target genes. ADARs also have effects on miRNAs independent of their catalytic activity.

2 MiRNA Biogenesis and Function

Biogenesis of miRNAs is regulated through sequential steps in the nucleus and cytoplasm (Kim et al. 2009; Krol et al. 2010; Siomi and Siomi 2010). Pri-miRNAs are generally transcribed from the genome by RNA polymerase II as 5′-capped and polyadenylated transcripts of several kilobases in length. This makes them similar to precursor messenger RNAs (pre-mRNAs). In fact, some pri-miRNAs are harbored within pre-mRNAs. What sets pri-miRNAs apart from other RNA polymerase II transcripts is one or more palindromic sequences which fold up to form hairpins of ~70 nt in length.

These hairpins are recognized by the nuclear Drosha-DGCR8 complex. DGCR8 helps enzymatic Drosha bind its pri-miRNA substrates, and Drosha's RNase III domains cleave both hairpin strands about two turns of the helix away from the hairpin loop. This excises the ~65 nt precursor miRNA (pre-miRNA). Like most RNase III enzymes, Drosha leaves a 3′ ~2 nt overhang and a 5′ phosphate at its cleavage site. These function as identifiers of the pre-miRNA.

Following Drosha cleavage, the pre-miRNA is exported to the cytoplasm. Nuclear transport receptor Exportin-5 (Exp-5) binds dsRNA with 3′ overhangs and 5′ phosphates like those left by Drosha. By also binding to the GTP-bound trimeric GTPase Ran, Exportin-5 exports the pre-miRNA through the nuclear pore complex to the cytoplasm.

In the cytoplasm, the pre-miRNA is recognized by the Dicer-TRBP complex. TRBP helps Dicer bind the dsRNA region of the pre-miRNA. Dicer's PAZ domain binds the pre-miRNA 3′ overhang and 5′ phosphate, and its RNase III domains cleave off the hairpin loop. The distance between these domains measures the ~21 base pairs to be incorporated into the mature miRNA duplex. Like Drosha, Dicer leaves a 3′ overhang and 5′ phosphate at its cleavage site.

The two strands of the mature miRNA duplex are referred to as the 5p and 3p strands, depending on whether they originate from the 5′ or 3′ end of the pri-miRNA. One gets loaded onto RISC while the other is degraded. The choice of the strand loaded onto RISC, referred to as the *guide strand*, depends on the relative stabilities of the duplex ends. The strand with the less stable 5′ end usually

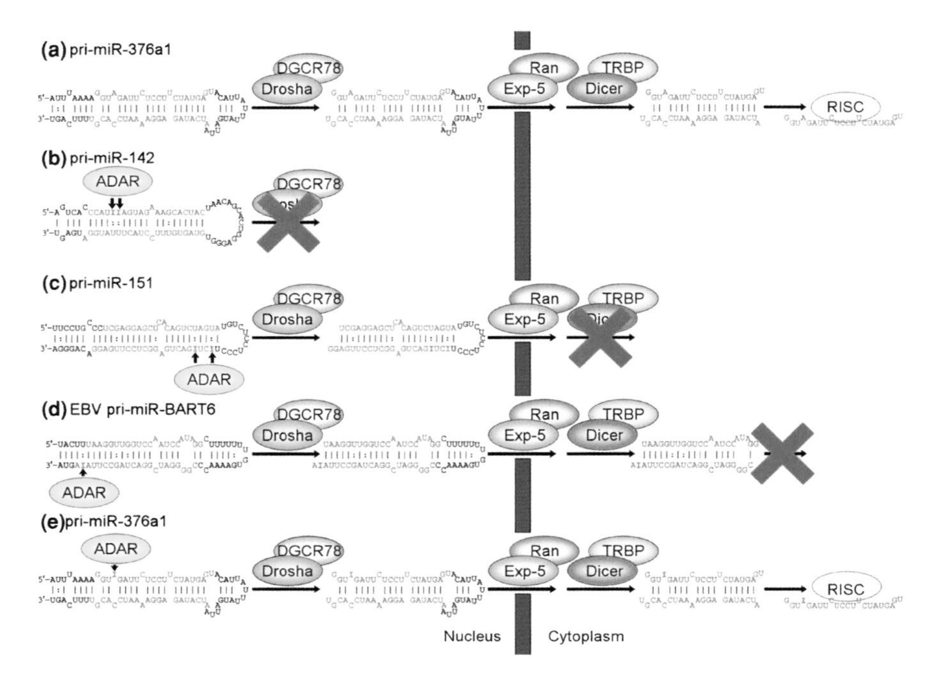

Fig. 1 Possible consequences of miRNA editing. **a** In the absence of editing, pri-miRNAs are processed by the Drosha-DGCR8 complex into pre-miRNAs, exported to the cytoplasm by Exportin-5 and processed by the Dicer-TRBP complex into mature miRNA duplexes. One strand of this duplex is then loaded onto the RISC complex. Editing can (**b**) prevent Drosha cleavage (Yang et al. 2006), **c** prevent Dicer cleavage (Kawahara et al. 2007a) or (**d**) prevent RISC loading (Iizasa et al. 2010). **e** If none of these steps are prevented by editing, a mature miRNA with an altered sequence can be expressed. This can cause redirection of target silencing (Kawahara et al. 2007b)

becomes the guide strand. For most miRNAs, one strand is predominantly chosen as the guide strand. This entire maturation process is illustrated in Fig. 1a.

Following miRNA loading, RISC proceeds to degrade or sequester mRNAs complementary to its guide strand. These transcripts are referred to as targets of the miRNA. Most of them contain several guide strand-complementary *target sites* in their 3′ untranslated regions (UTRs), although some also contain target sites in the 5′UTRs or coding regions. Perfect complementarity leads to degradation, while less perfect complementarity leads to sequestration. Prediction of target sites is difficult but can be done with enough accuracy to form the basis for validation experiments (Grimson et al. 2007). Guide strand nucleotides 2–8 have particular importance for the choice of target sites and are therefore referred to as the *seed sequence* (Grimson et al. 2007).

3 Editing of MiRNAs

3.1 Identification of Pri-MiRNA Editing Sites

Inosine base pairs with cytidine during reverse transcription, so it shows up as guanosine during sequencing of complementary DNA (cDNA). An A-to-I editing site can therefore be inferred by the presence of guanosine at a given position in some cDNA sequences but only adenosine in the corresponding genomic position (Bass 2002; Nishikura 2010).

3.1.1 Systematic Surveys for MiRNA Editing Sites

A-to-I editing of a pri-miRNA was first reported in 2004 (Luciano et al. 2004). PCR amplification and sequencing of the region surrounding the pri-miR-22 hairpin revealed editing at several positions in mouse brain and human brain, lung, and testis (Fig. 2a). However, editing frequency was very low; less than 10%. While the biological consequence of this editing is still unclear, its discovery demonstrated the existence of miRNA editing and sparked further studies.

In 2006, Blow et al. attempted to isolate, PCR amplify, and sequence all 231 human pri-miRNAs registered in miRBase and their corresponding genomic DNA from ten different tissues: adult human brain, heart, liver, lung, ovary, placenta, skeletal muscle, small intestine, spleen, and testis (Blow et al. 2006). They succeeded for 99 pri-miRNAs.

Six of these showed A-to-I editing for at least one tissue: pri-miR-99a, pri-miR-151 (Fig. 2b), pri-miR-379 (Fig. 2c), pri-miR-223, pri-miR-376a1 (Fig. 2d) and pri-miR-197. One further editing site was located in a novel pri-miRNA-like hairpin that has since been annotated as pri-miR-545 in miRBase. Finally, the survey identified five uridine-to-cytidine (U-to-C) conversions (in pri-miR-133a1, pri-miR-371, pri-miR-144, pri-miR-451, pri-miR-215, and pri-miR-194-1). The origin of these U-to-C conversions is uncertain since there is no known enzyme capable of catalyzing the U-to-C conversion. Since the method used was unable to distinguish between reverse complements, they might represent transcription and editing of the opposite strand. These results indicate editing in about 6% of miRNAs. However, this might be an underestimation. The survey did not detect editing in pri-miR-22 and might therefore also have missed other pri-miRNAs edited at low levels.

In 2008, these results were extended by a second systematic survey (Kawahara et al. 2008). Known editing sites indicated that adenosines are particularly prone to editing when located in UAG motifs. For this reason, Kawahara et al. decided to focus on the 257 pri-miRNAs in miRBase at the time with such UAG motifs, with at most a mismatch at either the U or G, in their hairpin stems. They isolated total RNA from human brain and successfully amplified and sequenced 209 of these

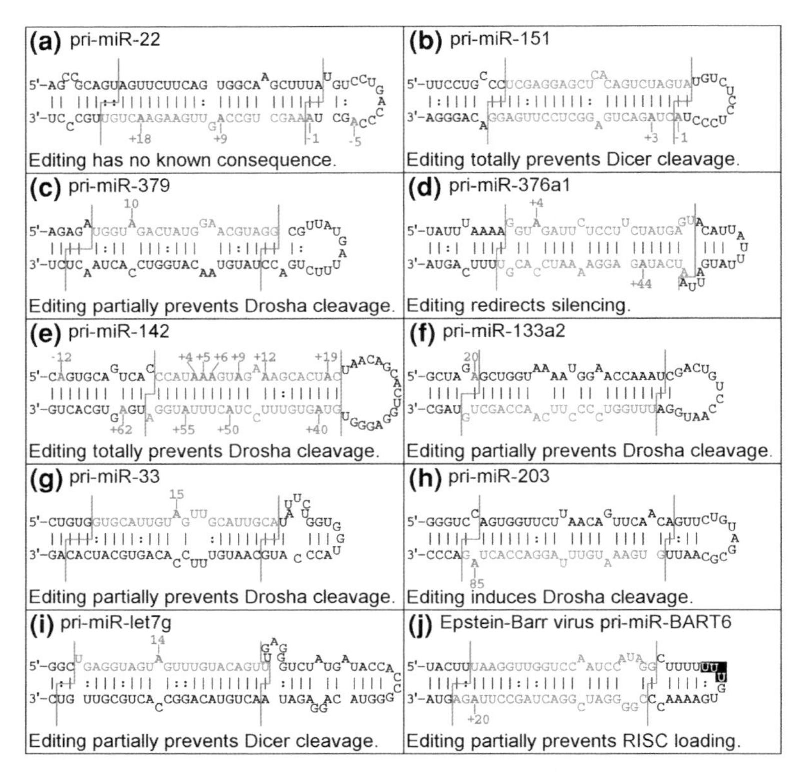

Fig. 2 Some known edited pri-miRNAs. The figure shows the secondary structures of some known edited pri-miRNAs and lists the consequences of their editing. Editing sites are highlighted in red and their positions are given. Positions in pri-miR-22, pri-miR-151, pri-miR-376a1, pri-miR-142, and pri-miR-BART6 are labeled based on schemes established by Luciano et al. 2004; Kawahara et al. 2007a, b; Yang et al. 2006; Iizasa et al. 2010, respectively. For the remaining pri-miRNAs, the first nucleotide of the pri-miRNA as given by miRBase is defined as position 1. Sequences commonly loaded onto RISC are highlighted in *green*. Drosha and Dicer cleavage sites are indicated by *red lines*. **a** Although human and mouse pri-miR-22 have identical sequences, they are edited at different nucleotides. Mouse pri-miR-22 is edited at positions −41, +1, and +2 (Luciano et al. 2004). **b** Editing of pri-miR-151 prevents Dicer cleavage (Kawahara et al. 2007a). **c** Editing of pri-miR-379 partially prevents pre-miR-379 Dicer cleavage (Kawahara et al. 2008). **d** Editing of pri-miR-376a1 redirects silencing (Kawahara et al. 2007b). **e** Editing of pri-miR-142 prevents Drosha cleavage and causes degradation by Tudor-SN (Yang et al. 2006). **f** Editing of pri-miR-133a2 partially prevents Drosha cleavage (Kawahara et al. 2008). **g** Editing of pri-miR-33 partially prevents Drosha cleavage (Kawahara et al. 2008). **h** Editing of pri-miR-203 editing increases the efficiency of Drosha cleavage (Kawahara et al. 2008). **i** Editing of pri-miR-let7 partially prevents pre-miR-let7 Dicer cleavage (Kawahara et al. 2008). **j** Two different sequence variants of EBV-encoded pri-miR-BART6 have been identified. One lacks the three boxed U residues. For the pri-miRNA without the UUU deletion, editing prevents RISC loading. For the pri-miRNA with the UUU deletion, editing prevents Drosha cleavage (Iizasa et al. 2010)

pri-miRNAs. This identified 43 edited UAG adenosines and 43 edited non-UAG adenosines in 47 pri-miRNAs.

From these results, a new estimate of the frequency of pri-miRNA editing was proposed. Forty-seven of the 209 sequenced pri-miRNAs containing UAG motifs were edited. This rate suggests another 11 edited pri-miRNAs among the 48 pri-miRNAs not successfully sequenced. Twenty of the 209 sequenced pri-miRNAs contained non-UAG editing sites. This rate suggests another 20 edited pri-miRNAs among the 217 pri-miRNAs excluded from this study because they did not contain UAG motifs. The conclusion is that ~79 or ~16% of the 474 pri-miRNAs known at the time were edited. Although the number of known pri-miRNAs is growing, that ~16% of pri-miRNAs are edited likely remains a valid estimate.

This represents the most extensive search for editing sites in pri-miRNAs to date; yet it covers only a fraction of the 1,048 human pri-miRNAs currently in miRBase. As explained in Sect. 3.2.6, the pri-miRNAs still unexamined could likely reveal interesting and yet undiscovered consequences of miRNA editing.

3.1.2 High-Throughput Sequencing of Mature MiRNAs

Since 2006, large-scale sequencing of small RNAs (<35 nt) has been carried out for a variety of organisms, cell lines, and tissues, and many of the sequences generated correspond to mature miRNAs (Ruby et al. 2006; Landgraf et al. 2007; Babiarz et al. 2008; Kuchenbauer et al. 2008; Morin et al. 2008; Suzuki et al. 2009; Chiang et al. 2010; Linsen et al. 2010; Schulte et al. 2010; Berezikov et al. 2011; Wulff et al. 2011). Mapping and comparing these sequences to their genomic origins could allow identification of sequence alterations representing A-to-I editing. However, this approach to editing site identification has its difficulties.

Mature miRNAs are only 19–22 nt long and may have 3′ non-templated nucleotide additions. Furthermore, certain miRNAs are derived from miRNA families whose members have high sequence similarities. These traits make them susceptible to cross-mapping, in which sequences originating from one locus are inadvertently mapped to another (de Hoon et al. 2010). For example, a 3′ non-templated adenosine addition can create a mature miRNA sequence that does not map perfectly to its origin. If there exists a second member of the same miRNA family whose mature miRNA contains this additional 3′ adenosine, cross-mapping is likely to occur. If the second miRNA also differs from the first by an A-to-G difference, cross-mapping could lead to the false identification of editing in the second miRNA.

To avoid this, some studies have applied stringent filters when mapping miRNA sequences (Chiang et al. 2010; Linsen et al. 2010; Berezikov et al. 2011). With these filters, the number of identified putative editing sites dwindles to only a handful. However, there are indications that this improves the reliability of the reported editing sites: previously known editing sites (Chiang et al. 2010; Linsen et al. 2010) and editing sites in UAG motifs (see Sect. 3.1.1) are overrepresented (Chiang et al. 2010). These miRNAs are good candidates for further investigation of redirection of miRNA silencing by editing (see Sect. 3.2.5).

An algorithm for preventing cross-mapping that does not discard many of its sequences by filtering has also been developed (de Hoon et al. 2010). First, one maps the sequences by the naïve method described above. From the number of sequences mapping to each genomic origin, one estimates its expression level. By the types of mismatches in the alignments to each genomic origin, one estimates its error profile. Then, one discards the original mapping results and repeats the mapping procedure. However, this time, sequences are not divided equally between potential genomic origins but are preferentially assigned to the sites with the higher expression levels and the error profiles better corresponding to the sequence in question. This procedure is repeated until convergence, and only then are the sequence reads assigned to their final genomic origins. Application of this algorithm to the FANTOM4 small RNA sequence library (Suzuki et al. 2009), which contains 236 mature miRNAs, revealed that editing can be confirmed with only one of these mature miRNA, miR-376c, at least in a human macrophage cell line examined in this study (de Hoon et al. 2010).

3.2 Consequences of Pri-MiRNA Editing

MiRNA biogenesis and function are easily affected by single nucleotide exchanges (Gottwein et al. 2006; Sun et al. 2009), so editing can have critical effects on the life of a miRNA. It appears that most instances of miRNA editing prevent some step in the miRNA's maturation (Kawahara et al. 2008). However, some instances of editing have no known effect on, or even facilitate, maturation (Kawahara et al. 2008). This can lead to expression of edited mature miRNAs (Kawahara et al. 2007b).

The reason for the generally antagonistic effect of editing on miRNA maturation is not clear. One possibility is that most pri-miRNA sequences have evolved to be efficiently processed into mature miRNAs, and any alteration of the miRNA sequence might interfere with this ability. Another possibility is that ADARs have evolved to antagonize miRNA biogenesis. Alternatively, both these statements could be true. Furthermore, how often miRNA editing serves a significant biological function and how often it is an off-target effect of ADARs remains poorly understood.

3.2.1 Effect on Drosha Cleavage

Drosha cleavage is the only miRNA maturation step unable to rely on a 3′ overhang and 5′ phosphate for miRNA recognition. Instead, the Drosha-DGCR8 complex must recognize the pri-miRNA hairpin's sequence and structure (Kim et al. 2009; Krol et al. 2010; Siomi and Siomi 2010). Both of these traits can be altered by A-to-I editing.

It is therefore not surprising that miRNA editing can prevent Drosha cleavage (Fig. 1b), as was first demonstrated for pri-miR-142 (Yang et al. 2006). Transfection of HEK293 with a plasmid expressing pri-miR-142 with or without A-to-G changes at editing sites +4, +5, +40 and +50 (Fig. 2e) showed that the A-to-G changes caused accumulation of pri-miR-142 and depletion of pre- and mature miR-142. Both ADAR1p110 and/or ADAR2 seem to be responsible for this editing depending on the individual editing site. In vitro processing of pri-miR-142—either wild-type, containing the four A-to-G changes or edited in vitro by ADAR1p110 and ADAR2—by recombinant Drosha-DGCR8 and Dicer-TRBP indicated that both A-to-G and A-to-I changes prevented Drosha cleavage. The same in vitro assay using pri-miRNA-142 with A-to-G changes only at the +4 and +5 sites or only at the +40 site revealed that editing at the +4 and +5 sites were enough to cause this effect while editing at the +40 site was not. Finally, endogenous miR-142-5p expression was found to be higher in the spleen of ADAR1 null mice (2.5-fold) and ADAR2 null mice (3.3-fold) as well as in the thymus of ADAR2 null mice (3.0-fold) compared to wild-type mice. All these data support the conclusion that pri-miR-142 editing prevents its Drosha cleavage (Yang et al. 2006).

Similar results have since been found for other pri-miRNAs. In vitro processing of six randomly selected human edited pri-miRNAs suggested that Drosha cleavage is affected by editing of five of the six (Kawahara et al. 2008). Editing of pri-miR-133a2 (Fig. 2f), pri-miR-33 (Fig. 2h), and pri-miR-379 (Fig. 2c) partially prevented Drosha cleavage. In contrast, editing increased Drosha cleavage very slightly for pri-miR-197 and substantially for pri-miR-203 (Fig. 2g). For pri-miR-let7 g (Fig. 2i), Drosha cleavage was unaffected.

Pri-miR-BART6 is an edited miRNA encoded by the Epstein-Barr Virus (EBV) genome (Iizasa et al. 2010). Its sequence can vary between EBV-infected cell lines. Pri-miR-BART6 from Daudi Burkitt lymphoma or nasopharyngeal carcinoma C666-1 cells have a UUU deletion in its hairpin loop that is not found in pri-miR-BART6 from lymphoblastoid GM607 cells (Fig. 2j). An A-to-G change at the editing site of this pri-miR-BART6, but not of the one from GM607 cells, absolutely prevents in vitro processing by the Drosha-DGCR8 complex (Iizasa et al. 2010).

Interestingly, electrophoretic mobility shift assays (EMSA assays) indicate that binding of DGCR8 to the pre-miR-BART6 with the UUU deletion is unaffected by the A-to-G change (Iizasa et al. 2010). This suggests that it is Drosha cleavage, and not binding, that editing affects. The proximity of the editing site to the Drosha cleavage site might therefore be important for effects of editing on Drosha cleavage. This idea is supported by the positions of editing sites in pri-miR-142, pri-miR-133a2, pri-miR-379, pri-miR-203 and pri-miR-BART6, but not pri-miR-33 (Fig. 2).

3.2.2 Degradation of Inosine-Containing Pri-MiRNAs

An RNase activity specific for inosine-containing dsRNA was first observed in 2001 (Scadden and Smith 2001). In vitro editing of three dsRNAs by ADAR2 led to editing at about 40% adenosines. Subsequent incubation with *Xenopus* oocyte and HeLa nuclear, but not cytoplasmic, extracts resulted in their degradation. The activity did not degrade single-stranded RNAs or dsRNAs not edited. Interestingly, it also did not degrade dsRNAs where inosines were exchanged for guanosines, which indicated that it could distinguish inosine-containing from guanosine-containing RNA.

The number of degradation products formed from the three edited dsRNAs varied. One, the sequence of exons 2 and 3 of the rat α-tropomyosin gene without an in-between intron, was cleaved at only a single site to give two degradation products. The cleavage site was in the sequence of four consecutive wobble pairs I•U I•U U•I I•U. Mutating the third base pair to I•U prevented cleavage, indicating that inosine is required on both strands. The second RNA, from polyoma virus, gave four degradation products. The final RNA, chloramphenicol acetyl transferase, gave a large number of discrete degradation products. Based on the sequences of these RNAs, the number of degradation products seems to increase with the number of I•U/I•U stretches. The activity depends on alternating I•U and I•U wobble pairs (Scadden and Smith 2001) and increases with the number of consecutive such pairs (Scadden 2005).

Tudor staphylococcal nuclease (Tudor-SN), a member of RISC, became identified as the enzyme responsible for this inosine-specific RNase activity (Scadden 2005). Four proteins from *Xenopus laevis* oocyte extracts were found to bind an affinity matrix containing dsRNA with I•U wobble pairs but not to either of two affinity matrices containing dsRNA with G•U wobble pairs. Of these, the identity of only one could be determined: Tudor-SN. The other three proteins still remain unidentified. Recombinant Tudor-SN was shown to specifically cleave inosine-containing RNA, indicating that Tudor-SN is indeed the responsible agent (Yang et al. 2006).

Exactly how Tudor-SN can distinguish I•U from G•U wobble pairs remains unclear. Inosine only differs from guanosine by its lack of an exocyclic amine group pointing into the dsRNA minor groove. G•U wobble pairs are also slightly more stable than I•U wobble pairs as determined by melting temperature and susceptibility to single-stranded RNases (Bass and Weintraub 1988). Possibly, Tudor-SN can sense the presence or absence of the guanine amine group by probing the minor grove or by binding the base during breathing of the RNA wobble structure.

A biological function was suggested for Tudor-SN early on: disposal of viral dsRNA (Scadden and Smith 2001). The interferon-inducible ADAR1p150 is upregulated by various viral infections, and it localizes to the cytoplasm where it edits viral dsRNAs to high degrees (Bass 2002; Nishikura 2010). These edited viral dsRNAs could be suitable substrates for Tudor-SN. However, the only proven biological target of Tudor-SN to date is the edited pri-miR-142 (Yang et al. 2006).

As described in Sect. 3.3, editing of pri-miR-142 by ADAR1p110 or ADAR2 prevents its Drosha cleavage. It is therefore not surprising that co-transfection of HEK293 cells with a plasmid expressing pri-miR-142 and a plasmid express ing either ADAR1p110 or ADAR2 reduces mature miR-142-5p and -3p levels compared to cells transfected with the pri-miR-142 plasmid alone

(Yang et al. 2006). The more surprising result is that this does not cause accumulation of edited pri-miR-142 RNAs (Yang et al. 2006). Pri-miR-142 accumulation does take place when a plasmid expressing pri-miR-142 with A-to-G changes at editing sites is used. These results suggest that edited pri-miR-142 is degraded by an RNase specific for inosine-containing RNA, which immediately calls to mind Tudor-SN (Yang et al. 2006). Indeed, in vitro editing of pri-miR-142 made it increasingly susceptible to in vitro degradation by purified recombinant Tudor-SN. Furthermore, treating HEK293 cells over-expressing pri-miR-142 and either ADAR1p110 or ADAR2 with the specific competitive inhibitor of Tudor-SN 2′-deoxythimidine 3′,5′-bisphosphate (pdTp) led to accumulation of edited pri-miR-142 (Yang et al. 2006).

Edited pri-, and possibly pre-, miRNAs therefore seem to be important targets of Tudor-SN. Their degradation by Tudor-SN further adds to the antagonistic effect of ADAR editing on the miRNA pathway.

3.2.3 Effect on Dicer Cleavage

Pri-miR-151 (Fig. 2b) can be edited at two sites by ADAR1p110 and p150. This editing has been shown to completely block pre-miR-151 Dicer cleavage (Fig. 1c) (Kawahara et al. 2007a). Sequencing of more than 50 clones of mature miR-151-3p RNAs derived from human amygdala, mouse cerebral cortex, and mouse lung showed that none were edited. In contrast, all pre-miR-151 molecules detected in human amygdala and mouse cerebral cortex were completely edited at the +3 site (Fig. 2b). This discrepancy suggested that edited pre-miR-151 was prevented from being processed into mature miR-151 either due to inhibition of export to the cytoplasm or Dicer processing. In vitro processing of pre-miR-151 by the Dicer-TRBP complex was prevented by an A-to-G change at the −1 site, +3 site, or both. This indicates that Dicer cleavage is the step prevented by editing. Nearly identical K_d values for binding of Dicer-TRBP complex to unedited and singly or doubly edited pre-miR-151 RNAs, as estimated by EMSA assays, indicate that Dicer cleavage, not binding, is inhibited by editing. This might not be surprising since at least part of Dicer's affinity for its substrate relies on the ~2 nt 3′ overhang and 5′ phosphate. Apart from partial prevention of pre-let-7 g Dicer cleavage by ADAR2 editing (Kawahara et al. 2008), pre-miR-151 remains the only example of editing that inhibits Dicer cleavage.

3.2.4 Effect on RISC Loading

Following Dicer cleavage, one of the miRNA duplex strands is loaded onto RISC as the guide strand while the other is degraded. Which strand follows which fate depends on the relative stabilities of their 5′ ends. The strand with the less stable 5′ end is predominantly loaded onto RISC. Also, this step can be prevented by editing. The only currently known example of such prevention (Fig. 1d) is the editing of pri-miR-BART6 (Fig. 2j), a miRNA encoded by the EBV genome (Iizasa et al. 2010). When not edited, this pri-miRNA matures into miR-BART6-5p, which silences transcripts encoding Dicer.

As explained in Sect. 3.3, the sequence of the pri-miR-BART6 can vary. For the "wild type" sequence found in lymphoblastoid GM607 cells, without the UUU deletion (Fig. 2j), editing does not affect maturation to miR-BART6-5p. Yet an A-to-G change at pri-miR-BART6′s editing site reduces its ability to downregulate Dicer protein levels, as shown by transfecting HEK293 cells with plasmids expressing pri-miR-BART6. The transfection with the A-to-G change caused 25% suppression, while transfection without the change caused 60% suppression, of Dicer protein levels. miR-BART6-5p levels were the same in either case (Iizasa et al. 2010).

Incubation of edited or unedited pre-miR-BART6 with Dicer-TRBP complex and RISC presumably leads to Dicer cleavage and RISC loading. Addition of miR-BART6-5p targets allows the extent of RISC loading to be assessed by target cleavage. Interestingly, ~3-fold more targets are cleaved when unedited pre-miR-BART6 is used compared to when edited pre-miR-BART6. This indicates that editing prevents miR-BART6-5p RISC loading. Target of unedited or edited miR-BART6-3p are not significantly cleaved whether edited or unedited pre-miR-BART6 is used. This indicates that editing does not make miR-BART6-3p get loaded onto RISC instead of miR-BART6-5p, but rather prevents RISC loading altogether.

Editing of pri-miR-BART6, with or without the UUU deletion (Sect. 3.3), therefore, prevents miR-BART6 from silencing transcripts of Dicer. The consequent upregulation of Dicer levels in turn causes global upregulation of miRNA levels, which could severely affect the state of the cell. The effect on expression of genes associated with EBV lytic infection or state of latency has been investigated by antagomir inhibition of miR-BART6-5p in C666-1 cells (Iizasa et al. 2010). EBNA1, detected in type I, II, and III latency, was unaffected. EBNA2, an oncogene essential in B lymphocyte transformation, plays a central role in type III latency by upregulating promoters of latent EBV genes. Deficiency in EBNA2 is associated with type I and II latency. The antagomir treatment upregulated EBNA2 by ~5-fold. LMP1 is associated with type III latency and is weakly expressed in type I and II latency. The antagomir also upregulated LMP1 by ~2-fold. Furthermore, Zta and Rta, essential for initiation of the lytic EBV infection cycle, were both upregulated by 2–3-fold. Finally, the effect of the antagomir treatment on transcription from EBV promoters was also investigated. Transcription from Cp and Wp, characteristic of type III latency, was upregulated by 5.4-fold and 11-fold, respectively. Transcription from Qp, used in type I or II latency, was completely abolished. The consequences of downregulating Dicer expression using a short-hairpin RNA expressing vector on protein levels and promoter activities were the opposite of those observed upon miR-BART6-5p antagomir inhibition. The overall trend of these observations indicates that upregulation of Dicer by pri-miR-BART6 editing induces a shift away from type I and II latency towards the more immunoresponse-prone type III latency and promotes entry into the lytic EBV infection cycle.

3.2.5 Effect on Mature MiRNA Function

Following loading onto RISC, the miRNA guides RISC to sequester or degrade its complementary mRNAs (Kim et al. 2009; Krol et al. 2010; Siomi and Siomi 2010). Edited miRNAs can also be loaded onto RISC if editing did not prevent their maturation. Since editing alters the base pairing properties of the miRNA, it can also redirect its silencing. This is especially the case when editing takes place within the seed sequence, or nucleotides 2–8, of the miRNA, which is particularly important for target selection. The only proven case of redirection of silencing is editing of pri-miRNAs of the human and mouse miR-376 cluster (Kawahara et al. 2007b). Although sequencing of mature miRNAs (see Sect. 3.1.2) indicates expression of other mature miRNAs with edited sequences, the possible redirection of their silencing has not yet been examined.

Members of the miR-376 cluster are transcribed as one transcript from which each is excised as a separate pre-miRNA. They share high sequence similarity and most contain at least two editing sites: those corresponding to editing sites +4, located within the 5′ seed sequence, and +44, located within the 3′ seed sequence, of human cluster member miR-376a1 (Figs. 1e and 2d). Experiments using ADAR1 knockout mouse embryos or ADAR2 knockout cortex showed that the +4 site in pri-miR-376a is edited primarily by ADAR2 while the +44 site in pri-miR376b and pri-miR376c is edited primarily by ADAR1. Prediction of targets for unedited and edited human miR-376a1-5p or mouse miR-376a-5p (they have the same sequence) indicated different target sets with only minor overlap (Fig. 3).

The reliability of this prediction was examined by the random selection and experimental verification of three targets predicted for only edited miR-376a-5p (PRPS1, ZNF513 and SNX19) and three targets predicted for only unedited miR-376a-5p (TTK, SFRS11, and SLC16A1). HeLa cells were co-transfected with miR-376a-5p RNAs and luciferase reporter constructs containing in their 3′UTRs the predicted target sites of one of the six above targets. Co-transfection using miR-376a-5p with an adenosine at the +4 sites repressed luciferase activity only for the constructs containing the targets of the unedited miR-376a-5p. Co-transfection using miR-376a-5p with a guanosine or inosine at the +4 sites repressed luciferase activity only for the constructs containing the targets of the edited miR-376a-5p.

One of the targets specific to the miR-376a edited at the +4 site by ADAR2 was phosphoribosyl pyrophosphate synthetase 1 (PRPS1), an essential housekeeping enzyme involved in purine metabolism and the uric acid synthesis pathway. The importance of tight control of PRPS1 expression is demonstrated by an X-chromosome-linked human disorder characterized by gout and neurodevelopmental impairment with hyperuricemia due to a 2–4-fold increase in PRPS1 levels. ADAR2 knockout mice had PRPS1 levels and uric acid levels both

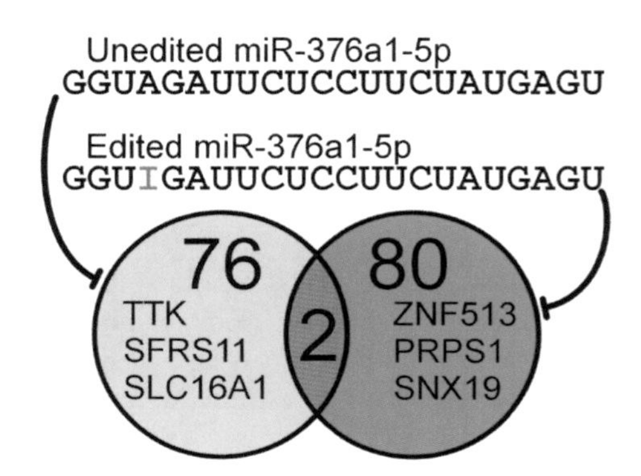

Fig. 3 Relationship between targets of edited and unedited pri-miR-376a-5p. The Venn diagram illustrates the relationship between the 78 mRNA targets predicted for unedited human pri-miR-376a1-5p and the 82 mRNA targets predicted for human pri-miR-376a1-5p edited at the +4 site. Experimentally confirmed targets are plotted (Kawahara et al. 2007b)

upregulated ~2-fold in the cortex. By contrast, the levels of TTK, a target of unedited miR-376a-5p, in the cortex were not affected. Nor were PRPS1 levels or uric acid levels in the liver. The latter is consistent with the observation that half of pri-miR-376a +4 site adenosines were edited in wild-type mouse cortex, while pri-miR-376a +4 editing was barely detected in wild-type mouse liver. This shows that ADAR2 influences uric acid levels in a tissue-specific manner by redirection of miRNA silencing (Kawahara et al. 2007b).

3.2.6 Other Potential Consequences

Because most known consequences of pri-miRNA editing are only exemplified by one or a few known editing sites, it is likely that there exist consequences still exemplified by none. Further studies of pri-miRNA editing could identify these. For example, it is possible that pri-miRNA editing could alter Drosha or Dicer cleavage sites, prevent export to the cytoplasm or change what strand gets loaded onto RISC.

How the Drosha-DGCR8 complex chooses its cleavage site is still not entirely clear. Sequence and structure changes created by A-to-I editing could possibly shift the Drosha cleavage site. If a mature miRNA is produced, it could have a shifted seed sequence and consequently an altered set of target mRNAs.

Dicer functions as a ruler and cleaves the pre-miRNA about 21 base pairs away from the Drosha cut. Editing could alter the stability of the miRNA and therefore the number of base pairs that fit in the length measured by Dicer. This could shift the Dicer cleavage site. If the 3′ strand is loaded onto RISC, it would have a shifted seed sequence. A-to-G changes are associated with shifted 5′ ends in mature miRNA sequences (see Sect. 3.1.2). This indicates that editing might indeed shift Drosha and Dicer cleavage sites.

Nuclear export of pre-miRNAs is carried out by Exportin-5. The structure of Exportin-5 indicates that the major requirement for RNA export is some degree of double-strandedness ending in a 3′ overhang (Okada et al. 2009). The sequence and exact dsRNA structure are less important. It might therefore be difficult for editing to prevent miRNA export by Exportin-5.

Upon RISC loading, the miRNA duplex strand with the less stable 5′ end is more likely to be chosen as the guide strand. Editing near either of the duplex ends could change their relative stability. Thus, editing might affect which strand is loaded onto RISC. This could be significant because the two strands could target completely different sets of mRNAs.

3.3 *Possibility of Editing of Pre-MiRNAs*

The fraction of pri-miR-151 transcripts edited is ~41% in human amygdala and ~29% in mouse cerebral cortex (Kawahara et al. 2007a). Surprisingly, both these numbers increase to 100% for pre-miR-151 (Kawahara et al. 2007a). One explanation for this discrepancy presents itself. Editing prevents Dicer processing of pre-miR-151, which could lead to accumulation of edited pre-miR-151. However, pre-miR-151 can be edited in vitro by both ADAR1p110 and ADAR1p150 (Kawahara et al. 2007a). Taken together, these observations suggest that pre-miR-151 might be edited by ADAR1p110 in the nucleus and/or by ADAR1p150 in the cytoplasm.

4 Editing of MicroRNA Target Sites

Since single nucleotide changes in miRNA seed sequences can redirect silencing (see Sect. 3.2.5), it is conceivable that editing of target sequences could do the same. This seems especially plausible since both editing and target sites localize mostly to UTRs. However, it is difficult to investigate this possibility because the number of potential target sites is at least two orders of magnitude larger than the number of miRNAs (Grimson et al. 2007).

The earliest attempt at investigating this notion was therefore bioinformatics-based (Liang and Landweber 2007). Aligning ~28,000 putative editing sites and predicted miRNA target sites identified ~300 editing sites that could disrupt, and ~200 editing sites that could perfect, seed sequence matches in 3′UTRs. However, requiring miRNA expression and target site editing to take place in the same tissue left only two seed sequence matches disrupted by editing. There are several possible explanations for this low number. First, target site prediction relies on cross-species conservation while most known editing sites reside within primate-specific *Alu* sequences. Overlapping target and editing sites are therefore less likely to have been discovered. Second, target sites might be under evolutionary pressure to not be affected by editing. Third, dsRNA regions are required for editing but often sterically hinder miRNA targeting (Grimson et al. 2007).

However, one case of a seed sequence match perfected by editing has been verified (Borchert et al. 2009).

Examination of $\sim$12,000 putative editing sites revealed that $\sim$3,000 that could create a perfect seed sequence complement. Two hundred and fifty-eight putative editing sites in the sequence context 5′-CCUGUAA-3′ created a perfect match to the miR-513 seed sequence. Two hundred and fifty-two putative editing sites in the sequence context 5′-AAUCCCA-3′ created a perfect match to a seed sequence common to miR-769-3p and miR-450b-3p. Interestingly, $\sim$190 sites contained the 12 nt motif 5′-CCUGUAAUCCCA-3′, which becomes complementary to both seed sequences when edited.

Co-transfection of HEK293 cells with vectors expressing miR-513 or miR-796 hairpins as well as a luciferase transcript harboring the 12 nt motif led to a $\sim$50% reduction in luciferase activity when an A-to-G change was introduced in the 12 nt motif. One gene harboring this motif in its 3′UTR is DFFA (DNA fragmentation factor alpha—or ICAD). Cloning of DFFA from NB7 and HEK293 cells revealed that these motifs are edited in NB7, but not in HEK293, cells. Co-transfection of HEK293 cells with vectors expressing miR-513 or miR-796 hairpins as well as a luciferase transcript with the 3′UTR of DFFA as cloned from NB7 or HEK293 cells led to repression of luciferase activity only when the 3′UTR originated from NB7 cells. Finally, overexpression of miR-796 in NB7, but not HEK293, cells caused $\sim$60% reduction in endogenous DFFA levels (Borchert et al. 2009).

These results indicate that editing can cause mRNAs to come under the regulation of miRNA in a tissue-specific manner. How common this phenomenon is and whether editing can also prevent miRNA targeting, remains to be seen.

5 Editing-Independent Effects of ADARs

While ADARs have been mostly studied for their ability to edit RNAs, they also have functions independent of their catalytic activity. For example, they affect the small interfering RNA (siRNA) pathway by sequestering siRNAs (Yang et al. 2005). Recently, it was demonstrated that ADARs can similarly sequester miRNAs (Heale et al. 2009).

In vitro processing of human pri-miR-376a2 to pre-miR-376a2 by HeLa nuclear extract is prevented by addition of purified ADAR2, but not by A-to-G changes at the pri-miR-376a2 +4 or +44 editing sites. Furthermore, co-transfection of HEK293T cells with constructs expressing pri-miR-376a2, a luciferase transcript with miR-376a2 target sites and a catalytically inactive ADAR2 mutant caused $\sim$2-fold more luciferase activity than co-transfection with the pri-miRNA and luciferase constructs alone. The same result also held for pri-miR-376a1, but not for pri-miR-let7a1 or when the ADAR2 mutant was exchanged for ADAR1p110 (Heale et al. 2009).

These data indicate that ADAR2, but not ADAR1p110, is able to sequester pri-miR-376a1 and -376a2, but not -let7a1. ADAR2's superior ability to sequester miRNAs might explain its ability to downregulate ADAR1 editing of pri-miR-376b and -376c at the +44 position (Kawahara et al. 2007b). How many pri-miRNAs are similarly sequestered and whether there exist pri-miRNAs also sequestered by ADAR1p110 remains to be seen. Because ADARs might bind to a larger subset of miRNAs than they edit, many miRNAs might be affected by ADARs solely in an editing-independent manner.

6 Conclusions

MiRNAs have immense importance on the state of the cell. For example, abnormal miRNA expression is associated with cancer (Ryan et al. 2010). ADARs in turn, exert great influence on the miRNA phenomenon by editing or sequestering the miRNAs. In these ways, A-to-I editing of miRNAs also exerts influence on human health. For example, editing of EBV miRNAs seems to be connected to the virus moving to immunoresponse-prone types of latency or to lytic infection (Iizasa et al. 2010).

Great progress has been made in elucidating the effects of ADARs on miRNA in recent years. We now know of an extensive number of miRNA editing sites, various effects of their editing, editing of one miRNA target sequence, and editing-independent effects of ADARs.

Yet there are reasons to believe that there are many more facets to the interplay between ADAR and miRNAs that remain to be elucidated. For example, the study which identified Tudor-SN as a protein specifically binding to inosine-containing dsRNAs also showed the existence of other such proteins, but their identities have not yet been determined. Furthermore, there are various effects of miRNA editing that one might expect could happen, but which have never yet been observed. Finally, some of the effects of miRNA editing that we do know of, like inhibition of RISC loading or redirection of silencing targets, are only exemplified by one known miRNA editing site each. This indicates that we have still only learned of a random sample of a larger number of possible kinds of effects. The interplay between A-to-I editing and miRNAs is therefore likely to remain a growing field in the foreseeable future.

References

Babiarz JE, Ruby JG, Wang Y, Bartel DP, Blelloch R (2008) Mouse ES cells express endogenous shRNAs, siRNAs, and other Microprocessor-independent, Dicer-dependent small RNAs. Genes Dev 22:2773–2785

Bass BL (2002) RNA editing by adenosine deaminases that act on RNA. Annu Rev Biochem 71:817–846

Bass BL, Weintraub H (1988) An unwinding activity that covalently modifies its double-stranded RNA substrate. Cell 55:1089–1098

Berezikov E, Robine N, Samsonova A, Westholm JO, Naqvi A, Hung JH, Okamura K, Dai Q, Bortolamiol-Becet D, Martin R, Zhao Y, Zamore PD, Hannon GJ, Marra MA, Weng Z,

Perrimon N, Lai EC (2011) Deep annotation of *Drosophila melanogaster* microRNAs yields insights into their processing, modification, and emergence. Genome Res 21:203–215

Blow MJ, Grocock RJ, van Dongen S, Enright AJ, Dicks E, Futreal PA, Wooster R, Stratton MR (2006) RNA editing of human microRNAs. Genome Biol 7:R27

Borchert GM, Gilmore BL, Spengler RM, Xing Y, Lanier W, Bhattacharya D, Davidson BL (2009) Adenosine deamination in human transcripts generates novel microRNA binding sites. Hum Mol Genet 18:4801–4807

Chiang HR, Schoenfeld LW, Ruby JG, Auyeung VC, Spies N, Baek D, Johnston WK, Russ C, Luo S, Babiarz JE, Blelloch R, Schroth GP, Nusbaum C, Bartel DP (2010) Mammalian microRNAs: experimental evaluation of novel and previously annotated genes. Genes Dev 24:992–1009

de Hoon MJ, Taft RJ, Hashimoto T, Kanamori-Katayama M, Kawaji H, Kawano M, Kishima M, Lassmann T, Faulkner GJ, Mattick JS, Daub CO, Carninci P, Kawai J, Suzuki H, Hayashizaki Y (2010) Cross-mapping and the identification of editing sites in mature microRNAs in high-throughput sequencing libraries. Genome Res 20:257–264

Gottwein E, Cai X, Cullen BR (2006) A novel assay for viral microRNA function identifies a single nucleotide polymorphism that affects Drosha processing. J Virol 80:5321–5326

Grimson A, Farh KK, Johnston WK, Garrett-Engele P, Lim LP, Bartel DP (2007) MicroRNA targeting specificity in mammals: determinants beyond seed pairing. Mol Cell 27:91–105

Heale BS, Keegan LP, McGurk L, Michlewski G, Brindle J, Stanton CM, Caceres JF, O'Connell MA (2009) Editing independent effects of ADARs on the miRNA/siRNA pathways. EMBO J 28:3145–3156

Iizasa H, Wulff BE, Alla NR, Maragkakis M, Megraw M, Hatzigeorgiou A, Iwakiri D, Takada K, Wiedmer A, Showe L, Lieberman P, Nishikura K (2010) Editing of Epstein-Barr virus-encoded BART6 microRNAs controls their dicer targeting and consequently affects viral latency. J Biol Chem 285:33358–33370

Jin Y, Zhang W, Li Q (2009) Origins and evolution of ADAR-mediated RNA editing. IUBMB Life 61:572–578

Kawahara Y, Zinshteyn B, Chendrimada TP, Shiekhattar R, Nishikura K (2007a) RNA editing of the microRNA-151 precursor blocks cleavage by the Dicer-TRBP complex. EMBO Rep 8:763–769

Kawahara Y, Zinshteyn B, Sethupathy P, Iizasa H, Hatzigeorgiou AG, Nishikura K (2007b) Redirection of silencing targets by adenosine-to-inosine editing of miRNAs. Science 315:1137–1140

Kawahara Y, Megraw M, Kreider E, Iizasa H, Valente L, Hatzigeorgiou AG, Nishikura K (2008) Frequency and fate of microRNA editing in human brain. Nucleic Acids Res 36:5270–5280

Kim VN, Han J, Siomi MC (2009) Biogenesis of small RNAs in animals. Nat Rev Mol Cell Biol 10:126–139

Krol J, Loedige I, Filipowicz W (2010) The widespread regulation of microRNA biogenesis, function and decay. Nat Rev Genet 11:597–610

Kuchenbauer F, Morin RD, Argiropoulos B, Petriv OI, Griffith M, Heuser M, Yung E, Piper J, Delaney A, Prabhu AL, Zhao Y, McDonald H, Zeng T, Hirst M, Hansen CL, Marra MA, Humphries RK (2008) In-depth characterization of the microRNA transcriptome in a leukemia progression model. Genome Res 18:1787–1797

Landgraf P, Rusu M, Sheridan R, Sewer A, Iovino N, Aravin A, Pfeffer S, Rice A, Kamphorst AO, Landthaler M, Lin C, Socci ND, Hermida L, Fulci V, Chiaretti S, Foa R, Schliwka J, Fuchs U, Novosel A, Muller RU, Schermer B, Bissels U, Inman J, Phan Q, Chien M, Weir DB, Choksi R, De Vita G, Frezzetti D, Trompeter HI, Hornung V, Teng G, Hartmann G, Palkovits M, Di Lauro R, Wernet P, Macino G, Rogler CE, Nagle JW, Ju J, Papavasiliou FN, Benzing T, Lichter P, Tam W, Brownstein MJ, Bosio A, Borkhardt A, Russo JJ, Sander C, Zavolan M, Tuschl T (2007) A mammalian microRNA expression atlas based on small RNA library sequencing. Cell 129:1401–1414

Liang H, Landweber LF (2007) Hypothesis: RNA editing of microRNA target sites in humans? RNA 13:463–467

Linsen SE, de Wit E, de Bruijn E, Cuppen E (2010) Small RNA expression and strain specificity in the rat. BMC Genomics 11:249

Luciano DJ, Mirsky H, Vendetti NJ, Maas S (2004) RNA editing of a miRNA precursor. RNA 10:1174–1177

Morin RD, O'Connor MD, Griffith M, Kuchenbauer F, Delaney A, Prabhu AL, Zhao Y, McDonald H, Zeng T, Hirst M, Eaves CJ, Marra MA (2008) Application of massively parallel sequencing to microRNA profiling and discovery in human embryonic stem cells. Genome Res 18:610–621

Nishikura K (2010) Functions and regulation of RNA editing by ADAR deaminases. Annu Rev Biochem 79:321–349

Okada C, Yamashita E, Lee SJ, Shibata S, Katahira J, Nakagawa A, Yoneda Y, Tsukihara T (2009) A high-resolution structure of the pre-microRNA nuclear export machinery. Science 326:1275–1279

Ruby JG, Jan C, Player C, Axtell MJ, Lee W, Nusbaum C, Ge H, Bartel DP (2006) Large-scale sequencing reveals 21U-RNAs and additional microRNAs and endogenous siRNAs in *C. elegans*. Cell 127:1193–1207

Ryan BM, Robles AI, Harris CC (2010) Genetic variation in microRNA networks: the implications for cancer research. Nat Rev Cancer 10:389–402

Scadden AD (2005) The RISC subunit Tudor-SN binds to hyper-edited double-stranded RNA and promotes its cleavage. Nat Struct Mol Biol 12:489–496

Scadden AD, Smith CW (2001) Specific cleavage of hyper-edited dsRNAs. EMBO J 20:4243–4252

Schulte JH, Marschall T, Martin M, Rosenstiel P, Mestdagh P, Schlierf S, Thor T, Vandesompele J, Eggert A, Schreiber S, Rahmann S, Schramm A (2010) Deep sequencing reveals differential expression of microRNAs in favorable versus unfavorable neuroblastoma. Nucleic Acids Res 38:5919–5928

Siomi H, Siomi MC (2010) Posttranscriptional regulation of microRNA biogenesis in animals. Mol Cell 38:323–332

Sun G, Yan J, Noltner K, Feng J, Li H, Sarkis DA, Sommer SS, Rossi JJ (2009) SNPs in human miRNA genes affect biogenesis and function. RNA 15:1640–1651

Suzuki H, Forrest AR, van Nimwegen E, Daub CO, Balwierz PJ, Irvine KM, Lassmann T, Ravasi T, Hasegawa Y, de Hoon MJ, Katayama S, Schroder K, Carninci P, Tomaru Y, Kanamori-Katayama M, Kubosaki A, Akalin A, Ando Y, Arner E, Asada M, Asahara H, Bailey T, Bajic VB, Bauer D, Beckhouse AG, Bertin N, Bjorkegren J, Brombacher F, Bulger E, Chalk AM, Chiba J, Cloonan N, Dawe A, Dostie J, Engstrom PG, Essack M, Faulkner GJ, Fink JL, Fredman D, Fujimori K, Furuno M, Gojobori T, Gough J, Grimmond SM, Gustafsson M, Hashimoto M, Hashimoto T, Hatakeyama M, Heinzel S, Hide W, Hofmann O, Hornquist M, Huminiecki L, Ikeo K, Imamoto N, Inoue S, Inoue Y, Ishihara R, Iwayanagi T, Jacobsen A, Kaur M, Kawaji H, Kerr MC, Kimura R, Kimura S, Kimura Y, Kitano H, Koga H, Kojima T, Kondo S, Konno T, Krogh A, Kruger A, Kumar A, Lenhard B, Lennartsson A, Lindow M, Lizio M, Macpherson C, Maeda N, Maher CA, Maqungo M, Mar J, Matigian NA, Matsuda H, Mattick JS, Meier S, Miyamoto S, Miyamoto-Sato E, Nakabayashi K, Nakachi Y, Nakano M, Nygaard S, Okayama T, Okazaki Y, Okuda-Yabukami H, Orlando V, Otomo J, Pachkov M, Petrovsky N et al (2009) The transcriptional network that controls growth arrest and differentiation in a human myeloid leukemia cell line. Nat Genet 41:553–562

Wulff BE, Sakurai M, Nishikura K (2011) Elucidating the inosinome: global approaches to adenosine-to-inosine RNA editing. Nat Rev Genet 12:81–85

Yang W, Wang Q, Howell KL, Lee JT, Cho DS, Murray JM, Nishikura K (2005) ADAR1 RNA deaminase limits short interfering RNA efficacy in mammalian cells. J Biol Chem 280:3946–3953

Yang W, Chendrimada TP, Wang Q, Higuchi M, Seeburg PH, Shiekhattar R, Nishikura K (2006) Modulation of microRNA processing and expression through RNA editing by ADAR deaminases. Nat Struct Mol Biol 13:13–21

Nuclear Editing of mRNA 3′-UTRs

Ling-Ling Chen and Gordon G. Carmichael

Abstract Hundreds of human genes express mRNAs that contain inverted repeat sequences within their 3′-UTRs. When expressed, these sequences can be promiscuously edited by ADAR enzymes, leading to the retention of mRNAs in nuclear paraspeckles. Here we discuss how this retention system can be used to regulate gene expression.

Contents

L.-L. Chen (✉)
State Key Laboratory of Molecular Biology,
Institute of Biochemistry and Cell Biology,
Shanghai Institutes for Biological Sciences,
Chinese Academy of Sciences, 320 Yueyang Road,
200031 Shanghai, China
e-mail: linglingchen@sibcb.ac.cn

G. G. Carmichael
Department of Genetics and Developmental Biology,
University of Connecticut Stem Cell Institute,
University of Connecticut Health Center,
400 Farmington Avenue, Farmington, CT 06030-6403, USA
e-mail: carmichael@nso2.uchc.edu

Current Topics in Microbiology and Immunology (2012) 353: 111–121
DOI: 10.1007/82_2011_149

Published Online: 3 July 2011

1 Double Stranded RNA in the Nucleus

Double stranded RNA (dsRNA) can be formed in two distinct ways: via sense–antisense transcription followed by intermolecular RNA strand annealing or via the intramolecular hybridization of inverted repeat sequences within transcripts. While sense–antisense annealing in *trans* has not been clearly documented for endogenous mammalian RNAs in either the nucleus or the cytoplasm, it is commonly found cytoplasmically as a byproduct or intermediate in the replication of RNA virus genomes. The intramolecular annealing of inverted repeats is far more common in mammalian cells, and most of this type of dsRNA is found in the nucleus.

The cellular response to dsRNA structures depends not only on the subcellular location of the duplexes, but also on their length. In the cytoplasm, dsRNA is rare and its presence in that compartment is generally associated with viral infection. Cytoplasmic dsRNA often triggers the interferon signaling pathway and leads to nonspecific inhibition of gene expression (Wang and Carmichael 2004). The response to dsRNA is quite different in the nucleus. Nuclear dsRNAs often serve as substrates for A-to-I editing by members of the ADAR enzyme family, which are ubiquitously expressed in higher eukaryotes (Bass 2002). Depending on the length and quality of the RNA duplex, editing can be either site-selective or promiscuous. Editing directed by short dsRNA structures can be site-specific, leading to coding changes in mRNAs. This type of editing is directed to specific adenosines imbedded in favorable secondary structures, often involving an unpaired adenosine (Kallman et al. 2003; Wong et al. 2001). In contrast, long dsRNA regions (at least 25–30 bp in length) (Bass and Weintraub 1988; Nishikura 1992) are edited promiscuously, with up to half of their adenosines being changed to inosines (Bass 2002). In long perfect duplexes up to 50% of the A's on each strand are edited in a rather promiscuous fashion, except for a 5′-neighbor preference for A or U (Polson and Bass 1994). The resulting RNAs contain I–U base pairs which render the RNA duplexes unstable (Bass and Weintraub 1988).

Promiscuous editing in the nucleus has been shown to lead to the association of inosine-containing RNAs with a protein complex containing the factor $p54^{nrb}$, which interacts strongly with inosine-containing RNAs (Zhang and Carmichael 2001). This in turn leads to nuclear retention, thus perhaps providing a quality control system to prevent the inappropriate export of some mRNAs with altered codons. $p54^{nrb}$ is an abundant and highly conserved multifunctional nuclear protein that binds both DNA and RNA and has been reported to be involved in a number of processes, including transcription initiation (Yang et al. 1993), pre-mRNA splicing (Dong et al. 1993; Kameoka et al. 2004), transcription elongation (Kaneko et al. 2007) as well as nuclear retention of edited RNAs (Zhang and Carmichael 2001). $p54^{nrb}$ not only forms homodimers, but can also heterodimerize with the related proteins PTB-associated splicing factor (PSF) and paraspeckle protein 1α (PSP1α) (Akhmedov and Lopez 2000; Fox et al. 2005; Myojin et al. 2004; Zhang et al. 1993). Evidence that these proteins together may participate in

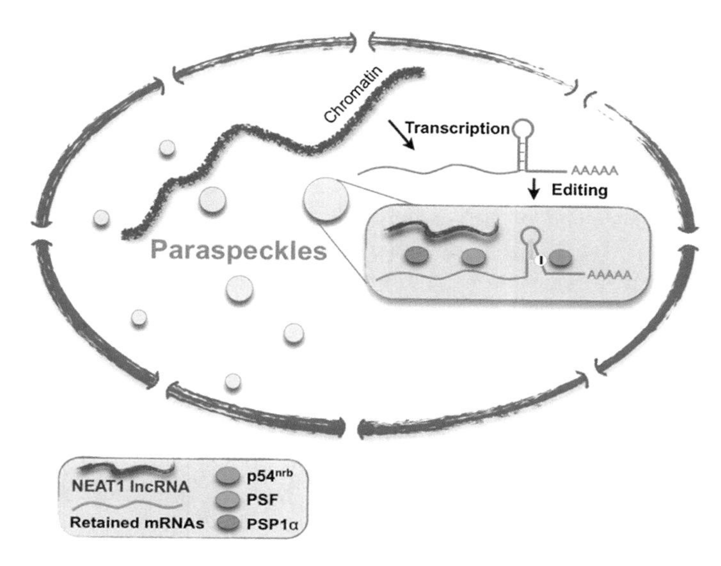

Fig. 1 A common fate of nuclear mRNAs with inverted repeats in their 3′-UTRs. Transcription and pre-mRNA splicing sometimes generate mRNAs with inverted repeats, generally IR*Alus*. The dsRNA structures are substrates for promiscuous ADAR editing and edited RNAs are bound by $p54^{nrb}$ and sequestered in nuclear paraspeckles. Paraspeckles are organized by a long noncoding RNA, NEAT1 and contain $p54^{nrb}$, PSF, PSP1α and other proteins (See text for details)

the nuclear retention of edited RNAs comes from the observation that these proteins and RNAs colocalize to specific subnuclear structures called paraspeckles (see Fig. 1)(Chen and Carmichael 2009; Chen et al. 2008; Fox et al. 2002; Prasanth et al. 2005), which are cell-cycle-regulated subnuclear domains (Fox et al. 2005). Several groups have recently independently shown that paraspeckle assembly and integrity depends on the expression of an abundant noncoding RNA called NEAT1 (or Menε/β in the mouse) (Bond and Fox 2009; Chen and Carmichael 2009; Clemson et al. 2009; Mao et al. 2011; Sasaki et al. 2009; Sunwoo et al. 2009). Importantly, the nuclear retention of mRNAs containing edited inverted repeats in their 3′-UTRs also appears to require not only the expression of $p54^{nrb}$ and other paraspeckle proteins, but also paraspeckles themselves and NEAT1 RNA (Chen and Carmichael 2009).

2 Editing in mRNA 3′-UTRs

As mentioned above, dsRNA structures in the nucleus are readily edited. But how often are such molecules produced and how often are they edited? By far the most common inverted repeat structures in nuclear transcripts in human cells are comprised of a single class of highly abundant short interspersed nuclear repetitive

DNA elements (SINEs), the *Alu* elements. About 45% of the human genome is composed of repetitive and transposable elements, which include SINEs, LINEs and retrotransposons (Lander et al. 2001). The vast majority of human SINEs are *Alu* elements. There are up to 1.4×10^6 copies of these 300 bp elements in the genome, corresponding to more than 1 *Alu* about for every 3,000 bp of genomic DNA (DeCerbo and Carmichael 2005). *Alu* elements are very similar to one another in sequence and are not randomly distributed throughout the genome, but tend to be concentrated in gene rich regions, generally within noncoding segments of transcripts, such as in introns and untranslated regions (Versteeg et al. 2003). Due to their abundance, the average human pre-mRNA contains more than 16 *Alu* elements (DeCerbo and Carmichael 2005). Also, since they are randomly oriented with respect to gene transcription units, a large fraction of human genes express transcripts that generate inverted *Alu* repeat structures (IR*Alu*s) which can serve as efficient substrates for ADAR editing. Indeed, in searches of cDNA and EST databases for clusters of A-to-G changes as indicators of editing, it was found that the more than 90% of A-to-I editing in humans is within *Alu* elements (Athanasiadis et al. 2004; Blow et al. 2004; Kim et al. 2004; Levanon et al. 2004). There are now many thousands of instances of *Alu* editing in the human transcriptome (Athanasiadis et al. 2004; Barak et al. 2009; Blow et al. 2004; Greenberger et al. 2010; Kim et al. 2004; Levanon et al. 2004; Osenberg et al. 2010; Sakurai et al. 2010). Since *Alu* elements within coding regions would disrupt open reading frames, most IR*Alu*s are found within introns. Thus, although RNA hairpin structures are likely formed and edited during transcription and pre-mRNA processing, most are rapidly degraded and do not contribute significantly to gene expression. Of more interest are those that remain in mRNAs following splicing. Our bioinformatic analysis identified 333 human genes with IR*Alu*s within their 3′-UTRs, and a number of these *Alu*s have already been shown to be susceptible to promiscuous editing (Chen et al. 2008). If these edited mRNAs are retained in paraspeckles, how are they be expressed and regulated?

3 Regulation of Gene Expression of Genes Containing IR*Alu*s

Figure 2 offers several ways in which genes encoding IR*Alu*s in their 3′-UTRs might regulate their expression. Here we show three typical genes with IR*Alu*s, METTL7A, NICN1 and LIN28. Each of these genes expresses multiple transcripts that differ in their 3′-UTR structures. Each appears to express at least one isoform in which the IR*Alu*s structure is not present. First, some isoforms appear to have 3′-UTR sequences removed, either by pre-mRNA splicing or by some other mechanism. We see this phenomenon for many of the genes containing IR*Alu*s in their 3′-UTRs (Chen et al. 2008; and data not shown). This is consistent with a recent claim that *Alu* elements are often cleaved at both ends of the inverted repeat region, followed by rejoining of the two parts of the transcript on both sides (Osenberg et al. 2009). While it is not yet clear whether such cleavage is general,

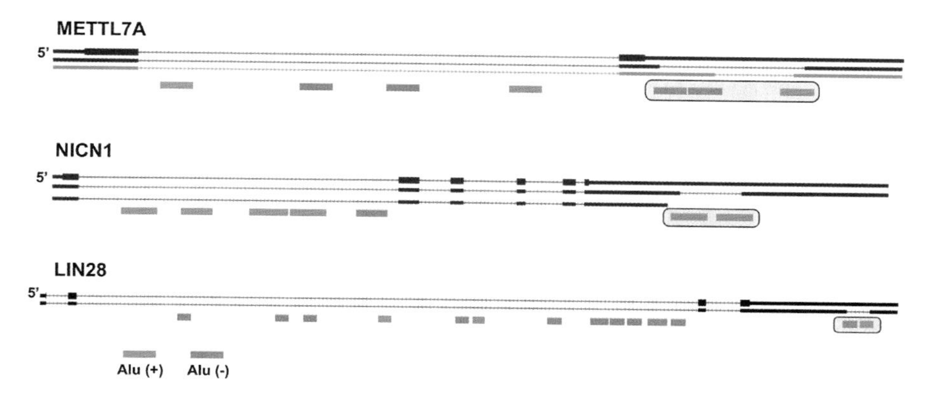

Fig. 2 Several illustrative genes containing IR*Alu*s. Exons are depicted by *solid lines*, with coding exons thicker. Positions of *Alu* elements are shown and their different orientations are in different colors. Note that each of these genes contains intronic *Alu*s as well as IR*Alu*s in the 3′-UTRs (marked by *shaded boxes*). Also note that NICN1 has alternative polyadenylation signals and each gene may express RNAs that cannot form inverted repeats

or whether many cDNA sequences annotated this way result from reverse transcriptase artifacts during cloning (Chen et al. 2008), such cleavage and rejoining could generate mRNAs that escape the retention pathway.

Second, alternative polyadenylation could alter gene expression or regulation. NICN1 has two alternative polyadenylation signals. Choice of the upstream signal results in an mRNA completely lacking IR*Alu*s (Fig. 3). In fact, the majority of genes with IR*Alu*s in their 3′-UTRs possess alternative polyadenylation signals of this sort (Chen et al. 2008). About half of all mammalian genes use alternative cleavage and polyadenylation to generate multiple mRNA isoforms differing in their 3′-UTRs (Beaudoing and Gautheret 2001; Edwalds-Gilbert et al. 1997; Lee et al. 2007; Zhang et al. 2005). Alternative poly(A) site choice may be tissue-specific for some genes (Zhang et al. 2005). For example, many alternative 3′-UTR mRNA isoforms were observed during T cell activation (Sandberg et al. 2008). This resulted not only in 3′-UTRs of different lengths, but containing distinct sequence elements such as microRNA binding sites. Thus, regulated poly(A) site choice could influence the retention or export of many mRNAs, and this could involve editing in the 3′-UTR regions of mRNAs. A related mechanism of regulation involves the expression of alternative 3′-UTRs by a combination of alternative splicing and alternative poly(A) signal usage. For example, caspase 8 and caspase 10 lie adjacent to one another on chromosome 2 and each gene can express mRNAs with either an upstream 3′-UTR that contains IR*Alu*s or a downstream 3′-UTR that does not.

IR*Alu*s are not always edited at the same nucleotides, and editing, while promiscuous, is quite variable in extent from transcript to transcript. Further, nuclear retention of transcripts containing IR*Alu*s is not absolute, and the significance of this observation is not yet clear (Chen et al. 2008). In situ hybridization studies showed that the majority of IR*Alu*s-RNA localized within the nucleus, but in a minority of cells completely cytoplasmic localization was seen. One explanation for this might

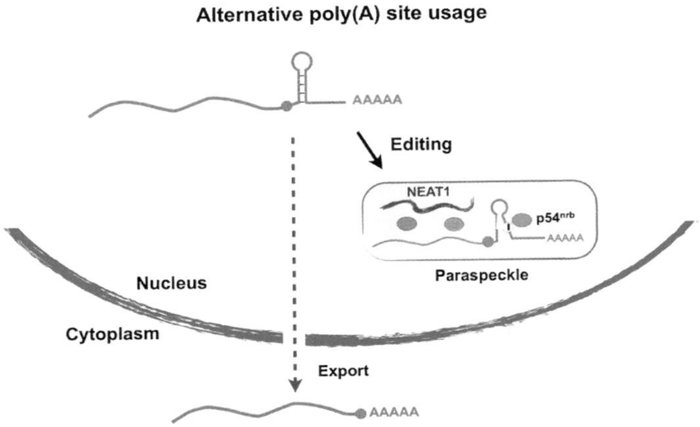

Fig. 3 A model for gene regulation via alternative poly(A) site usage. Many genes with IR*Alus* in their 3′-UTRs also have multiple polyadenylation signals. Choice of the upstream signal leads to efficient mRNA export to the cytoplasm, but choice of the downstream signal leads to editing and nuclear retention

be that the degree of A-to-I RNA editing of IR*Alus*-RNA determines its nuclear/cytoplasmic distribution. If editing were the primary cause of retention, then transcripts that lack inosines or contain only low level of inosines would be exported to the cytoplasm, whereas more highly edited RNA isoforms would be selectively retained in the nucleus. Thus, editing might serve to modulate gene expression of IR*Alus* containing RNAs by titrating the amount of mRNA that is allowed to reach the cytoplasm. We do not yet know whether different cells or tissues differ in their relative editing efficiencies of IR*Alus*, but it is well known that ADAR1 and ADAR2 are expressed in a tissue-specific and developmentally controlled manner in mammals. For instance, ADAR1 and ADAR2 can both promiscuously edit dsRNAs (Riedmann et al. 2008), and the expression and activity levels of both enzymes are higher in the brain (Bass 2002; Blow et al. 2004). Thus, genes with IR*Alus* in their 3′-UTRs may exhibit enhanced nuclear retention in the brain, where editing levels are highest. In addition, editing activity might differ between cancer cells and normal cells (Cenci et al. 2008; Paz et al. 2007), thus perhaps altering the nucleocytoplasmic export of some mRNAs. Further, ADAR activity is increased after inflammation, interferon treatment and immune stimulation (George and Samuel 1999; George et al. 2005; Liu et al. 1997; Rabinovici et al. 2001; Yang et al. 2003). Finally, we expect that in some cells, or under some conditions, the retention pathway may be overridden to allow export of mRNAs with IR*Alus*, whether or not they have been edited (Fig. 4a). This would be consistent with the observation that some mRNAs with structured or edited 3′-UTRs can be found in the cytoplasm, associated with polysomes (Chen and Carmichael 2009; Chen et al. 2008; Hundley et al. 2008; Hundley and Bass 2010).

Are long hairpins nuclear retention elements, even in the absence of editing? At this moment we cannot exclude the possibility that long hairpin structures that

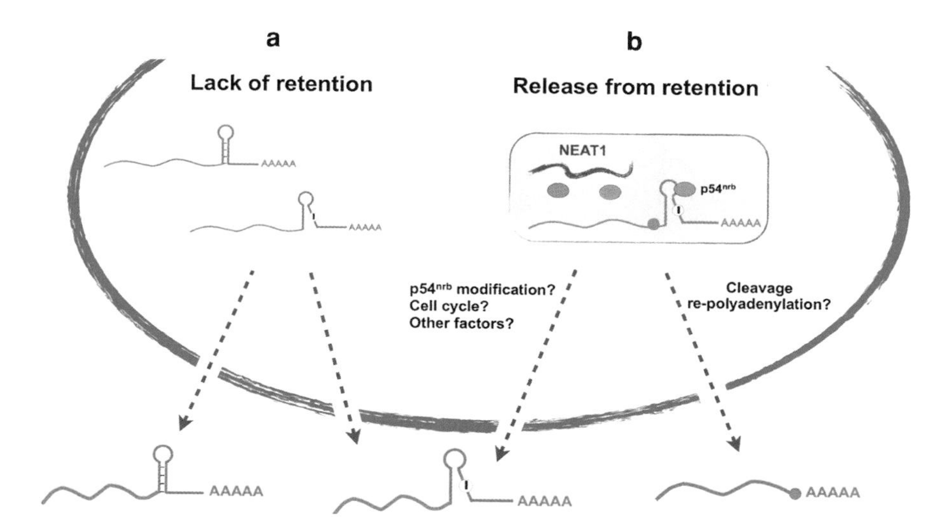

Fig. 4 Gene regulation by inverted repeats in 3′-UTRs. **a** In some cells or under some conditions (see text for details) mRNAs with hairpins may be exported efficiently to the cytoplasm, regardless of their editing status. **b** In cases where mRNAs with 3′-UTR hairpins are retained in paraspeckles, these mRNAs might be allowed to leave the nucleus upon appropriate molecular signal(s). Cleavage and possible re-polyadenylation has been demonstrated but the other mechanisms are speculative

formed by IR*Alu*s in mRNA 3′-UTRs might lead to paraspeckle localization and nuclear retention themselves, regardless of A-to-I editing. No studies have been reported yet on the fate of such mRNAs in cells that do not express ADAR1 and/or ADAR2.

4 Why are Some Edited mRNAs Retained in Paraspeckles, and How can They be Released?

Why would the cell retain some mRNAs in the nucleus while allowing the nucleocytoplasmic export of others? Can nuclear-sequestered mRNAs be released for export and, if they can, what mechanisms or signals are involved? What is the purpose of this retention system? While we do not yet know the answers to these questions, there are a number of possible clues and models. One model (Fig. 4b) is that for at least some mRNAs containing IR*Alu*s in their 3′-UTRs, editing and retention in paraspeckles allows the cell to store these mRNAs in the nucleus for rapid nucleocytoplasmic export upon the appropriate cellular signal. Thus, paraspeckles might serve as a nuclear repository for mRNAs that are not needed immediately, but which are available for very rapid mobilization. A good example of this type of regulation is the mouse CTN-RNA, which has been shown retained in the nucleus via its extended 3′-UTR which contains numerous A-to-I editing sites generated by inverted repeats of a murine SINE (Prasanth et al. 2005). Upon cell stress, a cleavage event occurs upstream of the 3′-UTR nuclear retention signal (the inverted repeats of a murine SINE). The truncated message is then transported efficiently to the cytoplasm for translation (Prasanth et al. 2005). The signaling pathway leading to cleavage and the mechanism of cleavage are not yet known, but may represent components of a new way in which cells respond to stress or other signals. It is also not known whether cleavage results in the addition of a new poly(A) tail to the mRNAs, though this might be expected owing to the presence of upstream polyadenylation signals in many mRNAs with IR*Alu*s in their 3′-UTRs. Finally, it has been reported that some virus infections may induce the expression of NEAT1 RNA in the mouse brain (Saha et al. 2006). Whether this helps or hurts viral life cycles is unresolved but is consistent with a role of the editing-associated nuclear retention system is cellular response to stress.

There are many other ways in which we can envision gene regulation by nuclear retention of mRNAs with edited 3′-UTRs. In addition to cleavage of the tethering sequences, it is possible that retained mRNAs can be released in some other way, such as by modification or degradation of one or more paraspeckle proteins, by the expression of some factor that overrides the retention mechanism, or by altering the expression or stability of NEAT1 RNA. It is not known how mRNA retention is affected throughout the cell cycle and we are only beginning to learn how this gene regulation system is regulated in different tissues and cells, and throughout development. Strikingly, in human embryonic stem cells ADAR editing is robust but paraspeckles are absent and NEAT1 is barely detectable (Chen and Carmichael 2009). In these cells, mRNAs with IR*Alu*s are efficiently exported to the cytoplasm. When induced to differentiate into trophoblasts, NEAT1 expression increased, along with the appearance of nuclear paraspeckles (Chen and Carmichael 2009). NEAT1 is similarly down-regulated in induced pluripotent stem cells (unpublished results), suggesting that pluripotency is somehow correlated with or connected to the lack of this editing and retention system.

Acknowledgments This work was supported by grant 0925347 from the National Science Foundation, grant 2011CBA01105 from the National Basic Research Program of China and awards from the State of Connecticut under the Connecticut Stem Cell Research Grants Program to L-L Chen and GG Carmichael. Its contents are solely the responsibility of the authors and do not necessarily represent the official views of the State of Connecticut, the Department of Public Health of the State of Connecticut, or Connecticut Innovations, Inc.

References

Akhmedov AT, Lopez BS (2000) Human 100-kDa homologous DNA-pairing protein is the splicing factor PSF and promotes DNA strand invasion. Nucleic Acids Res 28:3022–3030

Athanasiadis A, Rich A, Maas S (2004) Widespread A-to-I RNA editing of Alu-containing mRNAs in the human transcriptome. PLoS Biol 2:e391

Barak M, Levanon EY, Eisenberg E, Paz N, Rechavi G, Church GM, Mehr R (2009) Evidence for large diversity in the human transcriptome created by Alu RNA editing. Nucleic Acids Res 37:6905–6915

Bass BL (2002) RNA editing by adenosine deaminases that act on RNA. Annu Rev Biochem 71:817–846

Bass BL, Weintraub H (1988) An unwinding activity that covalently modifies its double-stranded RNA substrate. Cell 55:1089–1098

Beaudoing E, Gautheret D (2001) Identification of alternate polyadenylation sites and analysis of their tissue distribution using EST data. Genome Res 11:1520–1526

Blow M, Futreal PA, Wooster R, Stratton MR (2004) A survey of RNA editing in human brain. Genome Res 14:2379–2387

Bond CS, Fox AH (2009) Paraspeckles: nuclear bodies built on long noncoding RNA. J Cell Biol 186:637–644

Cenci C, Barzotti R, Galeano F, Corbelli S, Rota R, Massimi L, Di Rocco C, O'Connell MA, Gallo A (2008) Down-regulation of RNA editing in pediatric astrocytomas: ADAR2 editing activity inhibits cell migration and proliferation. J Biol Chem 283:7251–7260

Chen L-L, Carmichael GG (2009) Altered nuclear retention of mRNAs containing inverted repeats in human embryonic stem cells: functional role of a nuclear noncoding RNA. Mol Cell 35:467–478

Chen L-L, DeCerbo JN, Carmichael GG (2008) Alu element-mediated gene silencing. EMBO J 27:1694–1705

Clemson CM, Hutchinson JN, Sara SA, Ensminger AW, Fox AH, Chess A, Lawrence JB (2009) An architectural role for a nuclear noncoding RNA: NEAT1 RNA is essential for the structure of paraspeckles. Mol Cell 33:717–726

DeCerbo J, Carmichael GG (2005) SINES point to abundant human editing. Genome Biol 6:216–217

Dong B, Horowitz DS, Kobayashi R, Krainer AR (1993) Purification and cDNA cloning of HeLa cell p54nrb, a nuclear protein with two RNA recognition motifs and extensive homology to human splicing factor PSF and drosophila NONA/BJ6. Nucleic Acids Res 21:4085–4092

Edwalds-Gilbert G, Veraldi KL, Milcarek C (1997) Alternative poly(A) site selection in complex transcription units: means to an end? Nucleic Acids Res 25:2547–2561

Fox AH, Bond CS, Lamond AI (2005) P54nrb forms a heterodimer with PSP1 that localizes to paraspeckles in an RNA-dependent manner. Mol Biol Cell 16:5304–5315

Fox AH, Lam YW, Leung AK, Lyon CE, Andersen J, Mann M, Lamond AI (2002) Paraspeckles. a novel nuclear domain. Curr Biol 12:13–25

George CX, Samuel CE (1999) Human RNA -specific adenosine deaminase ADAR1 transcripts possess alternative exon 1 structures that initiate from different promoters, one constitutively active and the other interferon inducible. Proc Natl Acad Sci USA 96:4621–4626

George CX, Wagner MV, Samuel CE (2005) Expression of interferon-inducible RNA adenosine deaminase ADAR1 during pathogen infection and mouse embryo development involves tissue-selective promoter utilization and alternative splicing. J Biol Chem 280:15020–15028

Greenberger S, Levanon EY, Paz-Yaacov N, Barzilai A, Safran M, Osenberg S, Amariglio N, Rechavi G, Eisenberg E (2010) Consistent levels of A-to-I RNA editing across individuals in coding sequences and non-conserved Alu repeats. BMC Genomics 11:608

Hundley HA, Krauchuk AA, Bass BL (2008) C. elegans and H. sapiens mRNAs with edited 3′ UTRs are present on polysomes. RNA 14:2050–2060

Hundley HA, Bass BL (2010) ADAR editing in double-stranded UTRs and other noncoding RNA sequences. Trends Biochem Sci 35:377–383

Kallman AM, Sahlin M, Ohman M (2003) ADAR2 A–I editing: site selectivity and editing efficiency are separate events. Nucleic Acids Res 31:4874–4481

Kameoka S, Duque P, Konarska MM (2004) p54(nrb) associates with the 5' splice site within large transcription/splicing complexes. EMBO J 23:1782–1791

Kaneko S, Rozenblatt-Rosen O, Meyerson M, Manley JL (2007) The multifunctional protein p54nrb/PSF recruits the exonuclease XRN2 to facilitate pre-mRNA 3′ processing and transcription termination. Genes Dev 21:1779–1789

Kim DD, Kim TT, Walsh T, Kobayashi Y, Matise TC, Buyske S, Gabriel A (2004) Widespread RNA editing of embedded alu elements in the human transcriptome. Genome Res 14:1719–1725

Lander ES, Linton LM, Birren B, Nusbaum C, Zody MC, Baldwin J, Devon K, Dewar K, Doyle M, FitzHugh W et al (2001) Initial sequencing and analysis of the human genome. Nature 409:860–921

Lee JY, Yeh I, Park JY, Tian B (2007) PolyA_DB 2: mRNA polyadenylation sites in vertebrate genes. Nucleic Acids Res 35:D165–D168

Levanon EY, Eisenberg E, Yelin R, Nemzer S, Hallegger M, Shemesh R, Fligelman ZY, Shoshan A, Pollock SR, Sztybel D et al (2004) Systematic identification of abundant A-to-I editing sites in the human transcriptome. Nat Biotechnol 22:1001–1005

Liu Y, George CX, Patterson JB, Samuel CE (1997) Functionally distinct double-stranded RNA-binding domains associated with alternative splice site variants of the interferon-inducible double-stranded RNA-specific adenosine deaminase. J Biol Chem 272:4419–4428

Mao YS, Sunwoo H, Zhang B, Spector DL (2011) Direct visualization of the co-transcriptional assembly of a nuclear body by noncoding RNAs. Nat Cell Biol 13:95–101

Myojin R, Kuwahara S, Yasaki T, Matsunaga T, Sakurai T, Kimura M, Uesugi S, Kurihara Y (2004) Expression and functional significance of mouse paraspeckle protein 1 on spermatogenesis. Biol Reprod 71:926–932

Nishikura K (1992) Modulation of double-stranded RNAs in vivo by RNA duplex unwindase. Ann NY Acad Sci 660:240–250

Osenberg S, Dominissini D, Rechavi G, Eisenberg E (2009) Widespread cleavage of A-to-I hyperediting substrates. RNA 15:1632–1639

Osenberg S, Paz Yaacov N, Safran M, Moshkovitz S, Shtrichman R, Sherf O, Jacob-Hirsch J, Keshet G, Amariglio N, Itskovitz-Eldor J et al (2010) Alu sequences in undifferentiated human embryonic stem cells display high levels of A-to-I RNA editing. PLoS One 5:e11173

Paz N, Levanon EY, Amariglio N, Heimberger AB, Ram Z, Constantini S, Barbash ZS, Adamsky K, Safran M, Hirschberg A et al (2007) Altered adenosine-to-inosine RNA editing in human cancer. Genome Res 17:1586–1595

Polson AG, Bass BL (1994) Preferential selection of adenosines for modification by double-stranded RNA adenosine deaminase. EMBO J 13:5701–5711

Prasanth KV, Prasanth SG, Xuan Z, Hearn S, Freier SM, Bennett CF, Zhang MQ, Spector DL (2005) Regulating gene expression through RNA nuclear retention. Cell 123:249–263

Rabinovici R, Kabir K, Chen M, Su Y, Zhang D, Luo X, Yang JH (2001) ADAR1 is involved in the development of microvascular lung injury. Circ Res 88:1066–1071

Riedmann EM, Schopoff S, Hartner JC, Jantsch MF (2008) Specificity of ADAR-mediated RNA editing in newly identified targets. RNA 14:1110–1118

Saha S, Murthy S, Rangarajan PN (2006) Identification and characterization of a virus-inducible non-coding RNA in mouse brain. J Gen Virol 87:1991–1995

Sakurai M, Yano T, Kawabata H, Ueda H, Suzuki T (2010) Inosine cyanoethylation identifies A-to-I RNA editing sites in the human transcriptome. Nat Chem Biol 6:733–740

Sandberg R, Neilson JR, Sarma A, Sharp PA, Burge CB (2008) Proliferating cells express mRNAs with shortened 3' untranslated regions and fewer microRNA target sites. Science 320:1643–1647

Sasaki YT, Ideue T, Sano M, Mituyama T, Hirose T (2009) MENepsilon/beta noncoding RNAs are essential for structural integrity of nuclear paraspeckles. Proc Natl Acad Sci USA 106:2525–2530

Sunwoo H, Dinger ME, Wilusz JE, Amaral PP, Mattick JS, Spector DL (2009) MEN epsilon/beta nuclear-retained non-coding RNAs are up-regulated upon muscle differentiation and are essential components of paraspeckles. Genome Res 19:347–359

Versteeg R, van Schaik BD, van Batenburg MF, Roos M, Monajemi R, Caron H, Bussemaker HJ, van Kampen AH (2003) The human transcriptome map reveals extremes in gene density, intron length, GC content, and repeat pattern for domains of highly and weakly expressed genes. Genome Res 13:1998–2004

Wang Q, Carmichael GG (2004) Effects of length and location on the cellular response to double-stranded RNA. Microbiol Mol Biol Rev 68:432–452

Wong SK, Sato S, Lazinski DW (2001) Substrate recognition by ADAR1 and ADAR2. RNA 7:846–858

Yang JH, Luo X, Nie Y, Su Y, Zhao Q, Kabir K, Zhang D, Rabinovici R (2003) Widespread inosine-containing mRNA in lymphocytes regulated by ADAR1 in response to inflammation. Immunology 109:15–23

Yang YS, Hanke JH, Carayannopoulos L, Craft CM, Capra JD, Tucker PW (1993) NonO, a non-POU-domain-containing, octamer-binding protein, is the mammalian homolog of Drosophila nonAdiss. Mol Cell Biol 13:5593–5603

Zhang H, Lee JY, Tian B (2005) Biased alternative polyadenylation in human tissues. Genome Biol 6:R100

Zhang WW, Zhang LX, Busch RK, Farres J, Busch H (1993) Purification and characterization of a DNA-binding heterodimer of 52 and 100 kDa from HeLa cells. Biochem J 290:267–272

Zhang Z, Carmichael GG (2001) The fate of dsRNA in the nucleus. A p54(nrb)-containing complex mediates the nuclear retention of promiscuously A-to-I edited RNAs. Cell 106:465–475

Control of ADAR1 Editing of Hepatitis Delta Virus RNAs

John L. Casey

Abstract Hepatitis delta virus (HDV) uses ADAR1 editing of the viral antigenome RNA to switch from viral RNA replication to packaging. At early times in the replication cycle, the virus produces the protein HDAg-S, which is required for RNA synthesis; at later times, as result of editing at the amber/W site, the virus produces HDAg-L, which is required for packaging, but inhibits further RNA synthesis as levels increase. Control of editing during the replication cycle is essential for the virus and is multifaceted. Both the rate at which amber/W site editing occurs and the ultimate amount of editing are restricted; moreover, despite the nearly double stranded character of the viral RNA, efficient editing is restricted to the amber/W site. The mechanisms used by the virus for controlling editing operate at several levels, and range from molecular interactions to procedural. They include the placement of editing in the HDV replication cycle, RNA structural dynamics, and interactions of both ADAR1 and HDAg with specific structural features of the RNA. That HDV genotypes 1 and 3 use different RNA structural features for editing and control the process in ways related to these features underscores the critical roles of editing and its control in HDV replication. This review will cover the mechanisms of editing at the amber/W site and the means by which the virus controls it in these two genotypes.

J. L. Casey (✉)
Department of Microbiology and Immunology,
Georgetown University Medical Center, Washington, DC, USA
e-mail: caseyj@georgetown.edu

Current Topics in Microbiology and Immunology (2012) 353: 123–143
DOI: 10.1007/82_2011_146

Published Online: 6 July 2011

Contents

1 Introduction

1.1 Hepatitis Delta Virus

Hepatitis delta virus (HDV) is an important human pathogen that causes potentially severe acute and chronic hepatitis. It requires simultaneous infection with hepatitis B virus (HBV). The helper function provided by HBV is the envelope protein, HBsAg, which is required for the assembly and release of HDV particles, as well as the ability of these particles to attach to and infect hepatocytes, the primary targets of infection. Compared with those infected with HBV alone, individuals infected with both HDV and HBV experience more severe liver disease, including cirrhosis, hepatocellular carcinoma and liver failure. Although HDV depends on HBV, current licensed anti-HBV pharmaceuticals are ineffective for treatment of this virus because HBsAg expression remains high enough to support continued propagation. Approximately 15 million individuals worldwide are chronically infected with HDV.

Eight clades (genotypes) of HDV have been identified (Deny 2006). Most molecular studies have been conducted using clones of genotype 1, which is the most geographically widespread and the predominant genotype in Europe and North America. Genotype 3 has also been of interest because it is the most distantly related genetically to other genotypes ($\sim$40% divergence at the nucleic acid level) and because it is associated with the most severe HDV disease in northern South America (Casey et al. 1993).

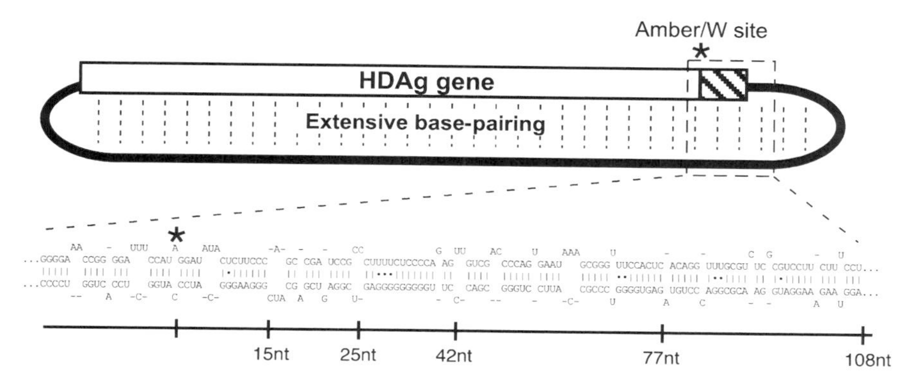

Fig. 1 The structure of the HDV genome. *Upper*: The *dark line* indicates the circular HDV genome, which forms an unbranched rod-like structure. Dashed vertical lines indicate base pairing between opposite sides of the circular RNA. The rectangle indicates the HDAg coding region; HDAg-S is encoded by the open portion of the *rectangle*; additional sequences added as a result of editing at the amber/W site and which form the C-terminus of HDAg-L are indicated by the striped portion. *Lower*: The predicted base pairing of a section of the antigenomic RNA that includes the amber/W site

1.2 Genome Structure of HDV

Hepatitis delta virus is unique in that it uses a helper virus to provide its envelope protein, but the molecular virology of HDV is arguably even more unusual. The approximately 1,700 nucleotide RNA genome is the smallest known to infect man. Yet, despite this small size, less than half the genome is devoted to encoding the sole viral protein, hepatitis delta antigen. To make up for the dearth of protein-coding information, the HDV replication cycle depends heavily on host factors and structural features of its RNA, including the unbranched rod-like structure of the entire genome, the double pseudo knotted ribozymes, and structures required for RNA editing. The genome and its replication intermediate, the antigenome, are circular RNAs that collapse into a characteristic unbranched rod-like structure in which about 70% of the nucleotides form base pairs (Fig. 1). This structure is essentially an elongated string of helices comprised of segments with 12 or fewer consecutive Watson–Crick base pairs interspersed with small internal bulges and loops, but no internal hairpins. In a two-dimensional representation of this structure (Fig. 1), the open reading frame for HDAg is on one side; sequences opposite the HDAg coding region in the circular RNA serve principally to form the unbranched rod structure. The unbranched structure plays several vital roles in the replication cycle. It is required for viral RNA replication (Casey 2002; Sato et al. 2004), binding to HDAg (Defenbaugh et al. 2009), the formation of viral particles (Chang et al. 1995; Lazinski and Taylor 1994) and for RNA editing and its regulation.

1.3 Hepatitis Delta Antigen

Hepatitis delta antigen (HDAg) is the sole protein encoded by HDV. This protein forms multimeric complexes and specifically binds HDV RNA in the unbranched rod-like structure. The two forms of HDAg produced during replication, HDAg-S and HDAg-L, differ structurally by the presence of an additional 19 or 20 amino acids at the C-terminus of HDAg-L. Functionally, they differ in two major respects. First, HDAg-S is required for RNA replication; HDAg-L not only does not support replication, but inhibits this process. Because of the multimeric nature of the protein, replication is highly sensitive to the relative levels of the two forms of HDAg: at an HDAg-L:HDAg-S ratio of 1:10 replication is reduced by as much as eight-fold (Chao et al. 1990; Xia and Lai 1992). The second functional difference between the two proteins is that HDAg-L is the limiting factor for the production of virus particles (Jayan and Casey 2005), which are formed via interactions between HBsAg and the C-terminal sequences unique to HDAg-L (Chang et al. 1991; Jenna and Sureau 1998).

1.4 The HDV Replication Cycle

The current model of HDV replication involves a double rolling circle (Fig. 2). The circular genome serves as a template for the synthesis of two RNAs: the mRNA for HDAg, and multimeric (concatameric) RNAs of the opposite sense. The latter are cleaved by internal ribozymes and ligated to form circular antigenomes, which subsequently serve as templates for the production of circular genomes in the same manner. One of the more remarkable aspects of this replication scheme is that it is accomplished by redirection of Pol II for HDV RNA synthesis (Chang et al. 2008). The specific mechanisms involved are not fully understood, but appear to involve HDAg-S binding Pol II (Yamaguchi et al. 2001, 2007). Related to the use of the host RNA polymerase, HDV replication occurs in the nucleus.

1.5 RNA Editing in HDV Replication

The HDV replication scheme, including the placement of editing, is shown in Fig. 2. Initially, the mRNA produced encodes HDAg-S. As replication proceeds, the antigenome RNA is edited such that the adenosine within the amber stop codon that terminates the HDAg-S ORF is deaminated to inosine. The codon is thus changed to tryptophan and the ORF is extended by an additional 19 or 20 amino acids to yield HDAg-L. In accord with the naming convention used for other editing sites that affect protein-coding regions, this editing site is referred to as the

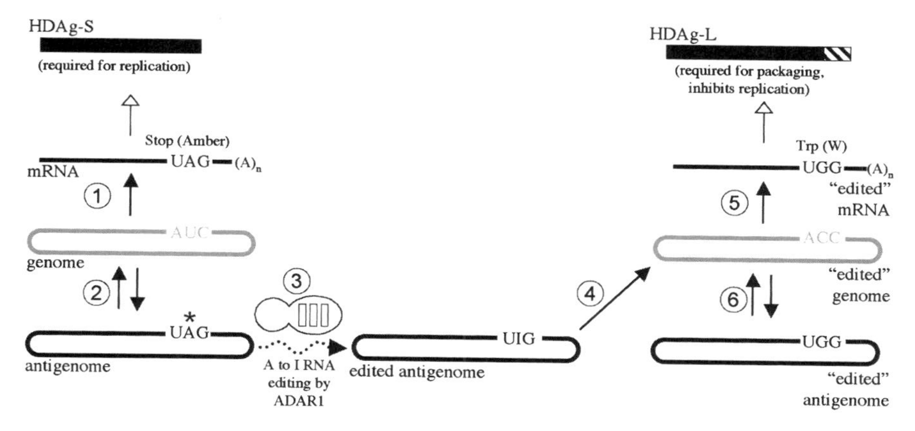

Fig. 2 The role of RNA editing in the HDV replication cycle. ① Genome RNA (*gray rounded rectangle*) serves as the template for synthesis of mRNA that is translated to produce HDAg-S, which is required for further RNA synthesis. ② Via rolling circle replication, the genome serves as the template for synthesis of the antigenome (*black rounded rectangle*) which also serves as the template for synthesis of additional genome RNA. ③ As replication proceeds, a fraction of antigenome RNAs is edited at the amber/W site by ADAR1 (*wavy dashed arrow*). ④ Edited antigenomes serve as the template for synthesis of "edited" genomes (*gray rounded rectangle with ACC*). ⑤ "Edited" genomes serve as templates for synthesis of mRNAs encoding HDAg-L, which is the limiting factor in virus packaging but also inhibits replication. ⑥ Simultaneously, "edited" antigenomes and additional "edited" genomes are synthesized via rolling circle replication. Thus, editing levels accumulate as replication proceeds. Note that the numbered scheme is intended to indicate an expanding repertoire of activities that persist as long as replication occurs rather than a stepwise progression in which earlier processes are terminated

amber/W site. As noted above, HDAg-S is required for viral RNA replication, whereas HDAg-L inhibits replication but is essential for virion morphogenesis. Thus, HDV uses RNA editing as a mechanism to switch from genome replication to packaging.

In cultured cells transfected with HDV cDNAs to initiate replication, editing, as well as HDAg-L production, increases from barely detectable levels 3 days post-transfection to around 25% by about 2 weeks (Casey et al. 1992; Chao et al. 1990; Sato et al. 2004; Wong and Lazinski 2002). In this manner, HDAg-L, which is the limiting component for production of HDV particles, accumulates and peaks after HDV RNA levels peak. The changes in editing levels are not due to changes in ADAR expression: no changes in ADAR1 levels were observed in cells transfected with an HDV replication expression construct or in HDV-infected liver (Jayan and Casey 2002b; Wong and Lazinski 2002). Rather, editing levels increase during virus replication due to the placement of editing in the virus RNA replication cycle. Unlike other editing substrates in which protein coding is affected, amber/W site editing does not occur on the HDV mRNA itself; instead, the substrate for editing is the antigenome, which is a replication intermediate (Fig. 2). Edited antigenomes serve as templates for the synthesis of genome RNAs with C at the corresponding position; these genome RNAs then serve as templates for the production of mRNAs encoding HDAg-L as well as additional "edited" antigenomes (Fig. 2). This scheme has several important consequences. First, the percentage of edited RNAs increases as long as replication occurs because edited RNAs are sustained via replication; moreover, additional editing events continue to convert unedited antigenomes to edited versions. Second, because HDAg-L inhibits viral RNA replication, but is the limiting factor for virion formation, virus replication and packaging are sensitive to the rate at which editing occurs. Finally, because edited genomes are packaged into virions but are not expected to be infectious, virus propagation could be negatively affected by excessive editing. It is therefore not surprising that HDV uses several specific mechanisms to ensure editing levels that are maximally productive.

2 Mechanism of HDV Amber/W Site Editing

2.1 Role of ADARs in Amber/W Site Editing

The relative roles of ADAR1 and ADAR2 in HDV RNA editing have been examined by several approaches, including siRNA knockdowns and analysis of expression levels in relevant cells, and comparison of the effects of site-directed mutagenesis of the RNA structure on editing by these enzymes. Both ADAR1 and ADAR2 are capable of efficiently editing the amber/W site. Early analyses found that *Xenopus* and *Drosophila* ADARs, which are similar to human ADAR1 and ADAR2, respectively, efficiently edited the amber/W site in the HDV antigenome RNA in vitro (Casey and Gerin 1995; Polson et al. 1996). Furthermore, when overexpressed, both human ADAR1 and human ADAR2 were found to efficiently edit the amber/W site in transfected cells (Jayan and Casey 2002a; Sato et al. 2001). However, the relative abilities of these two enzymes to edit the site have not been directly addressed; only hADAR1 has been examined for its ability to edit the RNA in vitro (Linnstaedt et al. 2006; Wong et al. 2003). In vitro, editing occurred efficiently under conditions that included only HDV RNA and purified enzyme, indicating that no additional factors are required (Polson et al. 1996; Wong et al. 2003).

Based on relative expression levels and on the effects of siRNA knockdown in cultured liver-derived cells, it is most likely that ADAR1 is responsible for editing at the amber/W site during the course of HDV infection. ADAR1 expression is 10- to 20-fold higher than ADAR2 in HDV-infected liver and in the liver derived cell line Huh-7 (Jayan and Casey 2002b; Wong and Lazinski 2002). Knockdown of ADAR1 expression reduced levels of editing in both reporter constructs and replicating genomes (Jayan and Casey 2002b; Wong and Lazinski 2002). Consistent with these results, in analyses of a series amber/W site mutations the patterns of editing activities by endogenous ADAR activity in Huh-7 cells better matched that obtained by co-transfected ADAR1 expression constructs (Casey and Gerin 1995; Polson et al. 1996; Sato et al. 2001). The ADAR1a splice form of ADAR1 is most abundant in infected liver and in Huh-7 cells and knockdown of this form in cultured cells reduced both editing and virus production (Jayan and Casey 2002b).

Given that interferon may be successful therapeutically for HDV-infected patients (Farci et al. 2004), and that HDV infection is capable of inducing production of cytokines (Jilbert et al. 1986), the relative activities of the constitutive and interferon-inducible forms of ADAR1 in amber/W site editing are of particular interest. Wong and Lazinski (2003) observed that the constitutive nuclear form of ADAR1 was more capable of editing the amber/W site than was the interferon-inducible form in the context of replication in cells transfected with ADAR expression constructs. Moreover, amber/W site editing was reduced by siRNA knockdown of both forms of ADAR1, whereas knockdown of the interferon-inducible form alone had no effect (Wong and Lazinski 2002). These observations are consistent with the higher expression level of the constitutive form of ADAR1 in the nucleus, where HDV RNA replication occurs. Wong and Lazinski (2002) did observe that the longer, inducible form of ADAR1 was responsible for the bulk of editing of an amber/W editing reporter RNA; however, this observation was probably a consequence of the use of a reporter mRNA with a long residence time in the cytoplasm. Thus, amber/W site editing in cultured cells is primarily due to the constitutive form ADAR1a, which is the most abundant form in the liver.

2.2 RNA Structure Requirements for Editing

The specific RNA structural features of ADAR substrates and their interactions with ADAR double-stranded RNA binding motifs and deaminase domains are described in detail elsewhere in this volume. A common feature is base pairing that flanks the editing site and extends at least about ~25 bp in one direction. In most cases, base pairing extends 3′ of sites and includes a limited number of mismatches, bulges and small internal loops that might help to properly position the deaminase domain (Ohman et al. 2000). As described below, the secondary structures required for amber/W site editing have been experimentally evaluated for two of the eight HDV genotypes–genotype 1, which is the most common, and genotype 3, which is the most distantly related genetically. In both instances, the amber/W site is present in a base-paired context, but the structures differ considerably. Compared with other editing sites, the overall size of the base-paired structure required for HDV genotype 1 and genotype 3 amber/W site editing may be larger.

2.2.1 The HDV Genotype 1 Amber/W Site

Site-directed mutagenesis has indicated that, for HDV genotype 1, amber/W site editing involves the unbranched rod structure (Casey et al. 1992; Sato et al. 2001), which is also required for HDV RNA replication (Casey 2002; Sato et al. 2004). This site occurs as an A–C mismatch pair in the midst of eight canonical Watson–Crick base pairs (Fig. 1). Both the A–C mismatch and the base pairs immediately surrounding the site have been shown to be critical for editing in cells and in vitro (Casey et al. 1992; Casey and Gerin 1995; Polson et al. 1996). A–C mismatches, which are found in some (but not all) other editing sites, have been found to maximize editing efficiency (Casey et al. 1992; Herb et al. 1996; Lomeli et al. 1994; Polson et al. 1996; Wong et al. 2001).

The role of base-paired regions outside the 8 bp immediately surrounding the genotype 1 amber/W site is not settled. Inspection of the RNA secondary structure downstream of the site indicates that it contains base-paired segments but is more frequently disrupted by bulges, mismatches and small internal loops than the region 3′ of other editing substrates (Fig. 1). The effects on amber/W site editing of site-directed mutations that either increase or decrease base pairing in this region raise the question of whether the quality of base pairing in this region is sufficient to play an important role in ADAR1 binding and activity (Jayan and Casey 2005; Sato et al. 2004). Mutations that improved base-pairing, particularly in the region 15–25 nt 3′ of the editing site, increased editing significantly (Jayan and Casey 2005; Sato et al. 2004). Moreover, mutations that further disrupted base pairing had little detectable effect on editing (Jayan and Casey 2005). These results suggest that base pairing in the region up to 25 nt 3′ of the amber/W site might not be sufficient to recruit ADAR1 to the editing site via interactions with the DRBMs.

This conclusion appears to be inconsistent with the results of Sato et al. (2001), who analyzed amber/W site editing using reporter constructs transfected with ADAR expression constructs. These authors observed editing using a minimal construct that contained just 24 nucleotides-principally the A–C mismatch and only the surrounding eight base pairs, and concluded, based on this and other data, that no additional secondary structure features were required. However, this interpretation is limited by the fact that ADAR1 was expressed at very high levels in the transfected cells. Several reports have indicated that such overexpression can alter the behavior of the enzyme. For example, Herbert and Rich (2001) found that overexpression of a form of ADAR1 lacking the double-stranded RNA binding domains exhibited levels of activity similar to that of the wild-type protein. We have observed a similar result for the genotype 1 amber/W site–overexpression of ADAR1 constructs lacking the DRBMs edited this site with efficiency approximately half that of wild type ADAR1 (Chen and Casey, unpublished). On the other hand, in vitro studies have shown that the DRBMs are essential for full enzymatic activity on both dsRNA substrates as well as substrates for site-specific editing (Liu et al. 1998; Liu and Samuel 1996). Overexpression of the protein also dramatically reduces editing specificity (Jayan and Casey 2002a). Overall, although the reasons for the discrepancy between the roles of the DRBMs in editing in cells and in vitro remains to be resolved, it is clear that results obtained in transfected cells expressing high levels of ADARs may need to be interpreted cautiously. Thus, the conclusions from Sato et al. (2001) are likely restricted by the high levels of ADAR1 expressed in the transfected cells and do not necessarily preclude a role for sequences and structures outside the immediate vicinity of the amber/W site for typical levels of endogenous ADAR1 expression.

More recent work indicates that the structure required for editing of the genotype I amber/W site may indeed be substantially larger than suggested by Sato et al. Inspection of the predicted secondary structure of the region 3′ of the amber/W site indicates that the highest degree of base pairing is located from 64 to 100 nt downstream (Fig. 1). Truncation to within 77 nt 3′ of the amber/W site diminished editing to less than one-third the level of full-length RNA and further reduction to 42 nt 3′ of the site led to nearly complete loss of activity (Chen and Casey, unpublished). This distal region includes 34 bp that are minimally disrupted by one mismatch pair and four asymmetric single nucleotide bulges. The role of sequences 5′ of the amber/W site has not been examined in detail. Base paired segments more than about 45 nt away from the 5′ side of the amber/W site are not required, at least when the base paired region extending about 100 nt 3′ of the site is present (Chen and Casey, unpublished).

Editing of the well-characterized GluR-B R/G site by ADAR2 requires a 25 bp hairpin 3′ of the site that contains two mismatch pairs (Jaikaran et al. 2002; Ohman et al. 2000); the predicted structures of other substrates for both ADAR1 and ADAR2 contain similar length contiguous base paired segments that contain few disruptions. The above results suggest that the HDV genotype 1 amber/W site may require a similar total amount of base pairing; however, unlike other substrates characterized thus far, the base pairing immediately flanking the amber/W site is not contiguous with other base paired segments that are required. A candidate model for the interaction of this structure with ADAR1 is that the DRBMs of ADAR1 bind within the base paired region between 64 and 100 nt 3′ of the amber/W site and position the deaminase domain at the editing site via a long range interaction that involves bending of the intervening partially dsRNA (Fig. 3). This model remains to be confirmed. No cellular substrates for site-specific adenosine deamination have yet been shown to use non-contiguous base paired segments. Perhaps such sites can be identified by expansion of current computational methods to include structures that include non-contiguous base paired segments.

2.2.2 The HDV Genotype 3 Amber/W Site

The secondary structure used by HDV genotype 3 for amber/W site editing differs from that used by genotype 1. Although genotype 3 also forms an unbranched rod structure that is required for replication, inspection of this structure indicated that the base pairing in the immediate vicinity of the amber/W site adenosine is much more disrupted than in genotype 1 (Fig. 4). In fact, the genotype 3 amber/W adenosine is not edited when the RNA is in the characteristic unbranched rod conformation (Casey 2002; Linnstaedt et al. 2006). The inability of this structure to be edited is due solely to the local disruption in base pairing around the amber/W site, because RNAs harboring site-directed mutations that produce a structure

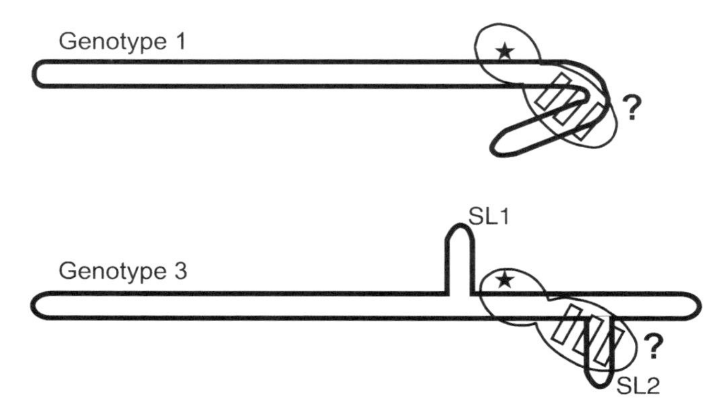

Fig. 3 Schematics showing hypothetical interactions between amber/W sites and ADAR1. In both cases, the DRBMs of ADAR1 (indicated by the *open rectangles*) may interact with base paired segments further removed from the editing site than in the GluR-B R/G site. For genotype 1, sequences more than 42 nt distal from the amber/W site, along the partially double stranded structure, appear to be involved. In genotype 3, sequences more than 25 nt away in the ~25 bp SL2 structure are required for editing in vitro

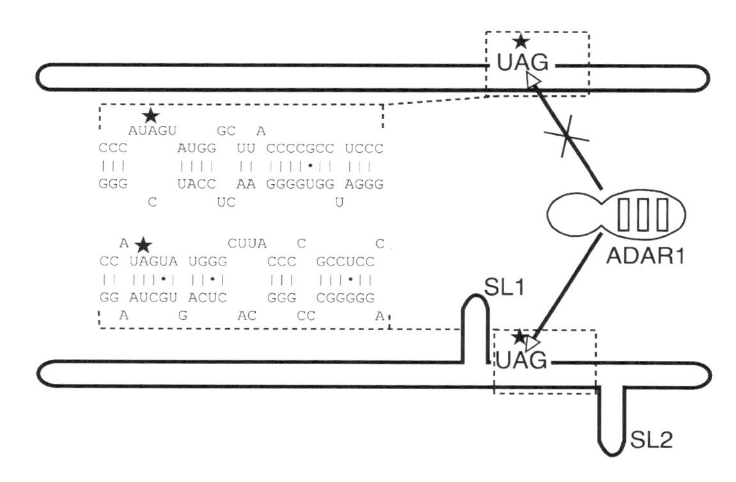

Fig. 4 Editing of the HDV genotype 3 amber/W site requires a branched structure. Schematics show the unbranched rod (*upper*) and branched structures (*lower*). The ~25 base pair stem loops SL1 and SL2 stabilize the branched structure. RNA secondary structures in the vicinity of the amber adenosine are shown for both the unbranched rod and branched structures. The unbranched rod structure is not a substrate for editing by ADAR1 due to the disrupted base pairing around the adenosine, which is indicated by the five-point star

similar to that of the genotype 1 amber/W site (an A–C mismatch pair flanked by 9 base pairs) are efficiently edited (Casey 2002). In addition to the unbranched rod structure, genotype 3 RNA is capable of forming an alternative branched structure in which the base pairing in the immediate vicinity of the editing site is increased (Fig. 4). This structure is efficiently edited by ADAR1 both in vitro and in cells (Casey 2002; Linnstaedt et al. 2006, 2009). In this branched structure, ca. 220 nt of the unbranched rod structure are rearranged to form two $\sim$25 bp stem-loops (SL1 and SL2) flanking a central base paired region that includes the amber/W site, which is itself base paired (Fig. 4).

The elements of the branched structure play different roles in editing of the genotype 3 amber/W site. Editing requires the central base paired region in the immediate vicinity of the amber/W site and is sensitive to sequence variations in this region in different HDV isolates (Linnstaedt et al. 2009). Perhaps, because the 28 nt mostly base-paired region 3′ of the genotype 3 amber/W site is disrupted by a 7 nt internal loop, and loops of more than 4 nt have been shown to affect editing of dsRNAs (Lehmann and Bass 1999), this region is not sufficient for editing in vitro (Linnstaedt and Casey, unpublished). Thus, like the genotype 1 amber/W site, the structural components of the genotype 3 site appear to extend over a larger segment of the RNA than for the GluR-B R/G site. Though SL1 stabilizes the branched structure required for editing, it does not participate in the editing reaction itself; removal of SL1 affects neither editing nor ADAR1 binding (Cheng et al. 2003). SL2, like SL1, stabilizes the branched structure, but may also play a direct role in editing. In vitro, SL2 is essential for ADAR1 binding and amber/W site editing of a miniaturized RNA that contains a limited amount of the unbranched rod structure beyond SL2 (Linnstaedt and Casey, unpublished). This result suggests a model for ADAR1 binding similar to that suggested for genotype 1 (Fig. 3). In this case, the binding of at least some of the DRBMs to SL2 may properly position the deaminase domain near the amber/W site. Conversely, in cells expressing longer RNAs that include more of the unbranched rod structure beyond SL2, removal of SL2 did not abolish editing (Cheng et al. 2003). Possibly, both SL2 and parts of the unbranched rod beyond SL2 can independently contribute to ADAR1 binding, perhaps via interactions with the DRBMs. Further analysis will be required to resolve this question.

3 Control of Editing

The roles of HDAg-L in the HDV replication cycle, the multimeric nature of HDAg and the susceptibility of replication intermediates with extensive dsRNA character (including both the antigenome and the similarly structured genome) to editing by ADAR1 necessitate that editing be controlled. This control occurs at several levels: (1) minimizing editing at sites other than the amber/W site; (2) maintaining the optimal rate of amber/W site editing to fit the timing of the viral replication cycle; and (3) preventing over-accumulation of editing.

3.1 Restriction of Editing to the Amber/W Site

ADAR1 and ADAR2 can extensively edit long (≥50 base-pairs) dsRNAs, in which up to 50% of adenosines may be deaminated. Clearly, promiscuous editing such as occurs on dsRNA would be deleterious to virus replication. Such highly promiscuous editing does not occur in HDV RNA in vitro (Polson et al. 1996), most likely because base-pairing in the HDV RNA unbranched rod structure is interrupted by frequent bulges, internal loops and mismatches, which have been shown to restrict editing on artificial dsRNA substrates (Lehmann and Bass 1999). However, because HDAg functions as a multimer, even moderate levels of non amber/W site editing could have deleterious effects by creating dominant negative inhibitors of replication. In fact, mutant forms of HDAg that arose as a result of spurious editing when ADAR1 or ADAR2 were overexpressed inhibited replication (Jayan and Casey 2002a). Thus, it is not surprising that promiscuous editing does not occur during typical HDV replication (Polson et al. 1998). In fact, the amber/W site is edited 600-fold more efficiently than the other 337 adenosines in the RNA (Polson et al. 1998).

There are three likely mechanisms by which HDV limits editing occurring at non-amber/W sites: (1) base pairing in the unbranched rod structure is frequently disrupted by internal bulges and loops that are likely to limit ADAR binding; (2) A–C mismatch pairs flanked by segments of canonical base pairing have been shown to be edited with the highest efficiency-the predicted secondary structure of genotype 1 RNA contains just three adenosines in such structures, one of which is the amber/W site; (3) the frequency of GA dinucleotides, which are strongly disfavored for editing (Polson and Bass 1994), is 60% higher than predicted based on a random distribution—indeed, G is the 5′ neighbor of 48% of adenosines and both of the non amber/W adenosines predicted to be present as A–C mismatch pairs have G as the 5′ neighbor.

It is important to note that, although the amount of editing that occurs at non-amber/W sites is very low relative to the amber/W site, such editing may occur at low levels during replication (Netter et al. 1995; Polson et al. 1998) and may contribute to the evolution of genetic changes in the virus that can affect the outcome of infection (Casey et al. 2006). Using a viroid-like model in which HDV RNA replication was decoupled from HDAg synthesis, Chang et al. (2005) observed A to G and U to C changes in the genome sequence that are consistent with editing by ADAR1 on the genome and antigenome, respectively. These changes accounted for 80% of all sequence changes that accumulated over the course of 1 year in this system. Similar results were found in experimentally infected woodchucks: more than two-thirds of sequence changes that accumulated during the course of HDV infection were consistent with ADAR activity (Casey et al. 2006; Netter et al. 1995). To be sure, none of the sequence modifications in these studies were specifically shown to be due to ADAR activity; however, the overwhelming bias in sequence changes towards transitions consistent with adenosine deamination strongly suggests that it plays an important role in the

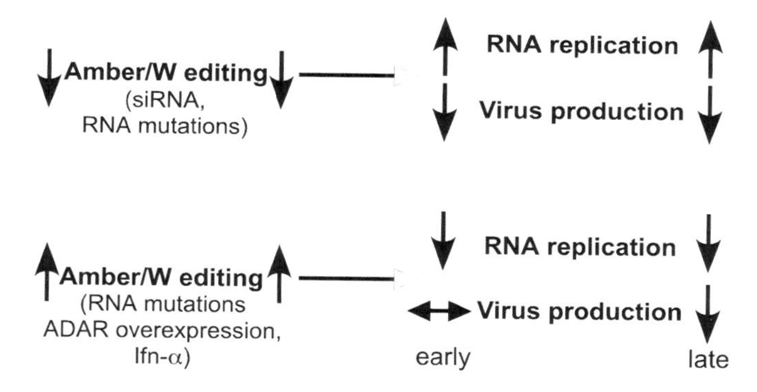

Fig. 5 Effects of experimental variations in amber/W editing activity on HDV replication and virion formation

majority. Significantly, in one of the woodchuck studies (Casey et al. 2006), some of the potentially ADAR-initiated sequence changes were associated with the development of chronic HDV infection, a serious health complication in humans. Thus, in addition to playing a specific role in the HDV replication cycle by modifying the amber/W site, ADAR activity may also provide a mechanism for virus evolution that can affect the course of infection.

3.2 Control of Editing at the Amber/W Site

Considering the roles of HDAg-S and HDAg-L in virus replication, the amount of amber/W site editing that occurs is likely to be critical. Indeed, experimental manipulation of editing levels affected both RNA replication and virus production in ways consistent with the roles of HDAg-S and HDAg-L in these processes (Fig. 5). Thus, the rate of editing must be correlated with levels of viral RNA to maximize virus production. Furthermore, because edited genomes are packaged into virions, high levels of editing would compromise virus viability. Virus particles containing genomes encoding HDAg-L would not only be incapable of establishing replication on cell entry, but, in the event of co-infection with virions encoding HDAg-S, would be likely to inhibit replication of otherwise competent genomes. Thus, control of both the rate and ultimate level of editing is important for HDV to maximize the yield of infectious virus.

Comparison of HDV editing with other editing substrates indicates that amber/W site editing occurs much less efficiently. Cellular substrates for editing that are pre-mRNAs are not only frequently edited at high levels ($\sim 100\%$ in some cases), but must be edited prior to processing in the nucleus. By contrast, amber/W site editing HDV is barely detectable 4 days following the initiation of HDV replication in transfected cells (Casey et al. 1992; Sato et al. 2004; Wong and Lazinski 2002), but increases gradually as replication proceeds.

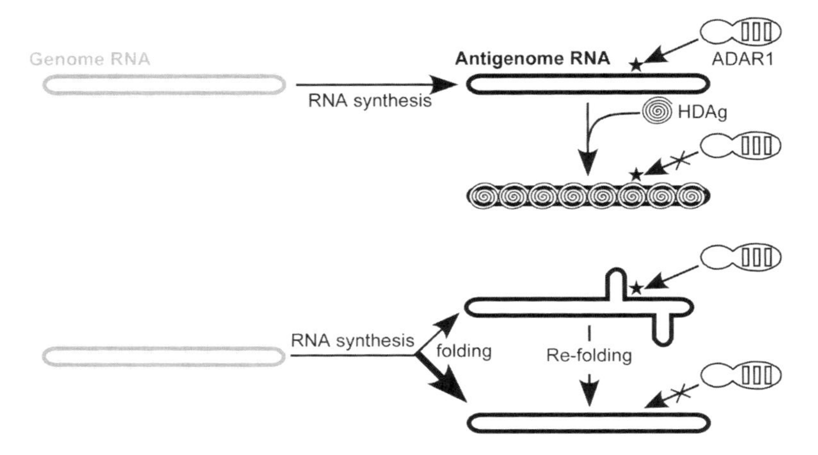

Fig. 6 Models for control of editing rates of the HDV genotype 1 and genotype 3 amber/W sites. Editing of the genotype 1 site (*upper*) involves the unbranched rod structure and is inhibited by HDAg, which binds this structure following synthesis. In this model, HDAg and ADAR1 compete for binding to the RNA. Editing of the HDV genotype 3 site (*lower*) requires a metastable branched structure that is formed only following transcription and eventually rearranges to the unbranched rod. Editing is limited by the fraction of RNA that folds into the branched conformation during transcription

In principle, the efficiency at which different adenosines are edited can be affected by a number of different mechanisms:

- ADAR expression levels
- Intrinsic activity of the editing substrate:
  - ADAR binding affinity
  - Accessibility of the targeted adenosine to the catalytic site after binding
- Substrate availability:
  - Alternative processing
  - Alternative folding
  - Binding by other factors

There is no evidence that editing rates, or efficiency, increase during the course of replication, nor do ADAR levels change (Sato et al. 2004; Wong and Lazinski 2002). The increased level of editing at later times during HDV replication is a consequence of the placement of editing in the replication cycle (Fig. 2). As discussed below, the intrinsic substrate activity of the amber/W site is sub-optimal for efficient editing in both genotype 1 and genotype 3 and this inefficiency contributes to the low editing levels observed. However, a corollary of reduced intrinsic activity of the amber/W site is decreased specificity for that site relative to other sites in the RNA. Thus, it is not surprising that both genotypes further decrease editing rates by limiting the availability of the RNA for editing by

ADAR1. That the specific mechanisms used to limit both the intrinsic editing activity and substrate availability differ for these genotypes underscores the critical need for the virus to restrict editing rates.

3.2.1 Sub-Optimal RNA Structures for Editing at the Amber/W Site

The K12 transcript of human herpes virus 8 (HHV-8) yields both a mRNA and miRNA that are edited at up to 76% under conditions of lytic virus replication (Gandy et al. 2007). Editing in vitro by ADAR1 of the genotype 1 amber/W site and the amber/W site in the genotype 3 RNA HDV_{PERU} is about threefold lower than that of the HHV-8 K12 editing site; editing of the genotype 3 HDV_{EC} amber/W site is even less efficient (Linnstaedt and Casey, unpublished). The features of the HDV amber/W sites responsible for the sub-optimal activity are different for genotype 1 and genotype 3. For genotype 1, the lack of sufficient base pairing within ~25 nt 3′ of the site limits activity. Increasing base pairing in this region increases both binding by ADAR1 (Chen and Casey, unpublished) and editing (Jayan and Casey 2005). In the branched structure used by HDV genotype 3, the amber/W site exists as an A–U base pair, which is edited less efficiently than the much more common A–C mismatch (Casey et al. 1992; Herb et al. 1996; Lomeli et al. 1994; Polson et al. 1996; Wong et al. 2001). This difference contributes to the sub-optimal activity of this site; changing to an A–C mismatch pair increases editing. Furthermore, additional sequence and structural variations within the region ~10 nt 3′ of the amber/W site contribute to differences in editing between the structures formed by Peruvian and Ecuadorian genotype 3 sequences (Linnstaedt et al. 2009).

3.2.2 Genotype 1—Control by Binding of HDAg to HDV RNA

While it is clear that the genotype 1 and genotype 3 amber/W sites are edited less efficiently than the HHV-8 K12 transcript in vitro, the differences in editing levels are not as dramatic as that seen in cells. Thus, the sub-optimal activity of the amber/W site cannot fully account for the lower efficiency of HDV editing. Although editing of HDV RNA 2–3 days following the initiation of replication is nearly undetectable, up to 40% of non-replicating genotype 1 RNA produced in transfected cells in the absence of HDAg are edited (Casey and Gerin 1995; Polson et al. 1998). In the presence of HDAg expression, editing of non-replicating RNA editing reporters is dramatically reduced (Polson et al. 1998; Sato et al. 2004). The levels of HDAg-S required for this inhibition are similar to those seen in cells replicating HDV RNA (Polson et al. 1998). Thus, HDAg prevents the rapid accumulation of editing early in the HDV genotype I replication cycle. HDAg has been shown to specifically bind the HDV RNA unbranched rod structure (Defenbaugh et al. 2009). Thus, inhibition likely occurs by binding of HDAg to HDV RNA and preventing access of ADAR1 (Polson et al. 1998; Sato et al. 2004) (Fig. 6).

Inhibition is not due to interference with ADAR1 activity per se; as discussed below, HDAg-S does not affect editing of genotype 3 RNA (Cheng et al. 2003). Moreover, recent work has shown that HDAg inhibits editing of HDV RNA by ADAR1 in vitro; editing of the K12 RNA, which is not bound by HDAg, is not affected. This mechanism of inhibition is advantageous with regard to the correlation between low editing activity and specificity. By binding the RNA, HDAg is likely to interfere with editing at non-amber/W sites at least as effectively, if not more so, than it interferes with editing at the amber/W site. Consistent with this idea, sequence analysis of cDNA clones indicated that non amber/W editing events were highly correlated with amber/W changes on the same RNA (Polson et al. 1998). Both HDAg-S and HDAg-L inhibit editing equally well, indicating that this mechanism for controlling editing is not directly responsive to editing levels.

3.2.3 Genotype 3—Control by RNA Folding Dynamics

HDV genotype 3 restricts editing rates by a completely different mechanism. Editing of the genotype 3 amber/W site is not inhibited by HDAg-S, either in transfected cells (Chen and Casey, unpublished) or in vitro (Cheng et al. 2003). The mechanistic explanation for this lack of inhibition is not yet clear, but may involve altered binding of HDAg to the branched structure required for editing. Rather than HDAg-binding, editing of the genotype 3 amber/W site is limited by the distribution of the RNA into several different folding conformations following synthesis. The branched structure required for editing of the genotype 3 amber/W site is less energetically stable than the unbranched rod structure and is therefore formed only immediately following transcription of the RNA as a metastable structure (Linnstaedt et al. 2006, 2009). In vitro, only a fraction of the RNA folds into the branched structure and, with time, the RNA converts from the branched to the unbranched structure (Linnstaedt et al. 2006). Because only the branched structure can be edited by ADAR1, the fraction of RNA that folds into this structure determines the amount of editing that can occur (Fig. 6). In support of this model, the predicted and observed in vitro folding tendencies of RNAs from two different HDV genotype 3 isolates correlated with the editing activities of these RNAs during replication (Linnstaedt et al. 2009).

3.3 Control of Final Editing Levels

As discussed above, genomes encoding HDAg-L accumulate during the course of HDV replication. Although these genomes are efficiently packaged into virions, they are not expected to be infectious because the encoded HDAg-L will not support RNA replication. There is no known mechanism employed by the virus to reverse the result of editing such that the genome could produce HDAg-S. Thus, the non-infectious genomes packaged into virions are, in a sense, a cost of editing to the virus. However, under conditions of high multiplicity of infection, the negative effects of packaging edited genomes may be far greater than the production of a population of viruses in which 25% are defective. During the acute stage of infection, viremia can approach 10^{12} particles per ml (Ponzetto et al. 1987); with a total of about 10^9 hepatocytes in the liver it is possible that cells will be infected by more than one virus particle. Under these conditions, genomes encoding HDAg-L could interfere with infection by HDAg-S-encoding genomes. Thus, it is likely important for the virus to not only prevent premature excessive editing by the mechanisms discussed above, but to also limit the total amount of editing that occurs, after sufficient HDAg-L has accumulated for producing virions.

Sato et al. (2004), demonstrated that HDAg-L controls its own synthesis by inhibiting HDV RNA replication and editing. This control appears to be intimately linked to the mechanisms by which editing occurs and by which editing rates are controlled for both genotypes 1 and 3. Because the antigenomic RNA is the substrate for editing, HDAg-L mRNA production from edited genomes requires two additional transcription reactions: synthesis of genome RNA, then synthesis of the mRNA. As levels of HDAg-L increase, these replication events are inhibited. But Sato et al. (2004) observed that editing levels were also controlled. To explain this result, it was suggested that newly synthesized RNAs are available to be edited by ADAR1 for a short time, but are rapidly bound by HDAg and become unavailable for editing thereafter (Sato et al. 2004). Thus, once RNA synthesis is stopped, amber/W editing also ceases.

Genotype 3 HDAg-L also inhibits replication of genotype 3 RNA (Casey and Gerin 1998). Although the study conducted by Sato et al. was performed using a genotype 1 clone, the model is also consistent with our understanding of how genotype 3 amber/W site editing occurs: editing of the genotype 3 amber/W site can only occur following folding of newly synthesized antigenome RNAs into the metastable branched structure (Linnstaedt et al. 2006, 2009). Once synthesis stops, editing will no longer occur because the RNA can no longer fold into the proper structure.

4 Relationship of HDV RNA Editing to Antiviral Responses

Interferon alpha (IFN-α) induces expression of a form of ADAR1 that can efficiently edit the amber/W site (Patterson and Samuel 1995; Wong et al. 2003) and HDV replication is very sensitive to editing levels. It is therefore important to consider the relationships between IFN-α treatment, editing levels and HDV RNA replication, particularly in light of the clinical use of IFN-α for HDV therapy. Although some results have been reported, this area is need of further study. Hartwig et al. (2004) showed that IFN-α treatment of Huh-7 cells led to an approximately two-fold increase in amber/W site editing and a marked decrease in total levels of HDAg expression 7 days post-transfection. These results appear to contradict an earlier study that found no effect of IFN-α on HDV replication (McNair et al. 1994). However, the lack of an effect in that work could be attributed to the use of a stably transfected cell line, in which HDV RNA levels could have been due to transcription from the integrated cDNA rather than viral RNA replication. In subsequent work, Hartwig et al. (2006) concluded that the increase in amber/W editing following IFN-α treatment is due to increased ADAR1-L expression. However, this conclusion needs further validation. The experimental system used did not fully knock down ADAR expression levels for the duration of HDV replication; thus, amber/W editing levels still rose nearly two-fold in cells treated with IFN-α even in the face of treatment with antisense oligonucleotides. Another unresolved question is how the extent of increased p150 expression observed by Hartwig et al. (2004) could account for the increase in amber/W editing. Even after IFN-α treatment, levels of ADAR1 p150 were lower than p110 (Hartwig et al. 2004), and Wong et al. (2003) previously showed that p110 is more effective editing the amber/W site during HDV replication. Thus, the extent to which the increase in amber/W site editing in IFN-α-treated cells is due to increased p150 expression, or to some other effect(s) of IFN-α treatment, remains to be determined.

Regardless of the specific process by which amber/W editing levels increase following IFN-α treatment, the results of Hartwig et al. (2004) suggest a potential mechanism of action of IFN-α therapy in infected patients that should be explored in greater detail. Do amber/W editing levels increase during therapy? Does sequence variability increase, as might be expected based on the effects of ADAR1 overexpression in transfected cells (Jayan and Casey 2002a)? How are HDV RNA replication levels affected? It is not clear whether to expect increased editing to produce decreases in HDV viremia, which is perhaps the simplest measure of antiviral success, because the connection between editing levels, HDAg-L expression and virus production might not be straightforward. In an analysis of genomes containing mutations that increased editing levels, it was found that virion production was not affected at early times, most likely because HDAg-L levels, which are limiting for virus production, remained constant because increased amber/W site editing counterbalanced decreased replication (Jayan and Casey 2005).

Related to the effects of IFN-α treatment are the questions of whether virus infection itself affects ADAR1 expression (and editing levels) by either inducing IFN-α expression (Jilbert et al. 1986) or by inhibiting IFN-α signaling pathways (Pugnale et al. 2009). Although levels of the interferon-inducible form of ADAR1 were reported to be the same in normal and HDV-infected liver (Jayan and Casey 2002b), it is possible that the analysis of bulk tissue samples would not have been sensitive to localized IFN-α expression near foci of infection. A recent report indicated that HDV may inhibit IFN-α signaling pathways (Pugnale et al. 2009). The mechanism by which this inhibition occurs has not been determined. Although this result appears to contradict the observed IFN-α induction of ADAR1 p150 in cells transfected with HDV (Hartwig et al. 2004), there are a number of differences in the experimental approaches (timing and dose of IFN-α treatment, timing of

HDV replication, accounting for untransfected cells); thus it remains unclear whether there is a controversy. Undoubtedly, there is much to be done to understand the interactions between HDV and IFN-α and the extent to which these interactions involve editing of the HDV RNA by ADAR1.

5 Perspective

Compared with cellular substrates for RNA editing, the HDV amber/W site displays both significant similarities and instructive differences. Although the base pairing in the immediate vicinity of sites is similar, ongoing analysis of the HDV amber/W site suggests that base paired segments further removed from the site (at least in a linear configuration) may be important, perhaps by binding the DRBMs of ADAR1. Whether such long-range interactions occur in cellular substrates is currently unknown. Determination of the specific interactions between ADAR1 elements and structural features in HDV RNA, including both the amber/W site itself and segments further removed, remains a critical goal. Possibly, understanding the interactions employed by HDV could expand current concepts of what constitutes a structure in cellular RNAs likely to be targeted by ADAR1. The mechanisms by which editing is controlled in different cellular environments and on different cellular RNAs has not been widely examined. However, it seems likely that control of editing will be just as important for cellular targets as for HDV. As control of editing of cellular substrates is analyzed in more depth, the mechanisms employed by HDV to control editing (including competition between ADARs and other proteins for the site, and RNA folding variability) will more than likely be among the mechanisms found to be used for these targets. Finally, the connections between ADAR1, IFN-α, and HDV editing are in need of further study, particularly in light of the therapeutic effects of IFN-α in HDV patients.

References

Casey JL (2002) RNA editing in hepatitis delta virus genotype III requires a branched double-hairpin RNA structure. J Virol 76:7385–7397

Casey JL, Gerin JL (1995) Hepatitis D virus RNA editing: specific modification of adenosine in the antigenomic RNA. J Virol 69:7593–7600

Casey JL, Gerin JL (1998) Genotype-specific complementation of hepatitis delta virus RNA replication by hepatitis delta antigen. J Virol 72:2806–2814

Casey JL, Bergmann KF, Brown TL, Gerin JL (1992) Structural requirements for RNA editing in hepatitis delta virus: evidence for a uridine-to-cytidine editing mechanism. Proc Natl Acad Sci USA 89:7149–7153

Casey JL, Brown TL, Colan EJ, Wignall FS, Gerin JL (1993) A genotype of hepatitis D virus that occurs in northern South America. Proc Natl Acad Sci USA 90:9016–9020

Casey JL, Tennant BC, Gerin JL (2006) Genetic changes in hepatitis delta virus from acutely and chronically infected woodchucks. J Virol 80:6469–6477

Chang FL, Chen PJ, Tu SJ, Wang CJ, Chen DS (1991) The large form of hepatitis delta antigen is crucial for assembly of hepatitis delta virus. Proc Natl Acad Sci USA 88:8490–8494

Chang MF, Chen CH, Lin SL, Chen CJ, Chang SC (1995) Functional domains of delta antigens and viral RNA required for RNA packaging of hepatitis delta virus. J Virol 69:2508–2514

Chang J, Gudima SO, Taylor JM (2005) Evolution of hepatitis delta virus RNA genome following long-term replication in cell culture. J Virol 79:13310–13316

Chang J, Nie X, Chang HE, Han Z, Taylor J (2008) Transcription of hepatitis delta virus RNA by RNA polymerase II. J Virol 82:1118–1127

Chao M, Hsieh SY, Taylor J (1990) Role of two forms of hepatitis delta virus antigen: evidence for a mechanism of self-limiting genome replication. J Virol 64:5066–5069

Cheng Q, Jayan GC, Casey JL (2003) Differential inhibition of RNA editing in hepatitis delta virus genotype III by the short and long forms of hepatitis delta antigen. J Virol 77:7786–7795

Defenbaugh DA, Johnson M, Chen R, Zheng YY, Casey JL (2009) Hepatitis delta antigen requires a minimum length of the hepatitis delta virus unbranched rod RNA structure for binding. J Virol 83:4548–4556

Deny P (2006) Hepatitis delta virus genetic variability: from genotypes I, II, III to eight major clades? Curr Top Microbiol Immunol 307:151–171

Farci P, Roskams T, Chessa L, Peddis G, Mazzoleni AP, Scioscia R, Serra G, Lai ME, Loy M, Caruso L, Desmet V, Purcell RH, Balestrieri A (2004) Long-term benefit of interferon alpha therapy of chronic hepatitis D: regression of advanced hepatic fibrosis. Gastroenterology 126:1740–1749

Gandy SZ, Linnstaedt SD, Muralidhar S, Cashman KA, Rosenthal LJ, Casey JL (2007) RNA editing of the human herpesvirus 8 kaposin transcript eliminates its transforming activity and is induced during lytic replication. J Virol 81:13544–13551

Hartwig D, Schoeneich L, Greeve J, Schutte C, Dorn I, Kirchner H, Hennig H (2004) Interferon-alpha stimulation of liver cells enhances hepatitis delta virus RNA editing in early infection. J Hepatol 41:667–672

Hartwig D, Schutte C, Warnecke J, Dorn I, Hennig H, Kirchner H, Schlenke P (2006) The large form of ADAR 1 is responsible for enhanced hepatitis delta virus RNA editing in interferon-alpha-stimulated host cells. J Viral Hepat 13:150–157

Herb A, Higuchi M, Sprengel R, Seeburg PH (1996) Q/R site editing in kainate receptor GluR5 and GluR6 pre-mRNAs requires distant intronic sequences. Proc Natl Acad Sci USA 93:1875–1880

Herbert A, Rich A (2001) The role of binding domains for dsRNA and Z-DNA in the in vivo editing of minimal substrates by ADAR1. Proc Natl Acad Sci USA 98:12132–12137

Jaikaran DC, Collins CH, MacMillan AM (2002) Adenosine to inosine editing by ADAR2 requires formation of a ternary complex on the GluR-B R/G site. J Biol Chem 277: 37624–37629

Jayan GC, Casey JL (2002a) Increased RNA editing and inhibition of hepatitis delta virus replication by high-level expression of ADAR1 and ADAR2. J Virol 76:3819–3827

Jayan GC, Casey JL (2002b) Inhibition of hepatitis delta virus RNA editing by short inhibitory RNA-mediated knockdown of Adar1 but not Adar2 expression. J Virol 76:12399–12404

Jayan GC, Casey JL (2005) Effects of conserved RNA secondary structures on hepatitis delta virus genotype I RNA editing, replication, and virus production. J Virol 79:11187–11193

Jenna S, Sureau C (1998) Effect of mutations in the small envelope protein of hepatitis B virus on assembly and secretion of hepatitis delta virus. Virology 251:176–186

Jilbert AR, Burrell CJ, Gowans EJ, Hertzog PJ, Linnane AW, Marmion BP (1986) Cellular localization of alpha-interferon in hepatitis B virus-infected liver tissue. Hepatology 6:957–961

Lazinski DW, Taylor JM (1994) Expression of hepatitis delta virus RNA deletions: cis and trans requirements for self-cleavage, ligation, and RNA packaging. J Virol 68:2879–2888

Lehmann KA, Bass BL (1999) The importance of internal loops within RNA substrates of ADAR1. J Mol Biol 291:1–13

Linnstaedt SD, Kasprzak WK, Shapiro BA, Casey JL (2006) The role of a metasj RNA secondary structure in hepatitis delta virus genotype III RNA editing. RNA 12:1521–1533

Linnstaedt SD, Kasprzak WK, Shapiro BA, Casey JL (2009) The fraction of RNA that folds into the correct branched secondary structure determines hepatitis delta virus type 3 RNA editing levels. RNA 15:1177–1187

Liu Y, Samuel CE (1996) Mechanism of interferon action: functionally distinct RNA-binding and catalytic domains in the interferon-inducible, double-stranded RNA- specific adenosine deaminase. J Virol 70:1961–1968

Liu Y, Herbert A, Rich A, Samuel CE (1998) Double-stranded RNA-specific adenosine deaminase: nucleic acid binding properties. Methods 15:199–205

Lomeli H, Mosbacher J, Melcher T, Hoger T, Geiger JR, Kuner T, Monyer H, Higuchi M, Bach A, Seeburg PH (1994) Control of kinetic properties of AMPA receptor channels by nuclear RNA editing. Science 266:1709–1713

McNair AN, Cheng D, Monjardino J, Thomas HC, Kerr IM (1994) Hepatitis delta virus replication in vitro is not affected by interferon-alpha or -gamma despite intact cellular responses to interferon and dsRNA. J Gen Virol 75:1371–1378

Netter HJ, Wu TT, Bockol M, Cywinski A, Ryu WS, Tennant BC, Taylor JM (1995) Nucleotide sequence stability of the genome of hepatitis delta virus. J Virol 69:1687–1692

Ohman M, Kallman AM, Bass BL (2000) In vitro analysis of the binding of ADAR2 to the pre-mRNA encoding the GluR-B R/G site. RNA 6:687–697

Patterson JB, Samuel CE (1995) Expression and regulation by interferon of a double-stranded-RNA- specific adenosine deaminase from human cells: evidence for two forms of the deaminase. Mol Cell Biol 15:5376–5388

Polson AG, Bass BL (1994) Preferential selection of adenosines for modification by double-stranded RNA adenosine deaminase. EMBO J 13:5701–5711

Polson AG, Bass BL, Casey JL (1996) RNA editing of hepatitis delta virus antigenome by dsRNA-adenosine deaminase. Nature 380:454–456

Polson AG, Ley HL 3rd, Bass BL, Casey JL (1998) Hepatitis delta virus RNA editing is highly specific for the amber/W site and is suppressed by hepatitis delta antigen. Mol Cell Biol 18:1919–1926

Ponzetto A, Hoyer BH, Popper H, Engle R, Purcell RH, Gerin JL (1987) Titration of the infectivity of hepatitis D virus in chimpanzees. J Infect Dis 155:72–78

Pugnale P, Pazienza V, Guilloux K, Negro F (2009) Hepatitis delta virus inhibits alpha interferon signaling. Hepatology 49:398–406

Sato S, Wong SK, Lazinski DW (2001) Hepatitis delta virus minimal substrates competent for editing by ADAR1 and ADAR2. J Virol 75:8547–8555

Sato S, Cornillez-Ty C, Lazinski DW (2004) By inhibiting replication, the large hepatitis delta antigen can indirectly regulate amber/W editing and its own expression. J Virol 78:8120–8134

Wong SK, Lazinski DW (2002) Replicating hepatitis delta virus RNA is edited in the nucleus by the small form of ADAR1. Proc Natl Acad Sci USA 99:15118–15123

Wong SK, Sato S, Lazinski DW (2001) Substrate recognition by ADAR1 and ADAR2. RNA 7:846–858

Wong SK, Sato S, Lazinski DW (2003) Elevated activity of the large form of ADAR1 in vivo: very efficient RNA editing occurs in the cytoplasm. RNA 9:586–598

Xia YP, Lai MM (1992) Oligomerization of hepatitis delta antigen is required for both the trans-activating and trans-dominant inhibitory activities of the delta antigen. J Virol 66:6641–6648

Yamaguchi Y, Filipovska J, Yano K, Furuya A, Inukai N, Narita T, Wada T, Sugimoto S, Konarska MM, Handa H (2001) Stimulation of RNA polymerase II elongation by hepatitis delta antigen. Science 293:124–127

Yamaguchi Y, Mura T, Chanarat S, Okamoto S, Handa H (2007) Hepatitis delta antigen binds to the clamp of RNA polymerase II and affects transcriptional fidelity. Genes Cells 12:863–875

Bioinformatic Approaches for Identification of A-to-I Editing Sites

Eli Eisenberg

Abstract The first discoveries of mammalian A-to-I RNA editing have been serendipitous. In conjunction with the fast advancement in sequencing technology, systematic methods for prediction and detection of editing sites have been developed, leading to the discovery of thousands of A-to-I editing sites. Here we review the state-of-the-art of these methods and discuss future directions.

Keywords A-to-I RNA editing · Alu repeats · Bioinformatics · Deep sequencing

Contents

Adenosine-to-inosine (A-to-I) RNA editing has the potential for a major diversification of the transcriptome beyond its genomic blueprint. This post-transcriptional modification of RNA is catalyzed by enzymes of the ADARs (adenosine deaminases that act on RNA) protein family, which bind double-stranded RNA

E. Eisenberg (✉)
Raymond and Beverly Sackler School of Physics and Astronomy,
Tel-Aviv University, 69978 Tel Aviv, Israel
e-mail: elieis@post.tau.ac.il

Current Topics in Microbiology and Immunology (2012) 353: 145–162
DOI: 10.1007/82_2011_147

Published Online: 13 July 2011

structures and deaminate targeted adenosines (A) within these structures into inosines (I). The inosines seem to be functionally equivalent to guanosines (Gs), and thus A-to-I editing affects downstream RNA processes, such as translation and splicing, resulting in different fates for the edited RNA molecules (Bass 2002; Nishikura 2010).

Twenty years ago, the first mammalian example for A-to-I RNA editing was reported—editing of an adenosine nucleotide within the coding sequence of the glutamate receptor subunit GluRB, resulting in a modified protein with a distinctive biochemical activity (Sommer et al. 1991). Despite much effort, only a handful of additional mammalian editing targets were found till 2003. On the other hand, a number of tantalizing hints suggested that editing is of high importance and wider scope: mice lacking ADARs die in utero or shortly after birth (Hartner et al. 2004; Higuchi et al. 2000; Wang et al. 2000). In addition, a number of neurological pathologies were linked to abnormal editing patterns, including epilepsy, brain tumors, amyotrophic lateral sclerosis (ALS), schizophrenia, depression and neuronal apoptosis following disruption of the blood flow to the brain (Brusa et al. 1995; Gurevich et al. 2002; Kawahara et al. 2004; Maas et al. 2001; Niswender et al. 2001; Paz et al. 2007; Wang et al. 2004). Most recently it was found that editing activity of ADAR1 is essential for hematopoiesis (Hartner et al. 2009; XuFeng et al. 2009). These phenotypes were not all explained by the few editing targets identified. Moreover, pioneering experimental work found that inosine exists in mRNA in large amounts (Paul and Bass 1998), much larger than could be accounted for by the small number of targets known at that time. Accordingly, the search for more targets continued and a variety of experimental methods to detect additional editing events and their levels were developed (Chateigner-Boutin and Small 2007; Chen et al. 2008b; Gallo et al. 2002; Lanfranco et al. 2009; Morse and Bass 1999; Ohlson et al. 2005; Ohlson and Ohman 2007; Sakurai et al. 2010; Suspene et al. 2008; Wong et al. 2009; Zilberman et al. 2009) with various levels of success.

1 Bioinformatic Screens

In principle, detection of editing sites should be straight-forward, analyzing cDNA sequencing data. Resembling the endogenous enzymes, most sequencing reactions also identify an edited adenosine "A" site within cDNA as a guanosine "G". Therefore, an A-to-G mismatch between a sequenced cDNA and its genomic reference is an indication of an A-to-I editing event. Naively, then, one has to only align the available cDNA data, including millions of publicly available ESTs and full-length RNAs, to the genome, and search for such A-to-G mismatches (Fig. 1). However, a simple application of this idea fails in reality due to the extremely low signal-to-noise ratio. The total fraction of mismatches between the genome and the expressed sequences amounts to 1–2% (Hillier et al. 1996). The main contributors for these discrepancies are then random sequencing errors in the expressed

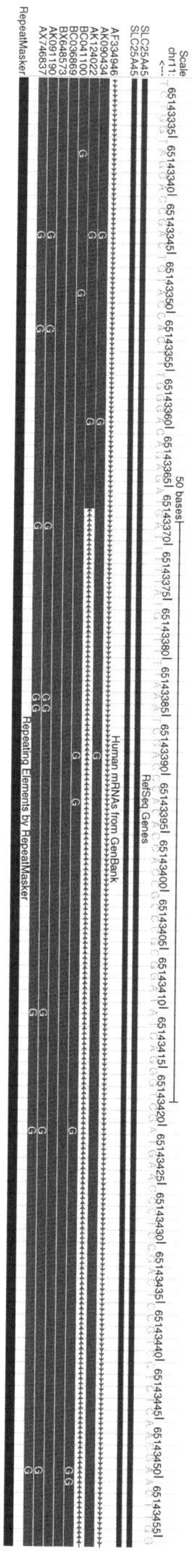

Fig. 1 *Editing traces in the public expressed sequences databases*. Evidence for editing in the 3′ UTR of the solute carrier family 25, member 45 (SLC25A45) gene as found by alignment of mRNAs to the reference genome observed in the UCSC genome browser. The mismatches are highlighted; all of them are A-to-G changes. The existence of the Alu repeat is indicated in the bottom of the screen-shot, in the Repeat-Masker panel. A number of inverted Alu repeats are located 3–4 kbp downstream the 3′ UTR

sequences, which alone account for 1–2 mismatches per 100 bp sequenced. Another important cause of variance between RNA and the genome includes genomic polymorphisms and somatic mutations that result in genomic differences between the different individuals, or the individual cells, contributing to the expressed sequences and the reference genome. In addition, misalignment of the RNA sequences to the genome is a major concern when dealing with repetitive regions of the genome.

The first discoveries of A-to-I editing sites resided all within the coding parts of mRNAs. These editing sites were shown to be functional—their editing results in a modified protein, with biochemical properties different than those of the unedited version. Studying these sites, it has been noticed that the genomic sequence surrounding them is highly conserved among species (Hoopengardner et al. 2003). This can be readily understood in terms of an additional evolutionary constraint: in addition to the sequence conservation against changes in the amino-acid coding information, the double-stranded RNA structure must be left intact in order to preserve the editing event. This constraint leads to higher conservation at the DNA level, and has proven to be a very useful signature of editing sites, to be employed in bioinformatic searches (Clutterbuck et al. 2005; Hoopengardner et al. 2003; Levanon et al. 2005a). The first study identified highly conserved regions and then used extensive sequencing to look for editing sites, resulting in 16 novel sites in *Drosophila melanogaster* and one in human (Hoopengardner et al. 2003). The conservation may be further used to sift through the mismatches in available cDNA data. Unlike editing sites, sequencing errors and genomic polymorphisms are not often shared between species. Note, however, that specific types of sequencing errors are not random but rather follow from a given pattern in the neighboring sequence (Zaranek et al. 2010), these would seem as 'conserved' between species. Focusing on mismatches that reoccur in different species allows one to find the few editing recoding sites among tens of millions of mismatches between ESTs/RNAs and the genome. This strategy was applied by a number of groups: looking for such conserved mismatches located in the exact same position in human and mouse resulted in a few additional A-to-I editing substrates (Levanon et al. 2005a; Ohlson et al. 2007; Sie and Maas 2009). The newly discovered sites are now under investigation in order to determine their biological function and regulation potential (Galeano et al. 2010; Hideyama et al. 2010; Kwak et al. 2008; Nicholas et al. 2010; Nishimoto et al. 2008; Riedmann et al. 2008; Rula et al. 2008). One might have used an additional characterization of the editing sites to further improve this analysis, namely the requirement for having a dsRNA structure at the editing site. However, based on the examples of editing sites known so far, it seems that the typical dsRNA structures are rather weak and hard to predict computationally (Bhalla et al. 2004). Interestingly, many of the novel editing sites appear in the SNP database (dbSNP), due to an erroneous interpretation of the variability among expressed sequences in these sites as a sign for a single-nucleotide polymorphism (SNP) (Eisenberg et al. 2005a). A careful analysis of dbSNP could result in more editing sites hidden as mis-annotated SNPs (Gommans et al. 2008).

2 Editing Within Repetitive Elements

In 1990s, experimental evidence for a significant amount of inosine in total RNA has emerged. In the decade to follow, we have witnessed an impressive growth in the number of known editing sites within the coding sequence (editing of which might modify the encoded protein), especially as deep-sequencing methods have been introduced in the past 2 years (see below). However, these are far from being able to account for the total inosine levels observed: the currently known editing sites within the coding region amount to about 400 sites. Thus, they represent roughly 1:150,000 of all nucleotides in exons. The editing efficiency is spread between 0 and 100%, with average efficiency less than 50%, so one expects not more than 1:300,000 inosine to adenosine ratio in total mRNA. This rough estimate is at odds with an observed ratio of 1:17,000nt in rat's brain (Paul and Bass 1998) and results showing up to one inosine per 2,000 nt in poly-adenylated mRNA from human brain (Blow et al. 2004). In addition, a number of clusters of editing events were found in non-coding regions, providing first hints for the significance of the non-coding RNA for the global A-to-I editing pattern (Morse and Bass 1999).

In 2004, three groups have devised computational methods for identifying such clusters, based on analysis of mismatches in otherwise almost perfect alignments of RNA (Athanasiadis et al. 2004; Eisenberg et al. 2005b; Kim et al. 2004; Levanon et al. 2004). The methods differ by the clustering criteria used and the statistical analysis employed. Remarkably, the three independent procedures resulted in highly similar results: A-to-G substitutions, which could arise from A-to-I editing events, account for more than 80% of the 12 possible types of mismatches in the selected set of transcripts. As this disparity in mismatches distribution is unlikely to occur for genomic polymorphisms and sequencing errors, it provides a clear signature of editing in tens of thousands of sites within the human transcriptome.

Editing events couple with splicing, thus they may occur in introns as well. However, computational approaches based on expressed sequences are obviously limited in their ability to detect editing within introns. Therefore, it is anticipated that the actual number of editing sites in the human genome is even much higher than the tens of thousands sites reported in the above works. Indeed, direct sequencing of human brain total RNA has revealed that up to 1 in 1,000 bp of the expressed regions are being edited, compared to only 1:2000 bp in poly-adenylated mRNA (Blow et al. 2004).

Virtually all clusters of editing sites are harbored within Alu repetitive elements (Levanon et al. 2004). Alu elements are short interspersed elements (SINEs), roughly 300 bp long each. Humans have about a million copies of Alu, accounting for ~10% of its genome (Lander et al. 2001). Since these repeats are so common, especially in gene-rich regions, pairing of two oppositely-oriented Alus located in the same pre-mRNA structure is likely. Such pairing produces a long and stable dsRNA structure, an ideal target for the ADARs. Alu repeats are primate specific

(Batzer and Deininger 2002), but other mammals have a similar number of different SINEs. For example, the number of rodent-specific SINEs in the mouse genome is larger than the number of Alu SINEs in humans, and they occupy a similar portion of the genome (7.6% in mouse, 10.7% in human) (Waterston et al. 2002). However, genome-wide analysis of the properties of these SINE repeats explains the order-of-magnitude difference in the global editing levels observed in measurements of total inosine abundance as well as in bioinformatic screens editing sites (Eisenberg et al. 2005b; Kim et al. 2004). It turns out that the shorter length and higher diversity of the mouse SINE repeats are responsible for this disparity in editing levels (Neeman et al. 2006). This global difference between human and other mammals such as mouse is intriguing, as it is generally believed that cellular mechanisms are generally conserved between human and mouse. However, the significance of this difference is not clear yet, as the role of editing in non-coding repeats is yet elusive.

3 Deep-Sequencing Approaches

Recent advancements in massively parallel sequencing technologies open a new era in analysis of genome to transcriptome discrepancies. The first bioinformatic works studied the publicly available transcription data in GenBank, a result of group-effort of hundreds of labs around the world. In comparison, it is possible today to produce a similar amount of data in a single 1000$ experiment. Thus, one could start sequencing whole transcriptomes in order to determine the full scope of RNA editing. However, observing a consistent discrepancy between the RNA sequence and the reference genome is not sufficient to prove the site to be an editing site. One must exclude the possibility of genomic diversity between the reference genome and the genome of the RNA source tissue. It then follows that identifying RNA editing sites requires sequencing of both genomic DNA and cDNA from the same source, or two RNA samples of a wild-type and editing-deficient mutant. Second, editing levels vary among tissues, and therefore one would need to repeat the experiment for a wide variety of tissues in order to obtain the full organism-wide repertoire of editing. Current technology still renders this kind of experiment quite expensive. Accordingly, current usage of deep-sequencing to look at transcriptome-wide editing is usually limited to a single tissue, or to a limited part of the transcriptome (e.g. micro-RNAs, or a specific gene(s) of interest) throughout a number of tissues or developmental stages. Such studies are currently conducted by several groups, and are expected to significantly increase the scope of known editing levels, and may even detect consistent RNA–DNA mismatches other than A-to-G, reflecting RNA modifications beyond the dominant (at least when one includes Alu repeats in the analysis) A-to-I editing.

In a first and pioneering work in this direction, Rosenberg et al. (2011) have implemented a deep-sequencing approach followed by DNA–RNA mismatches analysis to discover 32 novel targets of APOBEC1 C-to-U RNA editing, edited in epithelial cells from the small intestines of a mouse. As it was the case for A-to-I editing, while the first known example of APOBEC1 editing (apoB mRNA) resides in the coding region (Chen et al. 1987; Powell et al. 1987), transcriptome-wide analysis have revealed editing in the non-coding regime. The functional role of this extensive catalytic activity in non-coding parts of the transcripts is an open challenge.

As mentioned above, studying a wide variety of tissues using a straight forward deep-sequencing approach is still impractical. An alternative approach was demonstrated recently by Li et al. (2009b). Combining a computational approach together with a novel targeted sequencing technique, they aimed to get a transcriptome-wide editing profile in a multi-tissues experiment. A bioinformatic search used alignments of eight million human ESTs against the human reference genome, in the spirit of the older bioinformatic approaches. After the repetitive portion of the human genome and known genomic polymorphisms were removed, there remained ~60,000 mismatches, which potentially could signal edited sites. A targeted capture and sequencing approach was employed to specifically deep-sequence the predicted sites. For each of the predicted sites, a padlock probe (also known as molecular inversion probe) was designed for specific anchoring and amplification (Li et al. 2009a). All sites were simultaneously captured, amplified and sequenced using genomic DNA and cDNA from several different tissues (mainly brain), all derived from a single donor in order to rule out polymorphisms among populations. The pool of probes was hybridized to the DNA and cDNA in separate amplification reactions. The amplicons were sequenced, and the resulting sequences were scanned in order to identify A-to-G mismatches between the genomic DNA and the RNA-derived cDNA. This method allows for parallel sequencing of tens of thousands of suspected sites in a single reaction. It resulted in detection of hundreds of novel A-to-I editing sites residing out of repetitive elements. This technology can now be applied to study the hundreds of confirmed editing sites (instead of the 60,000 candidates) in a large panel of tissues. In particular, it provides a promising cost-effective approach to study in large scale possible associations between the editing profile and various pathologies.

Similarly, Enstero et al. (2010) have first identified ~2,500 conserved regions which form putative double-stranded RNA structures, and then used deep sequencing of only 45 regions that were considered particularly promising based on sequence conservation and the existence of A-to-G mismatches in the public databases. This study has resulted in ten new editing sites, eight of which recode codons. However, the editing efficiency of these sites was minute-0.6–2.4%.

In addition to the improvement in detection of RNA editing, the deep-sequencing technology allows for much better quantification of the editing level. Counting the number of edited and non-edited reads is easy enough. However, an important (often neglected) concern is the possibility of an alignment bias. Current deep-sequencing technology often results in short reads, with a non-negligible amount of sequencing errors (1–2%). Due to the large amount of reads, fast-alignment protocols must be used. These often allow only a small number of mismatches within the read, e.g., retaining only reads with up to two mismatches.

An edited read will appear as if it has a mismatch in the edited site. Therefore, only a single additional mismatch is allowed if the read is to be aligned at all, while unedited reads will be aligned even if exhibiting two mismatches. This creates a bias against edited reads, resulting in an apparently lower level of editing. Situation is even worse considering the fact that many editing sites appear in clusters, where editing of the different sites within the cluster is often positively correlated. A satisfactory algorithmic solution for this problem has not been reported yet. As such solution will become easier as technology is shifting toward producing longer reads.

Finally, let us mention that the same methods described here for identification of RNA editing are applicable for the study of DNA editing and somatic mutations. Bioinformatic approaches for these phenomena are only beginning to emerge (Zaranek et al. 2010), but are expected to increase as large deep-sequencing data is accumulating.

4 Structural and Sequence Determinants of A-to-I Editing

A-to-I editing is characterized by a puzzling specificity and selectivity. In some targets, such as the AMPA receptor gluR-B subunit in mice (Seeburg et al. 1998) and the E1 site within the Alu-based alternative exon in the NARF gene (Lev-Maor et al. 2007), 100% of the transcripts are being edited at a specific adenosine. In contrast, most sites in the coding region show only a partial editing. Looking at sites in Alu repeats, a seemingly random editing pattern is observed: virtually all adenosines are targeted with varying editing efficiency, but only a few are edited in any given clone of the transcript. However, it was recently shown that editing in Alu repeats is also highly reproducible: the variability among healthy individuals in editing level at a given site within a specific Alu repeat is much lower than the site-to-site differences (Greenberger et al. 2010). The wide range of efficiencies and the significant consistency between individuals call for a sequence and structural motifs that determine the editing efficiency of each site relative to others. The sequence and the resulting dsRNA structure formed by Alu vary significantly from site-to-site, but are shared by all samples. Sequence analysis of editing sites revealed a number of weak motifs: C and T are over-represented at the nucleotide 5′ to the editing site, while G is under-represented. At the nucleotide 3′ to the site, G is significantly over-represented (Kleinberger and Eisenberg 2010; Lehmann and Bass 2000; Melcher et al. 1996; Polson and Bass 1994; Riedmann et al. 2008). However, these alone cannot account for the observed tightly regulated editing profiles. Therefore, the question still stands: what controls the editing level at each given site?

Given the dependence of ADAR activity upon the formation of a double-stranded RNA structure, it is plausible that structural motifs also play a role. Indeed, also some evidence has been accumulated supporting this idea. The editing level in a given Alu repeat can be shown to correlate with the existence of a nearby and reversely-oriented repeat, in support of the paired-Alu model. Analysis of thousands of examples has shown that effective editing requires a distance of roughly 2000 bp or less between the two Alus. Furthermore, the level of editing increases with the number of reverse complement Alus present within this distance (Athanasiadis et al. 2004; Blow et al. 2004; Kim et al. 2004; Levanon et al. 2004). These characteristics of the editing pre-requisites are instrumental in devising future searches for editing targets in human and other organisms. Interestingly, edited adenosines within the dsRNA structure are paired with a "U" or a "C" in the reverse strand, meaning that editing either strengthens or weakens the dsRNA structure, but virtually never has a neutral effect on the dsRNA-pairing energy (Levanon et al. 2004). However, a detailed analysis comparing editing levels of specific inosines within an Alu repeat is still not available. A first step in this direction has been done recently (Kleinberger and Eisenberg 2010), but results are still far from being able to explain in full the variability in editing levels. Ideally, one would like to have a predictive model which, given the genomic sequence, will provide the relative efficiencies of editing for all adenosines in the given sequence.

5 Correlations Between Editing Sites

Many edited targets include a number of editing sites. Analysis of correlations between editing of neighboring sites might reveal details regarding ADAR binding and catalytic activity. A recent study (Enstero et al. 2009) has identified positive correlations between different editing sites, as far as 25 bp apart. These positive correlations might support a model in which ADAR is attracted to a specific 'strong' editing site, and then edits weaker sites in its vicinity. Indeed such weak 'satellite' editing sites have been observed in the vicinity of several editing sites. A more complex pattern is observed when one looks at site–site correlations after correcting for the whole-transcript editing affinity. Then, a rich pattern of positive and negative correlations is seen, including pair and triple correlations for editing sites as far as 150 bp apart (Paz-Yaacov et al. 2010). These intriguing results may suggest that as editing of one site changes locally the double-stranded binding energy, it might induce changes in the global structure, which in turn may enhance or diminish editing efficiency in remote sites.

6 RNA Editing and Micro-RNAs

The role of Alu editing is yet to be explored. Recent observations suggest that editing is involved in molecular mechanisms based on dsRNA structure, such as RNAi (Tonkin and Bass 2003) and miRNA (de Hoon et al. 2010; Kawahara et al.

2008; Kawahara et al. 2007b; Luciano et al. 2004). miRNAs are short non-coding RNAs, endogenously expressed in the living cell, that bind to mRNAs and induce suppression of translation, by either leading to degradation of the RNA or inhibiting translation. The primary sequence of the miRNAs processes in the nucleus by Drosha and then further processes in the cytoplasm by Dicer, resulting in a mature sequence ~21 nucleotides long. These short RNAs are binded by the RISC complex (Bartel 2004). RNA editing is potentially coupled to miRNAs throughout their life cycle. The biogenesis of miRNAs through Drosha and Dicer processing hinges upon their double-stranded RNA structure. As these stages, editing of these pre-miRNA (or pri-miRNA) and double-stranded RNAs might interfere with the proper production of mature miRNAs, or even result in modified mature miRNA sequences, exhibiting a different set of targets. Furthermore, miRNA targets are often present in the 3′ UTRs, regions heavily targeted by RNA editing. Thus, editing might influence miRNA targets, increasing or decreasing their affinity toward miRNA binding.

The full picture of the relationship between miRNAs and A-to-I RNA editing is still missing. However, a number of interesting results have emerged in recent years (Blow et al. 2006; Kawahara et al. 2008; Luciano et al. 2004). It has been demonstrated that both ADAR enzymes edit specific adenosines within pri-miRNAs in human and in viruses (Iizasa et al. 2010; Kawahara et al. 2007a; Kawahara et al. 2007b; Yang et al. 2006). In these cases, editing was reported to suppress the processing by Drosha or Dicer, or prevent loading onto the RISC complex, resulting in a depleted amount of mature miRNA. In some cases, mature miRNAs with an altered sequence have been reported.

Systematic searches for editing sites in miRNAs have not yielded a large number of sites (Chiang et al. 2010; de Hoon et al. 2010; Linsen et al. 2010). However, these studies have focused on rodents, and it is possible that the results for humans might be different. Nevertheless, the low number of editing sites in miRNAs is surprising. It seems to suggest that edited pre-miRNAs are degraded or otherwise prevented from maturation. Alternatively, some other mechanism might be responsible for protection of the miRNA sequences from editing. These questions are yet to be explored.

Finally, modulation of miRNA targets has been also considered. Targets of the miRNA contain a seven nucleotide sequence which complements the miRNA seed. Thus, editing of the miRNA in the seed region should modify its set of targets. Similarly, editing of a target recognition site could alter its binding to the miRNA. Two bioinformatic studies have assessed the scope of this phenomenon (Borchert et al. 2009; Liang and Landweber 2007), concluding that hundreds to thousands of target sites might be affected. In particular, two human miRNAs, miRNA-513 and miRNA-769-3p, target a common motif present in the abundant Alu sequence only when it is edited. Here too, further studies are required to elucidate the importance of target editing for the miRNA regulation process.

7 More functional Roles of RNA Editing of Inverted Repeats

The role of Alu editing is yet to be explored. Recent observations suggest that editing is involved in molecular mechanisms based on dsRNA structure, such as RNAi (Tonkin and Bass 2003) and miRNAs (see above). RNA editing was also shown to be involved in splicing regulation in several cases (Lev-Maor et al. 2007; Moller-Krull et al. 2008), notably the self-editing of ADAR2 (Rueter et al. 1999). It has been suggested that hyper-editing of repetitive elements might result in gene silencing (Wang et al. 2005) or in an anti-retro element defense mechanism (Levanon et al. 2005b). A possibility gathering support in recent years is the suggestion that heavily edited transcripts are retained in the nucleus throughout complexes containing p54nrb (non-POU domain containing, octamer-binding, NONO) (Zhang and Carmichael 2001), later identified as paraspeckles (Chen and Carmichael 2009). Indeed, later studies have shown that a single pair of reversely-oriented Alu repeats in the 3′ UTR of a reporting gene strongly represses its expression, in conjunction with a significant nuclear retention of the mRNAs. Nuclear retention was demonstrated in detail for the endogenous Nicolin 1 (NICN11) mRNA harboring inverted Alus in its 3′ UTR (Chen et al. 2008a) and for mouse Slc7a2 edited transcripts (Prasanth et al. 2005). However, another group (Hundley et al. 2008) has recently reported no effect of editing within the 3′ UTR on mRNA localization and translation of several *Caenorhabditis elegans* and human transcripts, suggesting that the retention phenomenon might be different in different cells types, or conditions.

Nuclear retention of hyperedited transcripts was first interpreted as a means of protection against abnormal transcripts (Zhang and Carmichael 2001). This is supported by the abundance of hyperediting clusters in splicing-defective transcripts (Kim et al. 2004). This idea is in line with a similar proposed mechanism, suggesting that an I-specific cleavage of RNAs can lead to the selective destruction of edited RNAs (Scadden and Smith 2001). However, a recent study (Prasanth et al. 2005) opened a new perspective on the way transcript localization and inosine-specific cleavage might contribute to cell function. It was shown that inverted repeats within the 3′ UTR of the mouse Slc7a2 gene form a hairpin dsRNA structure and are highly A-to-I edited. The mRNA is then retained in the nucleus, as a reservoir of mRNAs that can be rapidly exported to the cytoplasm upon cellular stress. It has been demonstrated that under stress conditions, the edited part is post-transcriptionally cleaved, removing the edited SINEs from the 3′ UTR. Consequently, the mRNA is exported to the cytoplasm, where it translates into a protein. It thus turns out that A-to-I hyperediting may serve as a powerful means of retaining in the nucleus mRNA molecules that are not immediately needed to produce proteins but whose cytoplasmic presence is rapidly required upon a physiologic stress. This model might provide an elegant functional role to the global editing phenomenon. Naturally, one wonders what the scope of this model is, and whether it is relevant to the thousands of hyper-edited human genes. Some support to this idea has been provided bioinformatically, showing that there are hundreds of transcripts in public databases exhibiting cleavage of an inverted repeat structure, as if they have been retained in the nucleus and then cleaved and released prior to sequencing (Osenberg et al. 2009). However, it should be pointed out that it is not clear yet whether the nuclear retention is editing-mediated or rather stems from the double-stranded RNA structure.

While editing was shown to be coupled with several regulation mechanisms, it is yet too early to call whether any of these regulation mechanisms is as widespread as Alu editing itself. Thus, Alu editing is a mystery still waiting to be solved.

8 Future Directions

Advances in editing detection methods have opened the door for studies comparing editing levels globally between different conditions, pathologies and developmental stages. First results reporting such difference between normal and tumor brain tissues (Paz et al. 2007), mouse brain developmental stages (Wahlstedt et al. 2009) and along stem-cell differentiation process (Osenberg et al. 2010) have been published recently. We expect to see many more such studies in the near future, which will help to clarify the scope of processes affected by A-to-I RNA editing.

Alu repetitive elements are unique to the primates, but the occurrence of repetitive elements in general is common to all metazoa. Applying the same methods for editing detection to other organisms has shown that editing in human is about 40 times higher as compared with mouse (Eisenberg et al. 2005b; Kim et al. 2004). A similar picture was observed when comparing with rat, chicken and fly (Eisenberg et al. 2005b). The high-editing level in humans is likely due to the fact that humans have only one dominant SINE, which is relatively well-conserved (only $\sim$12% divergence between an average Alu and the consensus). In comparison, mouse has four different SINEs, which are shorter and more divergent ($\sim$20% average divergence) (Neeman et al. 2006). This has lead naturally to the question of the relative abundance of editing in humans as compared to other primates. In a recent study (Paz-Yaacov et al. 2010), a two-fold higher level was observed in human compared to chimpanzee and rhesus monkeys, for a set of six genes in which no significant genomic differences occur among the three species. In addition, human-specific Alu repeats have been shown to be associated with neurological pathways and disorders.

The exceptional level of editing in the primate brain makes it tempting to suggest a role in primate evolution. The over-representation of editing in brain tissues and the association of aberrant editing with neurological diseases are consistent with a possible connection between editing and brain capabilities. One thus may speculate that the massive editing of brain tissues is responsible in part for the brain complexity. As this large-scale editing is a direct result of Alu abundance, it follows that if the above idea has any merit then the massive invasion of Alus into the primate genome had a major impact on primates' evolution (Barak et al. 2009; Britten 2010; Eisenberg et al. 2005b; Gommans et al. 2009; Mattick and Mehler 2008; Paz-Yaacov et al. 2010).

The recent identification of hundreds of non-repetitive human RNA editing sites may be followed by many more very soon. The volume of RNA-sequence data collected in a couple of years already surpasses that the total amount deposited in EST database in two decades. This increasingly growing amount of data will allow for more predicted RNA editing sites. The dbSNP has also grown as a result of recent genomic sequencing efforts, in particular the 1,000 genomes project, improving one's ability to filter rare SNPs. As sequencing cost continues to drop, a comprehensive approach to identifying all RNA editing sites will become possible by sequencing the entire transcriptomes as well as the exomes or genomes. In addition to A-to-I sites, the full scope of other types of RNA editing and modifications are surely going to be revealed by these efforts.

References

Athanasiadis A, Rich A, Maas S (2004) Widespread A-to-I RNA editing of Alu-containing mRNAs in the human transcriptome. PLoS Biol 2:e391

Barak M, Levanon EY, Eisenberg E, Paz N, Rechavi G, Church GM, Mehr R (2009) Evidence for large diversity in the human transcriptome created by Alu RNA editing. Nucleic Acids Res 37:6905–6915

Bartel DP (2004) MicroRNAs: genomics, biogenesis, mechanism, and function. Cell 116: 281–297

Bass BL (2002) RNA editing by adenosine deaminases that act on RNA. Annu Rev Biochem 71:817–846

Batzer MA, Deininger PL (2002) Alu repeats and human genomic diversity. Nat Rev Genet 3:370–379

Bhalla T, Rosenthal JJ, Holmgren M, Reenan R (2004) Control of human potassium channel inactivation by editing of a small mRNA hairpin. Nat Struct Mol Biol 11:950–956

Blow M, Futreal PA, Wooster R, Stratton MR (2004) A survey of RNA editing in human brain. Genome Res 14:2379–2387

Blow MJ, Grocock RJ, van Dongen S, Enright AJ, Dicks E, Futreal PA, Wooster R, Stratton MR (2006) RNA editing of human microRNAs. Genome Biol 7:R27

Borchert GM, Gilmore BL, Spengler RM, Xing Y, Lanier W, Bhattacharya D, Davidson BL (2009) Adenosine deamination in human transcripts generates novel microRNA binding sites. Hum Mol Genet 18:4801–4807

Britten RJ (2010) Transposable element insertions have strongly affected human evolution. Proc Natl Acad Sci USA 107:19945–19948

Brusa R, Zimmermann F, Koh DS, Feldmeyer D, Gass P, Seeburg PH, Sprengel R (1995) Early-onset epilepsy and postnatal lethality associated with an editing-deficient GluR-B allele in mice. Science 270:1677–1680

Chateigner-Boutin AL, Small I (2007) A rapid high-throughput method for the detection and quantification of RNA editing based on high-resolution melting of amplicons. Nucleic Acids Res 35:e114

Chen LL, Carmichael GG (2009) Altered nuclear retention of mRNAs containing inverted repeats in human embryonic stem cells: functional role of a nuclear noncoding RNA. Mol Cell 35:467–478

Chen SH, Habib G, Yang CY, Gu ZW, Lee BR, Weng SA, Silberman SR, Cai SJ, Deslypere JP, Rosseneu M et al (1987) Apolipoprotein B-48 is the product of a messenger RNA with an organ-specific in-frame stop codon. Science 238:363–366

Chen LL, DeCerbo JN, Carmichael GG (2008a) Alu element-mediated gene silencing. Embo J 27:1694–1705

Chen YC, Kao SC, Chou HC, Lin WH, Wong FH, Chow WY (2008b) A real-time PCR method for the quantitative analysis of RNA editing at specific sites. Anal Biochem 375:46–52

Chiang HR, Schoenfeld LW, Ruby JG, Auyeung VC, Spies N, Baek D, Johnston WK, Russ C, Luo S, Babiarz JE, Blelloch R, Schroth GP, Nusbaum C, Bartel DP (2010) Mammalian microRNAs: experimental evaluation of novel and previously annotated genes. Genes Dev 24:992–1009

Clutterbuck DR, Leroy A, O'Connell MA, Semple CA (2005) A bioinformatic screen for novel A-I RNA editing sites reveals recoding editing in BC10. Bioinformatics 21:2590–2595

de Hoon MJ, Taft RJ, Hashimoto T, Kanamori-Katayama M, Kawaji H, Kawano M, Kishima M, Lassmann T, Faulkner GJ, Mattick JS, Daub CO, Carninci P, Kawai J, Suzuki H, Hayashizaki Y (2010) Cross-mapping and the identification of editing sites in mature microRNAs in high-throughput sequencing libraries. Genome Res 20:257–264

Eisenberg E, Adamsky K, Cohen L, Amariglio N, Hirshberg A, Rechavi G, Levanon EY (2005a) Identification of RNA editing sites in the SNP database. Nucleic Acids Res 33:4612–4617

Eisenberg E, Nemzer S, Kinar Y, Sorek R, Rechavi G, Levanon EY (2005b) Is abundant A-to-I RNA editing primate-specific? Trends Genet 21:77–81

Enstero M, Daniel C, Wahlstedt H, Major F, Ohman M (2009) Recognition and coupling of A-to-I edited sites are determined by the tertiary structure of the RNA. Nucleic Acids Res 37:6916–6926

Enstero M, Akerborg O, Lundin D, Wang B, Furey TS, Ohman M, Lagergren J (2010) A computational screen for site selective A-to-I editing detects novel sites in neuron specific Hu proteins. BMC Bioinformatics 11:6

Galeano F, Leroy A, Rossetti C, Gromova I, Gautier P, Keegan LP, Massimi L, Di Rocco C, O'Connell MA, Gallo A (2010) Human BLCAP transcript: new editing events in normal and cancerous tissues. Int J Cancer 127:127–137

Gallo A, Thomson E, Brindle J, O'Connell MA, Keegan LP (2002) Micro-processing events in mRNAs identified by DHPLC analysis. Nucleic Acids Res 30:3945–3953

Gommans WM, Tatalias NE, Sie CP, Dupuis D, Vendetti N, Smith L, Kaushal R, Maas S (2008) Screening of human SNP database identifies recoding sites of A-to-I RNA editing. RNA 14:2074–2085

Gommans WM, Mullen SP, Maas S (2009) RNA editing: a driving force for adaptive evolution? Bioessays 31:1137–1145

Greenberger S, Levanon EY, Paz-Yaacov N, Barzilai A, Safran M, Osenberg S, Amariglio N, Rechavi G, Eisenberg E (2010) Consistent levels of A-to-I RNA editing across individuals in coding sequences and non-conserved Alu repeats. BMC Genomics 11:608

Gurevich I, Tamir H, Arango V, Dwork AJ, Mann JJ, Schmauss C (2002) Altered editing of serotonin 2C receptor pre-mRNA in the prefrontal cortex of depressed suicide victims. Neuron 34:349–356

Hartner JC, Schmittwolf C, Kispert A, Muller AM, Higuchi M, Seeburg PH (2004) Liver disintegration in the mouse embryo caused by deficiency in the RNA-editing enzyme ADAR1. J Biol Chem 279:4894–4902

Hartner JC, Walkley CR, Lu J, Orkin SH (2009) ADAR1 is essential for the maintenance of hematopoiesis and suppression of interferon signaling. Nat Immunol 10:109–115

Hideyama T, Yamashita T, Nishimoto Y, Suzuki T, Kwak S (2010) Novel etiological and therapeutic strategies for neurodiseases: RNA editing enzyme abnormality in sporadic amyotrophic lateral sclerosis. J Pharmacol Sci 113:9–13

Higuchi M, Maas S, Single FN, Hartner J, Rozov A, Burnashev N, Feldmeyer D, Sprengel R, Seeburg PH (2000) Point mutation in an AMPA receptor gene rescues lethality in mice deficient in the RNA-editing enzyme ADAR2. Nature 406:78–81

Hillier LD, Lennon G, Becker M, Bonaldo MF, Chiapelli B, Chissoe S, Dietrich N, DuBuque T, Favello A, Gish W, Hawkins M, Hultman M, Kucaba T, Lacy M, Le M, Le N, Mardis E, Moore B, Morris M, Parsons J, Prange C, Rifkin L, Rohlfing T, Schellenberg K, Marra M et al (1996) Generation and analysis of 280, 000 human expressed sequence tags. Genome Res 6:807–828

Hoopengardner B, Bhalla T, Staber C, Reenan R (2003) Nervous system targets of RNA editing identified by comparative genomics. Science 301:832–836

Hundley HA, Krauchuk AA, Bass BL (2008) C-elegans and H-sapiens mRNAs with edited 3′ UTRs are present on polysomes. RNA 14:2050–2060

Iizasa H, Wulff BE, Alla NR, Maragkakis M, Megraw M, Hatzigeorgiou A, Iwakiri D, Takada K, Wiedmer A, Showe L, Lieberman P, Nishikura K (2010) Editing of Epstein-Barr virus-encoded BART6 microRNAs controls their dicer targeting and consequently affects viral latency. J Biol Chem 285:33358–33370

Kawahara Y, Ito K, Sun H, Aizawa H, Kanazawa I, Kwak S (2004) Glutamate receptors: RNA editing and death of motor neurons. Nature 427:801

Kawahara Y, Zinshteyn B, Chendrimada TP, Shiekhattar R, Nishikura K (2007a) RNA editing of the microRNA-151 precursor blocks cleavage by the Dicer-TRBP complex. EMBO Rep 8:763–769

Kawahara Y, Zinshteyn B, Sethupathy P, Iizasa H, Hatzigeorgiou AG, Nishikura K (2007b) Redirection of silencing targets by adenosine-to-inosine editing of miRNAs. Science 315: 1137–1140

Kawahara Y, Megraw M, Kreider E, Iizasa H, Valente L, Hatzigeorgiou AG, Nishikura K (2008) Frequency and fate of microRNA editing in human brain. Nucleic Acids Res 36:5270–5280

Kim DD, Kim TT, Walsh T, Kobayashi Y, Matise TC, Buyske S, Gabriel A (2004) Widespread RNA editing of embedded Alu elements in the human transcriptome. Genome Res 14:1719–1725

Kleinberger Y, Eisenberg E (2010) Large-scale analysis of structural, sequence and thermodynamic characteristics of A-to-I RNA editing sites in human Alu repeats. BMC Genomics 11:453

Kwak S, Nishimoto Y, Yamashita T (2008) Newly identified ADAR-mediated A-to-I editing positions as a tool for ALS research. RNA Biol 5:193–197

Lander ES, Linton LM, Birren B, Nusbaum C, Zody MC, Baldwin J, Devon K, Dewar K, Doyle M, FitzHugh W, Funke R, Gage D, Harris K, Heaford A, Howland J, Kann L, Lehoczky J, LeVine R, McEwan P, McKernan K, Meldrim J, Mesirov JP, Miranda C, Morris W, Naylor J, Raymond C, Rosetti M, Santos R, Sheridan A, Sougnez C, Stange-Thomann N, Stojanovic N, Subramanian A, Wyman D, Rogers J, Sulston J, Ainscough R, Beck S, Bentley D, Burton J, Clee C, Carter N, Coulson A, Deadman R, Deloukas P, Dunham A, Dunham I, Durbin R, French L, Grafham D, Gregory S, Hubbard T, Humphray S, Hunt A, Jones M, Lloyd C, McMurray A, Matthews L, Mercer S, Milne S, Mullikin JC, Mungall A, Plumb R, Ross M, Shownkeen R, Sims S, Waterston RH, Wilson RK, Hillier LW, McPherson JD, Marra MA, Mardis ER, Fulton LA, Chinwalla AT, Pepin KH, Gish WR, Chissoe SL, Wendl MC, Delehaunty KD, Miner TL, Delehaunty A, Kramer JB, Cook LL, Fulton RS, Johnson DL, Minx PJ, Clifton SW, Hawkins T, Branscomb E, Predki P, Richardson P, Wenning S, Slezak T, Doggett N, Cheng JF, Olsen A, Lucas S, Elkin C, Uberbacher E, Frazier M et al (2001) Initial sequencing and analysis of the human genome. Nature 409:860–921

Lanfranco MF, Seitz PK, Morabito MV, Emeson RB, Sanders-Bush E, Cunningham KA (2009) An innovative real-time PCR method to measure changes in RNA editing of the serotonin 2C receptor (5-HT(2C)R) in brain. J Neurosci Methods 179:247–257

Lehmann KA, Bass BL (2000) Double-stranded RNA adenosine deaminases ADAR1 and ADAR2 have overlapping specificities. Biochemistry 39:12875–12884

Levanon EY, Eisenberg E, Yelin R, Nemzer S, Hallegger M, Shemesh R, Fligelman ZY, Shoshan A, Pollock SR, Sztybel D, Olshansky M, Rechavi G, Jantsch MF (2004) Systematic identification of abundant A-to-I editing sites in the human transcriptome. Nat Biotechnol 22:1001–1005

Levanon EY, Hallegger M, Kinar Y, Shemesh R, Djinovic-Carugo K, Rechavi G, Jantsch MF, Eisenberg E (2005a) Evolutionarily conserved human targets of adenosine to inosine RNA editing. Nucleic Acids Res 33:1162–1168

Levanon K, Eisenberg E, Rechavi G, Levanon EY (2005b) Letter from the editor: Adenosine-to-inosine RNA editing in Alu repeats in the human genome. EMBO Rep 6:831–835

Lev-Maor G, Sorek R, Levanon EY, Paz N, Eisenberg E, Ast G (2007) RNA-editing-mediated exon evolution. Genome Biol 8:R29

Li JB, Gao Y, Aach J, Zhang K, Kryukov GV, Xie B, Ahlford A, Yoon JK, Rosenbaum AM, Zaranek AW, LeProust E, Sunyaev SR, Church GM (2009a) Multiplex padlock targeted sequencing reveals human hypermutable CpG variations. Genome Res 19:1606–1615

Li JB, Levanon EY, Yoon JK, Aach J, Xie B, Leproust E, Zhang K, Gao Y, Church GM (2009b) Genome-wide identification of human RNA editing sites by parallel DNA capturing and sequencing. Science 324:1210–1213

Liang H, Landweber LF (2007) Hypothesis: RNA editing of microRNA target sites in humans? RNA 13:463–467

Linsen SE, de Wit E, de Bruijn E, Cuppen E (2010) Small RNA expression and strain specificity in the rat. BMC Genomics 11:249

Luciano DJ, Mirsky H, Vendetti NJ, Maas S (2004) RNA editing of a miRNA precursor. RNA 10:1174–1177

Maas S, Patt S, Schrey M, Rich A (2001) Underediting of glutamate receptor GluR-B mRNA in malignant gliomas. Proc Natl Acad Sci USA 98:14687–14692

Mattick JS, Mehler MF (2008) RNA editing, DNA recoding and the evolution of human cognition. Trends Neurosci 31:227–233

Melcher T, Maas S, Herb A, Sprengel R, Seeburg PH, Higuchi M (1996) A mammalian RNA editing enzyme. Nature 379:460–464

Moller-Krull M, Zemann A, Roos C, Brosius J, Schmitz J (2008) Beyond DNA: RNA editing and steps toward Alu exonization in primates. J Mol Biol 382:601–609

Morse DP, Bass BL (1999) Long RNA hairpins that contain inosine are present in Caenorhabditis elegans poly(A) + RNA. Proc Natl Acad Sci USA 96:6048–6053

Neeman Y, Levanon EY, Jantsch MF, Eisenberg E (2006) RNA editing level in the mouse is determined by the genomic repeat repertoire. RNA 12:1802–1809

Nicholas A, de Magalhaes JP, Kraytsberg Y, Richfield EK, Levanon EY, Khrapko K (2010) Age-related gene-specific changes of A-to-I mRNA editing in the human brain. Mech Ageing Dev 131(6):445–447

Nishikura K (2010) Functions and regulation of RNA editing by ADAR deaminases. Annu Rev Biochem 79:321–349

Nishimoto Y, Yamashita T, Hideyama T, Tsuji S, Suzuki N, Kwak S (2008) Determination of editors at the novel A-to-I editing positions. Neurosci Res 61:201–206

Niswender CM, Herrick-Davis K, Dilley GE, Meltzer HY, Overholser JC, Stockmeier CA, Emeson RB, Sanders-Bush E (2001) RNA editing of the human serotonin 5-HT2C receptor. Alterations in suicide and implications for serotonergic pharmacotherapy. Neuropsychopharmacol 24:478–491

Ohlson J, Ohman M (2007) A method for finding sites of selective adenosine deamination. Methods Enzymol 424:289–300

Ohlson J, Enstero M, Sjoberg BM, Ohman M (2005) A method to find tissue-specific novel sites of selective adenosine deamination. Nucleic Acids Res 33:e167

Ohlson J, Pedersen JS, Haussler D, Ohman M (2007) Editing modifies the GABA(A) receptor subunit alpha3. RNA 13:698–703

Osenberg S, Dominissini D, Rechavi G, Eisenberg E (2009) Widespread cleavage of A-to-I hyperediting substrates. RNA 15:1632–1639

Osenberg S, Paz Yaacov N, Safran M, Moshkovitz S, Shtrichman R, Sherf O, Jacob-Hirsch J, Keshet G, Amariglio N, Itskovitz-Eldor J, Rechavi G (2010) Alu sequences in undifferentiated human embryonic stem cells display high levels of A-to-I RNA editing. PLoS One 5:e11173

Paul MS, Bass BL (1998) Inosine exists in mRNA at tissue-specific levels and is most abundant in brain mRNA. Embo J 17:1120–1127

Paz N, Levanon EY, Amariglio N, Heimberger AB, Ram Z, Constantini S, Barbash ZS, Adamsky K, Safran M, Hirschberg A, Krupsky M, Ben-Dov I, Cazacu S, Mikkelsen T, Brodie C, Eisenberg E, Rechavi G (2007) Altered adenosine-to-inosine RNA editing in human cancer. Genome Res 17:1586–1595

Paz-Yaacov N, Levanon EY, Nevo E, Kinar Y, Harmelin A, Jacob-Hirsch J, Amariglio N, Eisenberg E, Rechavi G (2010) Adenosine-to-inosine RNA editing shapes transcriptome diversity in primates. Proc Natl Acad Sci USA 107:12174–12179

Polson AG, Bass BL (1994) Preferential selection of adenosines for modification by double-stranded RNA adenosine deaminase. Embo J 13:5701–5711

Powell LM, Wallis SC, Pease RJ, Edwards YH, Knott TJ, Scott J (1987) A novel form of tissue-specific RNA processing produces apolipoprotein-B48 in intestine. Cell 50:831–840

Prasanth KV, Prasanth SG, Xuan Z, Hearn S, Freier SM, Bennett CF, Zhang MQ, Spector DL (2005) Regulating gene expression through RNA nuclear retention. Cell 123:249–263

Riedmann EM, Schopoff S, Hartner JC, Jantsch MF (2008) Specificity of ADAR-mediated RNA editing in newly identified targets. RNA 14:1110–1118

Rosenberg BR, Hamilton CE, Mwangi MM, Dewell S, Papavasiliou FN (2011) Transcriptome-wide sequencing reveals numerous APOBEC1 mRNA-editing targets in transcript 3′ UTRs. Nat Struct Mol Biol 18:230–236

Rueter SM, Dawson TR, Emeson RB (1999) Regulation of alternative splicing by RNA editing. Nature 399:75–80

Rula EY, Lagrange AH, Jacobs MM, Hu N, Macdonald RL, Emeson RB (2008) Developmental modulation of GABA(A) receptor function by RNA editing. J Neurosci 28:6196–6201

Sakurai M, Yano T, Kawabata H, Ueda H, Suzuki T (2010) Inosine cyanoethylation identifies A-to-I RNA editing sites in the human transcriptome. Nat Chem Biol 6:733–740

Scadden ADJ, Smith CWJ (2001) RNAi is antagonized by A® I hyper-editing. EMBO Reports 2:1107–1111

Seeburg PH, Higuchi M, Sprengel R (1998) RNA editing of brain glutamate receptor channels: mechanism and physiology. Brain Res Brain Res Rev 26:217–229

Sie CP, Maas S (2009) Conserved recoding RNA editing of vertebrate C1q-related factor C1QL1. FEBS Lett 583:1171–1174

Sommer B, Kohler M, Sprengel R, Seeburg PH (1991) RNA editing in brain controls a determinant of ion flow in glutamate-gated channels. Cell 67:11–19

Suspene R, Renard M, Henry M, Guetard D, Puyraimond-Zemmour D, Billecocq A, Bouloy M, Tangy F, Vartanian JP, Wain-Hobson S (2008) Inversing the natural hydrogen bonding rule to selectively amplify GC-rich ADAR-edited RNAs. Nucleic Acids Res 36:e72

Tonkin LA, Bass BL (2003) Mutations in RNAi rescue aberrant chemotaxis of ADAR mutants. Science 302:1725

Wahlstedt H, Daniel C, Enstero M, Ohman M (2009) Large-scale mRNA sequencing determines global regulation of RNA editing during brain development. Genome Res 19:978–986

Wang Q, Khillan J, Gadue P, Nishikura K (2000) Requirement of the RNA editing deaminase ADAR1 gene for embryonic erythropoiesis. Science 290:1765–1768

Wang Q, Miyakoda M, Yang W, Khillan J, Stachura DL, Weiss MJ, Nishikura K (2004) Stress-induced apoptosis associated with null mutation of ADAR1 RNA editing deaminase gene. J Biol Chem 279:4952–4961

Wang Q, Zhang Z, Blackwell K, Carmichael GG (2005) Vigilins bind to promiscuously A-to-I-edited RNAs and are involved in the formation of heterochromatin. Curr Biol 15:384–391

Waterston RH, Lindblad-Toh K, Birney E, Rogers J, Abril JF, Agarwal P, Agarwala R, Ainscough R, Alexandersson M, An P, Antonarakis SE, Attwood J, Baertsch R, Bailey J, Barlow K, Beck S, Berry E, Birren B, Bloom T, Bork P, Botcherby M, Bray N, Brent MR, Brown DG, Brown SD, Bult C, Burton J, Butler J, Campbell RD, Carninci P, Cawley S, Chiaromonte F, Chinwalla AT, Church DM, Clamp M, Clee C, Collins FS, Cook LL, Copley RR, Coulson A, Couronne O, Cuff J, Curwen V, Cutts T, Daly M, David R, Davies J,

Delehaunty KD, Deri J, Dermitzakis ET, Dewey C, Dickens NJ, Diekhans M, Dodge S, Dubchak I, Dunn DM, Eddy SR, Elnitski L, Emes RD, Eswara P, Eyras E, Felsenfeld A, Fewell GA, Flicek P, Foley K, Frankel WN, Fulton LA, Fulton RS, Furey TS, Gage D, Gibbs RA, Glusman G, Gnerre S, Goldman N, Goodstadt L, Grafham D, Graves TA, Green ED, Gregory S, Guigo R, Guyer M, Hardison RC, Haussler D, Hayashizaki Y, Hillier LW, Hinrichs A, Hlavina W, Holzer T, Hsu F, Hua A, Hubbard T, Hunt A, Jackson I, Jaffe DB, Johnson LS, Jones M, Jones TA, Joy A, Kamal M, Karlsson EK et al (2002) Initial sequencing and comparative analysis of the mouse genome. Nature 420:520–562

Wong K, Lyddon R, Dracheva S (2009) TaqMan-based, real-time quantitative polymerase chain reaction method for RNA editing analysis. Anal Biochem 390:173–180

XuFeng R, Boyer MJ, Shen H, Li Y, Yu H, Gao Y, Yang Q, Wang Q, Cheng T (2009) ADAR1 is required for hematopoietic progenitor cell survival via RNA editing. Proc Natl Acad Sci USA 106:17763–17768

Yang W, Chendrimada TP, Wang Q, Higuchi M, Seeburg PH, Shiekhattar R, Nishikura K (2006) Modulation of microRNA processing and expression through RNA editing by ADAR deaminases. Nat Struct Mol Biol 13:13–21

Zaranek AW, Levanon EY, Zecharia T, Clegg T, Church GM (2010) A survey of genomic traces reveals a common sequencing error, RNA editing, and DNA editing. PLoS Genet 6:e1000954

Zhang Z, Carmichael GG (2001) The fate of dsRNA in the nucleus: a p54(nrb)-containing complex mediates the nuclear retention of promiscuously A-to-I edited RNAs. Cell 106: 465–475

Zilberman DE, Safran M, Paz N, Amariglio N, Simon A, Fridman E, Kleinmann N, Ramon J, Rechavi G (2009) Does RNA editing play a role in the development of urinary bladder cancer? Urol Oncol 29:21–26

ADARs: Viruses and Innate Immunity

Charles E. Samuel

Abstract Double-stranded RNA (dsRNA) functions both as a substrate of ADARs and also as a molecular trigger of innate immune responses. ADARs, adenosine deaminases that act on RNA, catalyze the deamination of adenosine (A) to produce inosine (I) in dsRNA. ADARs thereby can destablize RNA structures, because the generated I:U mismatch pairs are less stable than A:U base pairs. Additionally, I is read as G instead of A by ribosomes during translation and by viral RNA-dependent RNA polymerases during RNA replication. Members of several virus families have the capacity to produce dsRNA during viral genome transcription and replication. Sequence changes (A–G, and U–C) characteristic of A–I editing can occur during virus growth and persistence. Foreign viral dsRNA also mediates both the induction and the action of interferons. In this chapter our current understanding of the role and significance of ADARs in the context of innate immunity, and as determinants of the outcome of viral infection, will be considered.

Contents

C. E. Samuel (✉)
Department of Molecular, Cellular and Developmental Biology,
University of California, Santa Barbara, CA 93106, USA
e-mail: samuel@lifesci.ucsb.edu

Current Topics in Microbiology and Immunology (2012) 353: 163–195
DOI: 10.1007/82_2011_148

Published Online: 14 July 2011

1 Introduction

Double-stranded RNA (dsRNA), the substrate of adenosine deaminases acting on RNA (ADARs), also functions as an effector molecule that is sensed by, and modulates the activity of, protein components that mediate the antiviral innate immune response. ADAR activity was described initially in *Xenopus* oocytes as a dsRNA unwinding activity during antisense RNA studies (Bass and Weintraub 1987; Rebagliati and Melton 1987). But rather than unwinding the dsRNA, it is now known that the deamination of adenosine (A) to inosine (I) that occurred in duplex RNA led to destablization of dsRNA because I:U mismatch bp are less stable than the A:U bp (Bass and Weintraub 1988; Wagner et al. 1989; Serra et al. 2004). Quantitation of the reduced stability of dsRNA following conversion of an A:U bp into a I:U mismatch pair has been described, with an internal I:U pair approximately 2 kcal/mol less stable than an internal A:U pair (Strobel et al. 1994). Because I hydrogen bonds with C instead of U, inosine is decoded as G instead of A by ribosomes during translation and by viral RNA-dependent RNA polymerases during replication and transcription, thereby leading to changes in biologic processes that involve either RNA sequence- or structure-dependent interactions and functions (Bass 2002; Maas et al. 2003; Nishikura 2010; George et al. 2011).

The innate immune response represents an early line of host defense against viral pathogens and includes the induction of *interferon*, the first cytokine discovered. Interferon (IFN) derives its name from its potent biologic activity, the ability to *interfere* with virus growth (Isaacs and Lindemann 1957). DsRNA has a long history in the IFN field (Colby and Morgan 1971; Stewart 1979). Both naturally occurring dsRNA in the form of reovirus genome RNA (Tytell et al. 1967) and synthetic polyribonucleotide dsRNA in the form of poly rI:poly rC (Field et al. 1967) were identified over four decades ago as potent inducers of IFN, both in intact animals and in cultured cells. We now have considerable insight into the molecular mechanisms by which dsRNAs are sensed as foreign nucleic acid and activate signal transduction pathways leading to the induction of IFNs and proinflammatory cytokines (Fig. 1, *left*). Among the innate immune signaling sensors that are triggered by pathogen-derived dsRNA leading to the production of IFN are the endosomal toll-like receptor 3, the cytoplasmic RIG-like receptors, RNA polymerase III that acts as a cytoplasmic DNA sensor and produces dsRNAs, and

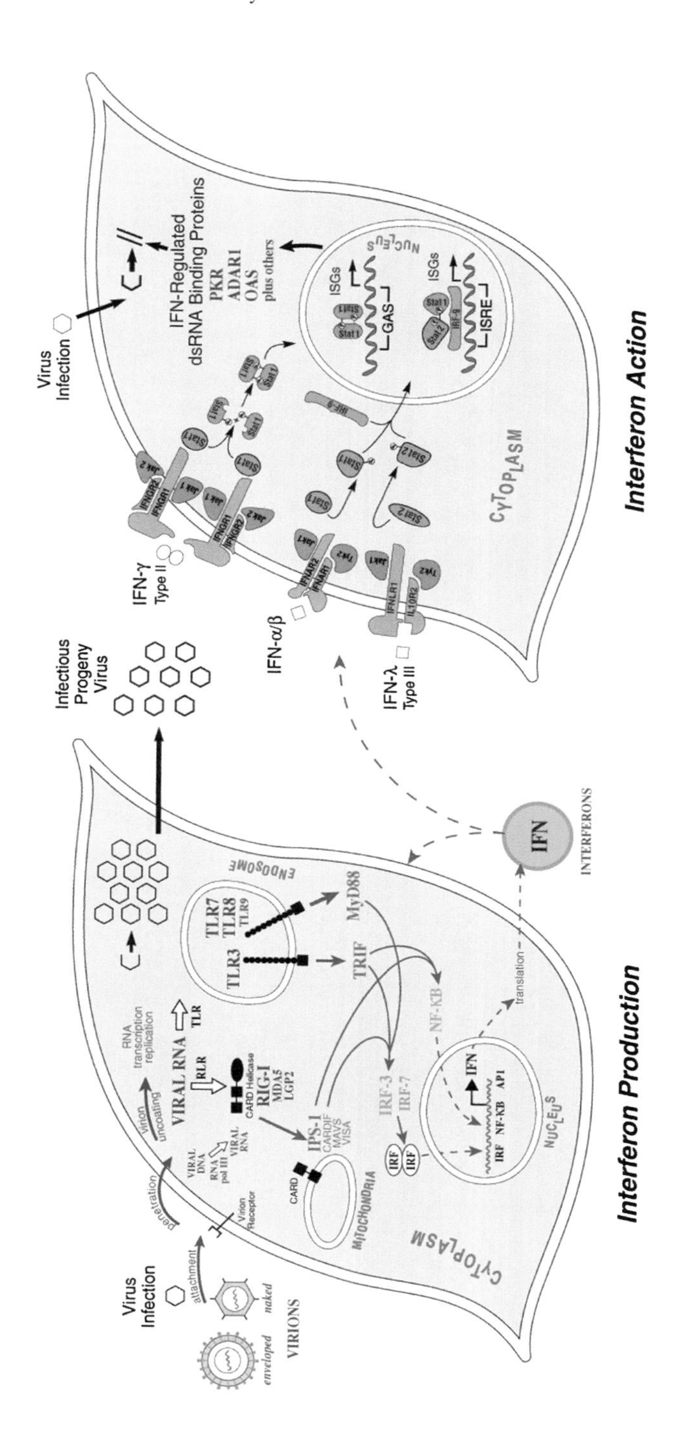
Virus Infection
VIRIONS
enveloped
naked
attachment
penetration
Virion Receptor
Virion uncoating
RNA transcription replication
VIRAL DNA
RNA pol III
VIRAL RNA
VIRAL RNA
RLR
TLR
CARD Helicase
RIG-I
MDA5
LGP2
IPS-1
CARDIF
MAVS
VISA
CARD
MITOCHONDRIA
TLR3
TLR7
TLR8
TLR9
ENDOSOME
TRIF
MyD88
NF-KB
IRF-3
IRF-7
IRF
IRF
IFN
IRF NF-KB AP1
NUCLEUS
translation
CYTOPLASM
IFN
INTERFERONS
Interferon Production
Infectious Progeny Virus
IFN-α/β
IFN-λ Type III
IFN-γ Type II
Virus Infection
IFN-Regulated dsRNA Binding Proteins
PKR
ADAR1
OAS
plus others
IFNGR1
IFNGR2
IFNAR1
IFNAR2
IFNLR1
IL10R2
Jak1
Jak2
Tyk2
Stat1
Stat2
IRF-9
ISGs
GAS
ISRE
NUCLEUS
CYTOPLASM
Interferon Action

◄ **Fig. 1** Signaling pathways activated by RNA leading to interferon production and action in virus-infected cells. *Left* following entry of virion particles by receptor-mediated fusion or endocytosis and uncoating, the transcription and replication of the viral genome can result in the production of viral RNA with double-stranded character (dsRNA) that is sensed as foreign RNA by pathogen recognition receptors (PRR) of the host. Among the cellular PRR sensors of foreign or non-self viral dsRNA are the RIG-family of cytosolic sensors (RIG-I, MDA5) that act through the mitochondrial adaptor protein IPS-1 (CARDIF, MAVS, VISA) and the endosomal toll-like receptor 3 (TLR3) sensor that acts through the TRIF adaptor protein. SsRNAs are sensed by TLR7 and TLR8, and CpG-rich dsDNA by TLR9, that act through MyD88. Foreign cytosolic dsDNA is also transcribed by RNA polymerase III to generate dsRNA that acts through IPS-1. The IPS-1, TRIF and MyD88 adaptors function downstream of their respective foreign nucleic acid PRRs to signal interferon regulatory factor IRF3 and NF-kB activation and subsequent transcriptional induction of type I interferons. *Right* interferon proteins initiate signaling by binding to their cognate receptors. The type I interferons that include the IFN-α subspecies and IFN-β act through a shared IFNAR receptor complex to activate the JAK1 and TYK2 protein kinases that lead to the activation of STAT1 and STAT2 signal transducers, which together with IRF9, form the trimeric ISGF3 complex that binds to the 13 bp ISRE DNA element to drive gene transcription. Among the genes induced by JAK-STAT-dependent IFN signaling are those that encode dsRNA-binding proteins including the dsRNA-dependent protein kinase PKR, the family of 2′, 5′-oligoadenylate synthetases activated by dsRNA, and the p150 isoform of ADAR1 that acts upon dsRNA (adapted from Samuel 2007)

PKR (Uematsu and Akira 2007; Yoneyama and Fujita 2007; O'Neill 2009; Kumar et al. 2011). DsRNAs are important not only as inducers of IFNs, but also as mediators of the actions of IFNs (Samuel 2001, 2007). Much has been learned about the IFN-induced gene products that, acting either alone or in combination with each other, inhibit virus growth in IFN-treated cells (Samuel 2001; Borden et al. 2007). Among these IFN-induced proteins with antiviral activity are two dsRNA-binding proteins that are enzymes activated by dsRNA (Fig. 1, *right*), the protein kinase PKR that phosphorylates translation initiation factor eIF-2 (Samuel 1979, 2001; Sadler and Williams 2007; Pindel and Sadler 2011) and the family of 2′, 5′-oligoadenylate synthetases that function via RNase L to degrade RNA (Chakrabarti et al. 2011). A third IFN-induced dsRNA binding protein, the inducible p150 form of ADAR1, utilizes dsRNA as its substrate (Toth et al. 2006; George et al. 2011).

The roles of ADAR proteins in viral infections, both antiviral and proviral, and the function of ADAR1 as a suppressor of innate immune responses including dsRNA-mediated protein activities, are reviewed in this chapter.

2 ADARs and Their Regulation

2.1 Genes

Three *Adar* genes have been characterized in mammals; they are designated *Adar1*, *Adar2* and *Adar3*. *Adar1* localizes to human chromosome 1q21 (Wang et al. 1995; Weier et al. 1995) and mouse chromosome 3F2 (Weier et al. 2000), and *Adar2* to human chromosome 21q22 (Mittaz et al. 1997) and mouse chromosome 10C1 (Slavov and Gardiner 2002). The *Adar1* and *Adar2* genes encode active deaminase enzymes that possess A–I RNA editing activity, whereas *Adar3* does not show demonstrable enzymatic activity (Bass 2002; Maas et al. 2003; Toth et al. 2006; Nishikura 2010; Samuel 2011). The *Adar1* gene transcripts possess 15 exons in the human (Liu et al. 1997) and the mouse (Hartner et al. 2004; Wang et al. 2004; George et al. 2005). Both the human and mouse *Adar2* genes also have 15 exons (Slavov and Gardiner 2002; Maas and Gommans 2009). Furthermore, multiple splice variants of both *Adar1* and *Adar2* transcripts have been described, in addition to a complex arrangement of multiple promoters that drive the expression of the *Adar1* gene (George et al. 2011).

A single major *Adar1* transcript of ~6.7 kb is detected by Northern analysis with RNA from human cells using probes derived from exons 2–15 (Patterson et al. 1995). While inducible by interferon and also following infection, a significant basal expression level of *Adar1* is observed both in cultured mammalian cells and animal tissues (Patterson et al. 1995; Shtrichman et al. 2002; George et al. 2005). The transcription of the *Adar1* gene occurs from at least three promoters, one of which is IFN inducible, in human (George and Samuel 1999a, b; Kawakubo and Samuel 2000) and mouse (George et al. 2005, 2008) cell lines. *Adar1* gene promoters drive the expression of transcripts with alternative exon 1 structures (exon 1A, 1B, 1C) that are spliced as far as is known to a common exon 2 junction. In addition to the three different forms of exon 1, splice variants of other exons, notably exon 7, are also observed (Toth et al. 2006; George et al. 2011).

Among the best-characterized transcriptional units are the two major alternative promoters responsible for the expression of two differently sized isoforms of ADAR1 protein, known generally as p110 and p150. The IFN inducible promoter drives transcription beginning with exon 1A that possesses a translation initiation codon (AUG1) and typically includes the alternative "b" form of exon 7 (George and Samuel 1999a, b; George et al. 2005). The major constitutively active promoter drives transcription beginning with exon 1B and includes the alternative "a" form of exon 7 that is 26 amino acids longer than exon 7b (Liu et al. 1997; George and Samuel 1999a, b; George et al. 2005). Translation initiation of the exon 1B-containing constitutively expressed transcripts begins at AUG296 within the unusually large exon 2. While the consensus open reading frame of the ADAR1 human cDNA is 1,226 amino acids (Kim et al. 1994; O'Connell et al. 1995; Patterson and Samuel 1995), because of the alternative splicing involving exons 1 and 7, multiple unique transcripts are produced of ~6.7 kb. The inducible exon 1A-containing transcript with 7b encodes the IFN-inducible p150 protein that is deduced to be 1,200 amino acids in size, whereas the constitutively expressed human p110 protein encoded by exon 1B-containing mRNA is deduced to be 931 amino acids (Toth et al. 2006; George et al. 2011).

The interferon-inducible *Adar1* gene promoter region possesses a consensus interferon-stimulated response element (ISRE). The *Adar1* ISRE is highly conserved between the inducible human (George and Samuel 1999b) and mouse (George et al. 2008) promoter sequences as well as with the ISRE elements present in the IFN inducible promoter of the mouse and human RNA-dependent protein kinase *Pkr* genes (Tanaka and Samuel 1994; Kuhen and Samuel 1997). The *Adar1* human and mouse gene ISRE elements are 12 bp and differ in only one position between each other, at the 3′-position which is either a T or C (GGAAA_CGAAAGT/C). The *Pkr* human and mouse gene ISRE elements are 13-bp and likewise differ in only the 3′ position from each other (GGAAAACGAAACT/A). ISRE elements enhance transcription in response to type I IFN treatment (Schindler et al. 2007; Randall and Goodbourn 2008). The IFN inducible *Adar1* promoter (George and Samuel 1999b; George et al. 2008), like the IFN inducible *Pkr* promoter (Tanaka and Samuel 1994; Kuhen and Samuel 1997), is TATA-less both in mouse and human cells. Transcriptional activation of genes by type I IFNs involve binding of the IFN (α/β) to the IFNAR receptor. Subsequent activation of JAK1 and TYK2 tyrosine kinases mediate activation of the STAT1 and STAT2 factors that together with IRF9 form the heterotrimeric ISGF3 complex that binds to the ISRE element to enhance gene expression (Fig. 1, *right*). As expected, IFN induction of the *Adar1*-encoded p150 protein is impaired in mouse embryo fibroblasts that are genetically deficient in either the IFNAR receptor, the JAK1 kinase or STAT2 (George et al. 2008). But, unexpectedly, IFN induction of p150 is independent of STAT1 in MEFs (George et al. 2008). The mechanism of STAT1-independent IFN induction of p150 ADAR1 is not established. Interestingly, the DNA editing enzyme APOBEC3G that catalyzes C–U deamination of first-strand retrovirus DNA likewise is reported to be induced in a STAT1-independent manner (Sarkis et al. 2006).

2.2 Proteins

The ADAR proteins possess multiple biochemical activities that include the ability to bind A-form dsRNA, Z-form dsDNA and Z-form dsRNA, and to deaminate adenosine in RNA substrates either in a highly selective manner at only one or a very few adenosines, or when the dsRNA substrate possesses extensive duplex character, then at multiple adenosine sites in a non-selective manner (Bass 2002; Maas et al. 2003; Toth et al. 2006; Nishikura 2010; George et al. 2011). The diversity of *Adar1* and *Adar2* transcripts with alternative exon structures suggests the existence of different ADAR1 and ADAR2 protein isoforms with potentially different activities that might include different substrate specificities, different subcellular localizations and different capacities to interact with protein partners or bind different nucleic acids. Among the best-characterized ADAR1 protein differences are those that arise from the alternative promoter usage and alternative splicing to generate the two size isoforms of ADAR1. The large or p150 form of ADAR1 is IFN inducible and localizes to both the cytoplasm and nucleus (Patterson and Samuel 1995; Poulsen et al. 2001; Strehblow et al. 2002). The small or p110 form of ADAR1 is constitutively expressed and is predominantly if not exclusively found in the nucleus (Patterson and Samuel 1995; Li et al. 2010). ADAR2 likewise is constitutively expressed and localizes exclusively to the nucleus, primarily in nucleoli (Desterro et al. 2003; Sansam et al. 2003). The catalytic and nucleic acid binding domain organization and function of the ADAR proteins are considered in depth elsewhere in this volume by Beal and Allain and their colleagues (Goodman et al. 2011; Barraud and Allain 2011). Briefly, for both the p110 and p150 isoforms of ADAR1, the deaminase catalytic domain is present in the C-terminal region and both size isoforms possess three copies of the dsRNA-binding motif (R, or dsRBD) positioned in the central region of p110 and p150 (Kim et al. 1994; O'Connell et al. 1995; Patterson and Samuel 1995; Liu and Samuel 1996; Liu et al. 1997). The dsRBD motifs of ADARs including ADAR1 are similar to the prototype dsRNA-binding motif first identified in PKR (McCormack et al. 1992, 1995; Fierro-Monti and Mathews 2000). The p150 protein, however, has an N-terminal extension of 295 amino acids compared to p110 ADAR1. The N-terminus of p150 includes a repeated domain ($Z\alpha$, $Z\beta$) with homology to a sequence present within the N-terminal half of the poxvirus E3L protein (Patterson and Samuel 1995) and this is now known as the Z-DNA binding domain (Herbert et al. 1997; Schwartz et al. 1999). Only one copy of the domain ($Z\beta$) is present in p110. Only more recently have insights been gained regarding the possible physiologic function of the Z-domains present in ADAR1. Interestingly, the $Z\alpha$ domain of ADAR1 p150 was shown to bind Z-RNA in addition to Z-DNA (Brown et al. 2000; Placido et al. 2007). The $Z\alpha$ domain of ADAR1 also forms stable complexes with ribosomes with high affinity in human cells, and a stoichiometric association blocks translation (Feng et al. 2011).

3 Innate Immunity Signal Transduction Pathways

3.1 Sensors and Adaptors of dsRNA Signaling

Viral nucleic acids produced in infected cells include double-stranded and 5′-triphospate-containing RNAs that are sensed as foreign or non-self nucleic acids, thereby triggering the innate immune response (Yoneyama and Fujita 2007, 2010; O'Neill 2009; Kumar et al. 2011; Nakhaei et al. 2009; Kawai and Akira 2010; Garcia-Sastre 2011). 5′-Capping and 2′-O methylation of RNAs, modifications characteristic of self or host mRNAs, generally prevent recognition by host cell cytoplasmic sensors. Among the cellular pattern recognition receptors (PRRs) that sense and respond to foreign viral RNAs are the cytoplasmic retinoic acid-inducible gene I (RIG-I)–like receptors (RLRs) and the membrane-associated Toll-like receptors (TLRs) as illustrated in Fig. 1 *left*. DsRNA is sensed by endosomal TLR3, whereas endosome-localized TLRs 7 and 8 detect single-stranded RNA and TLR9 detects CpG-rich unmethylated DNA. DsRNA also is sensed by RIG-I and MDA5 RLRs, with both the size of A-form dsRNA and the 5′-end structure of the RNA serving as contributing determinants for signaling via RIG-I or MDA-5 which are not functionally redundant. RLR signaling occurs via the mitochondrial IPS-1 adaptor, leading to the activation of IRF3 and NF-κB transcription factors, whereas TLR3 uses the TRIF adaptor, and TLR 7, 8 and 9 use the MyD88 adaptor. Cytoplasmic pathogen B-form DNA can be sensed by RNA pol III among other sensors, which gives rise to RNA subsequently detected by the RLR pathway.

Finally, the RNA-dependent protein kinase PKR functions also as a sensor of dsRNA in a manner that modulates the innate immune response leading to IFN production. Among the viruses where PKR plays a role in the induction of IFN are measles virus (MV), human parainfluenza virus (HPIV) and rotavirus. PKR enhances the induction of IFN-β by measles virus C and V deletion mutants by a pathway dependent upon the IPS-1 adaptor of the RLRs but not TRIF (McAllister et al. 2010). PKR activation also contributes to IFN-β induction via RLR signaling that follows the kinetics of dsRNA accumulation in human parainfluenza virus-infected cells (Boonyaratanakornkit et al. 2011), and the induction and secretion of IFN-β by rotavirus is likewise dependent upon PKR in addition to IPS-1and IRF3 (Sen et al. 2011).

3.2 *Suppression of Interferon Responses by ADARs*

A combination of findings from different lines of experimentation are consistent with the notion that ADAR1 functions as a suppressor of interferon responses. Studies including inducible *Adar1* gene disruption in mice (Hartner et al. 2009), stable knockdown of both p110 and p150 ADAR1 proteins in human cells in culture (Toth et al. 2009; Li et al. 2010), and transfection analyses with defined I:U-dsRNAs corresponding to hyperedited dsRNAs compared to control RNAs (Vitali and Scadden 2010), taken together, suggest that ADAR1 acts as a suppressor of IFN system responses. Given that poly(rI):poly(rC) is both a potent inducer of IFN (Stewart 1979; Yonemaya and Fujita 2007) and an activator of IFN-induced dsRNA-dependent enzymes including PKR and 2′,5′OAS (Samuel 2001), the finding that ADAR1 behaved as a suppressor of IFN responses was somewhat unexpected.

The inducible genetic disruption of *Adar1* p110 and p150 expressions in mice suggested that ADAR1 functions a suppressor of IFN signaling in addition to displaying an essential role for maintenance of both fetal and adult hematopoietic stem cells (Hartner et al. 2009). Loss of ADAR1 in hematopoietic stem cells led to rapid apoptosis and a global IFN response characterized by an upregulation of IFN-inducible transcripts as revealed by genome-wide transcriptome analyses. The gene expression signature of *Adar1*$^{-/-}$ cells showed similarities with the signatures of IFN-treated or virus-infected cells. Among the transcripts enhanced by ADAR1 deficiency were those for the PKR, RIG-I and TLR3, all of which are dsRNA sensors (Uemitsu and Akira 2007; Yonemaya and Fujita 2007; Samuel 2001), in addition to STAT1 and 2, IRF 1, 7 and 9, and Mx (Hartner et al. 2009). Human HeLa cells in culture made stably deficient in the expression of ADAR1 p110 and p150 proteins through the use of a short hairpin RNA-mediated knockdown strategy also revealed that apoptosis was enhanced in the ADAR1-deficient cells following infection with wild-type and V-deletion mutant MV (Toth et al. 2009). Furthermore, the level of PKR kinase activation and eIF-2α

phosphorylation was increased both in HeLa cells stably knocked down for ADAR1 (Li et al. 2010) and 293T cells transiently knocked down for ADAR1 (Nie et al. 2007). The enhanced apoptosis seen in the ADAR1-deficient cells following MV infection correlated with enhanced activation of the PKR kinase and IRF3 transcription factor, both key dsRNA sensors in the IFN system. These results suggest that the anti-apoptotic activity of ADAR1 in MV infected HeLa cells is achieved through suppression of pro-apoptotic and dsRNA-dependent activities illustrated by PKR and IRF3 (Toth et al. 2009). Likewise, MEF cells genetically deficient for the p150 isoform of ADAR1 and stably expressing the MV receptor displayed extensive virus-induced cytopathic effects following MV infection compared to wild-type MEF cells (Ward et al. 2011).

The mechanism by which ADAR1 modulates the expression of IFN inducible genes (characterized by elevated ISG expression in the absence of ADAR1) and the activation of dsRNA-dependent IFN system proteins (illustrated by the elevated activation both of the IFN inducible PKR kinase and the IRF3 transcription factor in the absence of ADAR1) is not yet fully elucidated. One possibility is that ADAR1 functions as a gatekeeper and modifies the structure of activator RNAs and, that when they are edited, the RNAs do not possess the structure sufficient to trigger the signaling and activation of the IFN response. But in the absence of ADAR, the integrity and amount of the structured RNA allows for induction and activation of the IFN system. Another possibility is that the A–I editing activity per se of ADAR1 is not required, but rather protein or nucleic acid binding interactions involving ADAR1 as a binding partner are sufficient, and these interactions become altered in the absence of ADAR1. I:U dsRNA, a mimic of hyperedited dsRNA, when transfected into HeLa cells suppresses the induction of IFN inducible genes and apoptosis by poly (rI);poly(rC) compared to control dsRNA (Vitali and Scadden 2010). Furthermore, transfection of I:U-dsRNA inhibited activation of IRF3 in parental HeLa cells (Vitali and Scadden 2010), consistent with the enhanced activation of IRF3 and increased apoptosis earlier seen in ADAR1-deficient HeLa cells following infection (Toth et al. 2009). As earlier mentioned, the PKR enhancement of IFN-β induction by virus infection requires the IPS-1 adaptor of the RIG-I/MDA5 signaling pathway that leads to IRF3 factor activation and subsequently to IFN induction (McAllister et al. 2010; Boonyaratanakornkit et al. 2011; Sen et al. 2011). That ADAR1 deficiency increases PKR activation (Toth et al. 2009; Li et al. 2010) and that I:U-dsRNA binds both RIG-I and MDA5 in a manner that may interfere with their sensing of foreign dsRNA (Vitali and Scadden 2010) suggests that multiple mechanisms may contribute to the ADAR1-mediated modulation of IRF3 activation status and apoptosis induction.

4 ADARs and Their Effects on Virus–Host Interactions

The profound importance of ADARs for normal development and neurophysiology in the absence of virus infection is illustrated in mice and flies by the phenotypes observed following genetic disruption of *Adar* genes or overexpression of ADAR

Table 1 Functions of ADARs in viral infections can be either anti- or pro- viral dependent upon the virus-host combination

Viral genome form and virus	Effect attributed to ADAR	References[a]
Negative-stranded RNA		
Measles virus	U–C, A–G; M RNA	Cattaneo et al. (1986, 1988);
		Wong et al. (1991)
		Schmid et al. (1992)
		Baczko et al. (1993)
	A–G; M RNA	Suspène et al. (2008, 2011)
	Protection against CPE; inhibition of PKR and IRF3 activation	Toth et al. (2009)
	Antiviral; protection against CPE	Ward et al. (2011)
Respiratory syncytial virus	A–G; gp G RNA	Rueda et al. (1994); Martínez and Melero (2002)
Parainfluenza virus 3	RNA 3′-region	Murphy et al. (1991)
Newcastle disease virus, canine distemper virus, Sendai virus	Protection against CPE	Ward et al. (2011)
Influenza virus	A–G; M1 RNA	tenOever et al. (2007)
	A–G; HA RNA	Suspène et al. (2011)
	Protection against CPE	Ward et al. (2011)
Vesicular stomatitis virus	A–G, U–C	O'Hara et al. (1984)
	Proviral; inhibition of PKR activation	Nie et al. (2007)
	No effect; inhibition of PKR activation	Li et al. (2011); Ward et al. (2011)
Rift Valley fever virus	A–G; L RNA	Suspène et al. (2008)
Lymphocytic choriomeningitis v.	A–G, U–C	Grande-Pérez et al. (2002)
	A–G, U–C; GP RNA	Zahn et al. (2007)
	No effect	Ward et al. (2011)
Hepatitis delta virus	Proviral; selective A–G, amber–W	Taylor (2003); Casey (2011)
Positive-stranded RNA		
Hepatitis C virus	Antiviral	Taylor et al. (2005)
Double-stranded RNA		
Orthoreovirus	No effect	Ward et al. (2011)
Retroviruses		
Human immunodeficiency virus 1	A–G TAR RNA	Sharmeen et al. (1991)
	Proviral; A–G RRE	Phuphuakrat et al. (2008)
	Proviral; inhibition of PKR activation	Clerzius et al. (2009); Doria et al. (2009, 2011)
Avian viruses: ALV and RAV-1	A–G	Felder et al. (1994); Hajjar and Linial (1995)

(continued)

Table 1 (continued)

Viral genome form and virus	Effect attributed to ADAR	References[a]
Double-stranded DNA		
Polyoma virus	A–G early/late RNA overlap	Kumar and Carmichael (1997); Gu et al. (2009)
Kaposi sarcoma-associated virus	Selective A–G K12 RNA	Gandy et al. (2007)
Epstein-Barr virus	A–G BART6 miR	Iizasa et al. (2010)

[a] See text for additional references

proteins as reviewed by Hartner and by Keegan and colleagues in other chapters in this volume (Hartner and Walkley 2011; Paro et al. 2011). Genetic knockout of *Adar1* in the mouse in a manner that disrupts expression of both p110 and p150 (Hartner et al. 2004, 2009; Wang et al. 2004; XuFeng et al. 2009) or only p150 (Ward et al. 2011) results in embryonic lethality between embryonic days 11.5 and 12.5. Genetic knockout of *Adar2* in the mouse, while not embryonic lethal, results in neurophysiological abnormalities (Higuchi et al. 2000), whereas overexpression of ADAR2 protein results in adult-onset obesity (Singh et al. 2007).

A number of studies have also revealed the importance of ADAR proteins and A–I editing during viral infections (Table 1). Somewhat surprising, the effect of ADARs on the virus–host interaction can be either antiviral or proviral, dependent upon the specific animal virus and mammalian host cell combination and the level and type of ADAR protein expression. The multiple roles played by ADARs will be illustrated in the following sections by considering examples of animal viruses from families with different genome organizations and subcellular sites of replication. These include viruses that have single-stranded RNA genomes of either negative, positive or ambisense coding organization; dsRNA genomes; and double-stranded DNA genomes (Knipe et al. 2007). The present understanding suggests that ADARs may act either directly on the virus by editing a viral RNA in a manner that impacts the outcome of the viral infection, or conceivably indirectly by editing a cellular RNA in a manner that alters a cellular product that subsequently impacts the interaction of the virus with the host. It is also possible that ADARs may function in an editing-independent manner, by altering protein or nucleic acid binding interactions, which subsequently affect the outcome of the viral infection. ADAR1 and ADAR2 have overlapping specificities with some substrates (Lehmann and Bass 2000). In those instances where the effect of ADAR protein function is exerted directly on the viral nucleic acid as implicated by viral sequence changes characterized by A–G (or U–C) substitutions, for viruses with exclusively a cytoplasmic localization for their multiplication, presumably the p150 isoform of ADAR1 would be the ADAR protein likely responsible; p150 ADAR1 is the only known cytoplasmic ADAR in mammalian cells (Bass 2002; Toth et al. 2006; Nishikura 2010; George et al. 2011). By contrast, for those viruses with a nuclear component to their multiplication that display A–G and U–C substitutions in the viral sequences, viruses which include dsDNA viruses and orthomyxoviruses among those considered below, then either ADAR1 size isoform or ADAR2 theoretically could be responsible for the editing events as all of these ADARs are nuclear proteins and enzymatically active deaminases (Bass 2002; Toth et al. 2006; Nishikura 2010; George et al. 2011).

4.1 Negative-Stranded RNA Viruses

Negative-stranded RNA viruses include those viruses with a single-stranded RNA genome that is the opposite sense, or complementary, to that of mRNA, and hence is not decoded directly by translation. Following infection, the viral genome of negative-stranded ssRNA viruses is transcribed by a virion-associated viral RNA-dependent RNA polymerase to produce mRNA for translation, and then replicated to produce full-length positive (and subsequently negative) sense ssRNA. Among the negative-stranded viruses are members of the *Paramyxoviridae*, *Orthomyxoviridae* and *Rhabdoviridae* families. The observations of A–G (U–C) nucleotide substitutions in viral RNA sequences determined for negative-stranded RNA viruses has a long history and has implicated ADAR-mediated A–I editing as a factor in the interaction of these viruses with their hosts (Cattaneo and Billeter 1992; Samuel 2011). Only more recently has evidence been obtained that directly establishes ADAR1 as a restriction factor affecting the outcome of interaction of negative-stranded viruses, particularly MV, with their hosts (Toth et al. 2009; Ward et al. 2011).

Paramyxoviridae and Orthomyxoviridae: Measles Virus is a member of the Morbillivirus genus of *Paramyxoviridae*. The enveloped virion of MV encloses a $\sim$16 kb negative-stranded ssRNA genome that consists of six genes, the N, M, F, H and L monocistronic genes and the polycistronic P/V/C gene. The co-transcriptional pseudo-templated G nucleotide insertion editing that occurs with MV to generate the V transcript with a frameshift compared to the P/C transcript, which is commonly referred to as "editing" by virologists, should not be confused with the nucleotide substitution editing by ADARs. While MV causes a typically acute infection spread by the respiratory route, a rare but serious complication is the subsequent persistent infection of the central nervous system known as subacute schlerosing panencephalitis (SSPE), a progressive fatal neurodegenerative disease (Moss and Griffin 2006; Oldstone 2009). It is in the context of SSPE where ADAR was initially implicated as causing hypermutations characterized by A–G substitutions (or U–C in the complementary strand). MV isolated from the central nervous system of SSPE patients differed from wild-type MV of acute infections. The novel genetic changes of the SSPE virus were characterized by extensive mutations affecting predominantly the MV matrix M gene and the fusion F gene (Cattaneo and Billeter 1992; Cattaneo et al. 1986; Wong et al. 1991, 1994; Schmid et al. 1992; Baczko et al. 1993). Most striking were biased hypermutations, primarily in the M gene where in one instance up to 50% of the U residues were converted to C. The observed M gene mutations are consistent with the defective production of viral matrix protein associated with SSPE persistent infection in the brain, and the lack of antibodies to M protein seen in SSPE patients (Hall et al. 1979; Liebert et al. 1986). Direct evidence that the hypermutated M gene of an SSPE strain virus contributes to the persistent infection and chronic CNS disease was obtained in studies with transgenic CD46 mice (Oldstone et al. 1999) and the Edmonston strain MV that causes acute infection, which was engineered by reverse genetics to replace the Edmonston M gene with the Biken SSPE strain M gene (Patterson et al. 2001).

Evidence for the editing of MV RNA in cell culture infections was obtained using a creative PCR-based strategy (3DI-PCR) to selectively amplify GC-rich RNAs, as would arise following ADAR editing which substitutes I(G) for A (or C for U in the complementary strand). With the 3DI-PCR approach the natural hydrogen-bonding rule is inversed by generating modified DNA that contains diaminopurine in place of adenine, and inosine in place of guanine, which allows for the selective amplification of GC-rich ADAR-edited RNAs by differential DNA denaturation. When the Schwarz strain of MV was grown on IFN sensitive MRC5 cells, MV sequences were obtained following 3DI-PCR amplification that had extensive A–G transitions characteristic of ADAR editing, whereas sequences amplified from MV grown on Vero cells that do not produce type I IFN α or β (and hence would not produce the induced p150 ADAR1) did not show the A–G transitions (Suspène et al. 2008). Using the 3DI-PCR approach it was also shown that a region of the M gene of the live attenuated measles vaccine virus and a region of the HA gene of inactivated seasonal influenza virus vaccine both possessed hypermutated sequences, with the mutation frequency greater for influenza compared to MV (Suspène et al. 2011). ADAR1 p150 earlier was implicated to edit influenza virus RNA, as A–G mutations were frequently found in influenza virus matrix M1 RNA isolated from the lung tissue of infected mice that possess an intact innate immune signaling system and expressing ADAR1. But mice genetically deficient in the IKKε kinase and defective in the induction of a class of IFN-induced genes including ADAR1 displayed infrequent A–G transitions compared to control mice (tenOever et al. 2007). Thus, sequence analyses both of MV, a member of the *Paramyxoviridae* that multiplies in the cytoplasm, and of influenza A virus, a segmented negative-strand virus that is a member of the *Orthomyxoviridae* that replicates and transcribes viral RNA in the nucleus (Samuel 2010a), reveal nucleotide substitutions characteristic of ADAR editing found under physiologic conditions where elevated levels of ADAR1 p150 would be expected (tenOever et al. 2007; Suspène et al. 2008, 2011). In addition, following infection with influenza A PR8 or Udorn strains, quantitative analysis of the nucleolar proteome also showed an increase in the nucleolar ADAR1 p110 and p150 levels (Emmott et al. 2010).

Loss of function approaches have provided direct evidence consistent with the conclusion that ADAR1 plays an important role as a host restriction factor for controlling MV host interactions (Toth et al. 2009; Ward et al. 2011). The effect of ADAR1 deficiency on the growth of the Moraten vaccine strain and virus-induced cell death was assessed with human HeLa cells made stably deficient in ADAR1 p110 and p150 through short hairpin RNA-mediated knockdown (Toth et al. 2009). The growth of MV mutants lacking expression of C or V accessory proteins was decreased in the ADAR1-deficient knockdown cells compared to ADAR1-sufficient cells. However, MV-induced apoptosis was enhanced in the ADAR1-

deficient cells. Furthermore, the C mutant-infected ADAR1-sufficient cells when ADAR1 did not protect against apoptosis, caspase cleavage of the p150 but not p110 protein was observed. The enhanced apoptosis observed in the ADAR1 knockdown cells correlated with enhanced activation of PKR and IRF3. To further evaluate the role of the p150 isoform in the host response to MV infection, mouse embryo fibroblast cells stably expressing the SLAM receptor for MV and genetically deficient in p150 were compared to wild-type (WT) MEFs following infection with the Edmonston vaccine strain (Ward et al. 2011). Deletion of the p150 isoform of ADAR1 increased susceptibility of MEF cells to MV infection. The p150 null MEFs but not the WT MEFs displayed extensive MV-induced CPE following infection. While at early times after infection the yield of infectious progeny was reduced in the p150 null compared to WT MEFs, at later times the yields were not statistically different. These results taken together indicate that, both in human HeLa and mouse MEF cells, ADAR1 plays an anti apoptotic role in the context of infection with MV vaccine strains (Toth et al. 2009; Ward et al. 2011).

The protection provided by the p150 isoform of ADAR1 against virus-induced cytopathic effects seen in MEF cells is not limited to MV of the Morbillivirus genus (Ward et al. 2011). Additional members of the *Paramyxoviridae* family including Newcastle disease virus (Avulavirus genus), Sendai virus (Respirovirus genus) and canine distemper virus (Morbillivirus genus) showed pronounced cytopathology in *Adar1 p150*$^{-/-}$ MEFs that was not observed in WT MEFs. The p150 isoform also protected against CPE in MEFs infected with the mouse-adapted influenza A WSN strain, a member of the *Orthomyxoviridae*. Reconstitution of the *Adar1 p150*$^{-/-}$ MEFs with mouse ADAR1 completely protected against development of CPE by the various myxoviruses. Taken together, these results indicate a general protective role of the p150 isoform of ADAR1 against infection with paramyxoviruses and orthomyxoviruses in MEFs (Ward et al. 2011). Three additional paramyxoviruses, human respiratory syncytial virus (RSV, Pneumovirus genus), human parainfluenza virus 3 (HPIV3, Respirovirus genus) and mumps virus (Rubulavirus genus) have also been shown to undergo mutation consistent with the A–I editing action of an ADAR (Murphy et al. 1991; Rueda et al. 1994; Martínez and Melero 2002; Amexis et al. 2002; Chambers et al. 2009). In the case of RSV, biased A–G substitutions that caused the loss of neutralization epitopes of the glycoprotein G were identified in G gene of antibody escape virus mutants, but not in the F gene glycoprotein (Rueda et al. 1994; Martínez and Melero 2002). With HPIV3, numerous A–G and U–C transitions were detected in the 3′-end of HPIV3 RNA recovered from long term persistently infected cells in culture (Murphy et al. 1991). ADAR catalyzed deamination has also been implicated in the sequence changes seen in the Jeryl Lynn vaccine strain of mumps virus (Amexis et al. 2002; Chambers et al. 2009).

Rhabdoviridae: Vesicular stomatitis virus (VSV), a member of the *Rhabdoviridae* family, has a negative-sense ssRNA genome of ~11 kb and multiplies in the cytoplasm by a scheme similar to MV (Knipe et al. 2007). However, in contrast to the enhanced virus-induced cytotoxicity effects mediated by ADAR1 deficiency seen for paramyxoviruses as illustrated by MV (Toth et al. 2009; Ward et al. 2011), ADAR1 deficiency did not affect the replication of VSV in the absence of IFN treatment (Li et al. 2010; Ward et al. 2011). Likewise, overexpression of either ADAR1 or ADAR2 did not significantly affect VSV growth even when the virus was passaged for ten rounds of growth in HEK 293 cells engineered to overexpress either p150 ADAR1 or ADAR2 (Li et al. 2010). The lack of a readily demonstrable effect on VSV multiplication of either ADAR1 overexpression or ADAR1 deficiency by genetic knockout or RNAi knockdown, is somewhat surprising given the similarities between VSV and MV in genome structure and replication mechanism. ADAR deficiency leads to enhanced virus-induced cytotoxicity in the case of MV (Toth et al. 2009; Ward et al. 2011), but not VSV (Li et al. 2010; Ward et al. 2011) which by comparison is more cytopathic than MV even in the presence of ADAR. However, two reports suggest that ADAR-mediated hypermutation may occur albeit at very low frequency of detection in mammalian or insect cells infected with rhabdoviruses. Clustered A–G transitions over a short A-rich sequence region have been described for a VSV defective interfering particle (O'Hara et al. 1984) and also for the *Drosophilia* sigma rhabdovirus PP3 gene region (Carpenter et al. 2009).

Bunyaviridae and *Arenaviridae*: *Bunyaviridae* and *Arenaviridae* have single-stranded RNA genomes, but they are segmented. Viruses in the family *Bunyaviridae* contain three ssRNA segments (S, M, L) whereas *Arenaviridae* members have two ssRNA segments (S, L). While most members of the *Bunyaviridae* are negative-stranded, phleboviruses including Rift Valley fever virus (RVFV) use an ambisense coding strategy for the S-segment that includes decoded sequence in both the negative and positive senses, as do members of the *Arenaviridae* family including lymphocytic choriomeningitis virus (LCMV) for the S and L segments (Knipe et al. 2007). Mutations characteristic of ADAR editing have been described in RVFV and LCMV infections (Grande-Pérez et al. 2002; Zahn et al. 2007; Suspène et al. 2008). The 3DI-PCR strategy was used to demonstrate that infection of MRC5 cells with an RVFV strain that lacks a functional NSs protein gave rise to RVFV genomes with extensive A–G editing of viral RNA, but when the infection was carried out in Vero cells only quasispecies variation was seen (Suspène et al. 2008). Vero cells are defective for IFN-α and β production, whereas MRC5 cells display a competent IFN response. RVFV with defective NSs protein was used, as wild-type NSs blocks IFN production and mediates degradation of the PKR kinase (Billecocq et al. 2004; Habjan et al. 2009).

LCMV has two ssRNA genome segments, S ($\sim$3.5 kb) and L ($\sim$7.2 kb), and among the LCMV S-segment gene products are the glycoprotein precursor that gives rise to GP1 and GP2. When the mutation rate and pattern of the GP region of genomic clones of LCMV were analyzed at early and late times after infection of L929 mouse fibroblast cells with the WE virus strain, the mutations seen did not exhibit any specific pattern at two days after infection. But by seven days after infection a distinct A–G mutation bias emerged that gave rise to nonfunctional glycoprotein at a high frequency (Zahn et al. 2007). The preference for A–G (U–C) substitutions was also seen in spleen tissue of LCMV infected mice with a higher rate of mutation found at late time points. LCMV did not antagonize ADAR1 activity, but rather infection upregulated ADAR1 p150 in L929 cells and mice (Zahn et al. 2007). LCMV replicates in the cytoplasm, and ADAR1 p150 is localized in part in the cytoplasm. These results taken together suggest that ADAR1 is antiviral in the context of LCMV infection at late times after infection. At early times after infection, however, no significant differences were observed between wild-type and ADAR1 p150$^{-/-}$ mutant MEFs in the growth of the LCMV clone 13 strain of virus (Ward et al. 2011).

Hepatitis delta virus: Hepatitis delta virus (HDV) possesses a circular ~1.7 kb ssRNA negative-sense RNA genome that forms a rod-like imperfect duplex that is extensively base-paired (Taylor 2003). HDV is a defective satellite of hepatitis B virus, and is dependent upon HBV helper function that provides the HBV envelope surface antigen as a component of the HDV particle. HDV genome replication takes place in the nucleus, and is dependent upon small delta antigen (HDAg-S). Late during replication large delta antigen (HDAg-L) is made that is required for HDV virion production. A–I editing of HDV RNA at the amber/W site, which converts an UAG termination codon into an UGG tryptophan (W) codon allows for the production of HDAg-L as reviewed by Casey in this volume (Casey 2011). While both ADAR1 and two are able to catalyze the editing of HDV RNA, ADAR1 is believed to be the enzyme primarily responsible for editing the amber/W site. Efficient editing of HDV RNA is restricted to the amber/W site despite the ~70% duplex character of HDV RNA. Under normal physiologic conditions A–I editing at the amber/W site is proviral.

Control of the rate and extent of A–I editing at the amber/W site however is essential for normal HDV replication (Casey 2011). Under conditions of ADAR1 overexpression, high levels of amber/W site editing leads to aberrant HDAg-L expression that aborts replication prematurely, and hence editing is antiviral. Elevated expression of p150 ADAR1 would be expected, for example, under conditions of IFN treatment as a therapy for HBV. IFNα induces ADAR1, increases editing and decreases HDV RNA replication (Patterson and Samuel 1995; Wong et al. 2003; Hartwig et al. 2004, 2006). However, HDV has also been described to impair IFN signaling (Pugnale et al. 2009) which would be anticipated to impair the induction of p150 ADAR1. The balance between pro- and anti-viral effects of ADAR editing of HDV RNA no doubt are determined by the combination of virus genotype, host cell and relative IFN induction and action levels achieved.

4.2 Positive-Stranded RNA Viruses

Positive-stranded RNA viruses include those viruses with positive-sense ssRNA genomes, that is, their genomic RNA is the polarity of mRNA and is decoded directly by translation to protein. Among the positive-stranded viruses is hepatitis C virus (HCV), a member of the *Flaviviridae* family. HCV possesses a ~9.6 kb positive-sense ssRNA genome enclosed within an enveloped virion that encodes a polyprotein precursor of ~3,000 amino acids which is processed by viral and cellular proteases to produce ten mature viral proteins (Knipe et al. 2007; Samuel 2010b).

HCV infection is a major global problem, with estimates of over 100 million individuals persistently infected, and an estimated ~350,000 deaths annually due to cirrhosis and heptacellular carcinoma (Perez et al. 2006). The treatment for chronic hepatitis infection is a combination therapy of pegylated type I IFN-α and ribavirin. Among the IFN inducible genes is ADAR1 p150. The site of HCV replication is the cytoplasm, and ADAR1 p150 localizes in part to the cytoplasm. Retrospective analysis of chronic hepatitis C virus-infected patients for responsiveness to treatment with IFN and ribavirin has revealed that patient genotype, in addition to the HCV virus genotype, viral load, and cirrhosis status are important factors in determining therapy responsiveness (Hwang et al. 2006; Welzel et al. 2009). And, ADAR1 is among the genes identified that associated with the responsiveness trait when DNA polymorphisms of responders and nonresponders of Taiwanese (Hwang et al. 2006) and European (Welzel et al. 2009) origins were analyzed. In addition to the IFN-inducible ADAR1, other IFN system components including the IFN receptor, JAK1 signaling kinase and IFN induced protein 44 were noted.

So far only limited mechanistic studies have been described to assess the potential role that ADAR1 plays directly on HCV replication. In Huh7 cells stably transfected with an HCV replicon, IFNα treatment was reported to inhibit replicon expression in part through the involvement of ADAR1 (Taylor et al. 2005). Adenovirus VAI RNA, identified initially as an IFN system antagonist of PKR (Kitajewski et al. 1986; Samuel 2001), also impairs ADAR1 activity in vitro (Lei et al. 1998). Inhibition of both PKR and ADAR1 by VAI RNA stimulated HCV replicon expression and decreased the amount of I-containing RNA found in replicon cells (Taylor et al. 2005). Consistent with the notion that ADAR1 was targeted by VAI RNA to inhibit HCV replicon expression, siRNA knockdown of ADAR1 was observed to stimulate replicon expression. While the HCV replicon system has provided a valuable approach to analyze HCV RNA replication (Appel et al. 2006), the availability of the JFH infectious virus and hepatoma cell culture system now makes possible mechanistic analyses of innate immune responses in HCV virus-infected cells (Lemon 2010). The infectious HCV virus cell culture system provides an approach to further assess the role of ADAR1 in the antiviral actions of IFN. Potential targets of ADAR1 relevant to the IFN response against HCV include microRNAs as well as the RNA-dependent protein kinase PKR. IFN modulation of cellular microRNAs has been reported as a component of the HCV antiviral response (Pedersen et al. 2007). HCV is known to use the abundant liver-specific miR122 to enhance replication (Skalsky and Cullen 2010; You et al. 2011), and ADAR is known to affect both the production and targeting of some cellular micro RNAs (Wulff and Nishikura 2011). Activation of the IFN-inducible PKR by HCV impairs cap-dependent translation and production of IFN response proteins (Garaigorta and Chisari 2009) as well as the production of IFN (Arnaud et al. 2010), and ADAR is known to function as a suppressor of PKR activation (Nie et al. 2007; Toth et al. 2009; Li et al. 2010).

4.3 Double-Stranded RNA Viruses

The only viruses with dsRNA genomes that are known to infect mammals are some members of the *Reoviridae* including the orthoreoviruses. Orthoreoviruses, commonly called reoviruses as they are the founding members of the family, are naked (non-enveloped) virions that possess a segmented dsRNA genome consisting of ten segments of fully complementary dsRNA that fall into three size classes; the total size of the 3 L, 3 M and 4 S-sized segments is ~23.5 kbp of dsRNA (Gomatos and Tamm 1963; Joklik 1981; Knipe et al. 2007). Replication of reoviruses occurs within the cytoplasm of the infected host. RNA transcription is by a virion core-associated dsRNA-dependent ssRNA polymerase that utilizes the dsRNA segments as templates for production of plus-sense transcripts that then serve as mRNA and also as templates for synthesis of progeny dsRNA during subviral particle morphogenesis. Despite over 20 kbp of naturally occurring dsRNA, relatively little is known regarding the roles, if any, that ADARs may play in reovirus replication and pathogenesis. What is clear is that reovirus multiplication occurs to high yields in many lines of cultured mammalian cells that possess both ADARs and also functional RNA interference machinery, both of which act on dsRNA. These observations suggest that the reovirus dsRNA is shielded from the host cell's defense machinery. Indeed, no free reovirus dsRNA is normally found in infected cells; as far as is known, reovirus dsRNA is present only enclosed in viral particles (Joklik 1981; Knipe et al. 2007). The yield of infectious Dearing strain reovirus in MEF cells genetically deficient for the cytoplasmic deaminase form, ADAR1 p150, is comparable to that seen in wild-type MEFs in the absence of IFN treatment; furthermore, type I IFN treatment reduces the yield of reovirus comparably in the *Adar1* p150$^{+/+}$ WT and p150$^{-/-}$ mutant cells (Ward et al. 2011).

4.4 Double-Stranded DNA Viruses

Double-stranded DNA viruses include viruses of the *Polyomaviridae* and *Herpesviridae* families. Their genomes are circular dsDNA of ~5 kbp in the case of polyoma virus and linear dsDNA of ~125–250 kbp dependent upon the specific herpesvirus. Transcription of viral genes in both cases occurs in the nucleus by the cellular RNA polymerase II machinery. ADAR-mediated RNA editing has been described for viral RNA transcripts of mouse polyoma virus (Liu et al. 1994; Kumar and Carmichael 1997; Gu et al. 2009) and two herpesviruses, Kaposi's sarcoma-associated herpesvirus (Gandy et al. 2007) and Epstein–Barr virus (Iizasa et al. 2010).

Polyomaviridae: Mouse polyoma virus, mPyV, is a small DNA virus in which the naked virion capsid encloses a single ~5 kbp molecule of circular dsDNA complexed with cellular histones to form a minichromosome. In mouse cells permissive for productive infection, the early and late regions of the viral genome are expressed bidirectionally, with early and late mRNAs transcribed from opposite strands of the DNA genome. Spliced early mRNAs encode three regulatory proteins that include large T antigen important for DNA replication. Spliced late mRNAs are expressed efficiently only after DNA replication and encode three capsid proteins VP1, VP2 and VP3 (Benjamin 2001; Knipe et al. 2007). Evidence has been presented that early RNA gene expression is regulated by sense-antisense interactions that result in extensive A–I editing of the early strand RNA transcripts at late times after infection (Liu and Carmichael 1993; Liu et al. 1994; Kumar and Carmichael 1997). Formation of dsRNA occurs in the regions of sense-antisense overlap of the early and late viral transcripts that includes overlap of the polyadenylation signals (Gu et al. 2009). Extensive A–G (I) sequence changes are seen in mPyV early strand RNAs present in the nucleus at late times after infection (Kumar and Carmichael 1997), consistent with the hyperediting action of an ADAR. Given the nuclear localization of mPyV transcription, either ADAR1 or ADAR2 both of which are nuclear enzymes could presumably be responsible for the biased hyperediting in the 3′-overlap region of mPyV RNAs. While transient knockdown of ADAR1 in NIH3T3 cells caused a defect in early-to-late switch suggesting mPyV infection is sensitive to ADAR1 protein levels (Gu et al. 2009), the availability of MEFs genetically null in *Adar1*, *Adar1 p150* and *Adar2* should permit the unequivocal identification of which ADAR edits mPyV RNA in the nucleus (Higuchi et al. 2000; Hartner et al. 2004, 2009; Wang et al. 2004; XuFeng et al. 2009; Ward et al. 2011). ADAR-catalyzed A–I editing of cellular mRNAs containing dsRNA structures in their 3′-UTRs, including inverted *Alu* repeats in the case of cellular mRNAs in human cells, may affect localization of the RNAs (Hundley and Bass 2010). However, for mPyV, it is not yet fully clear whether it is the formation of the dsRNA in the 3′-region of overlapping polyadenylation signals or the A–I editing *per se* of the dsRNA that is the critical determinant for regulation of mPyV RNA expression.

Herpesviridae: Kaposi's sarcoma-associated herpesvirus, KSHV or human herpesvirus 8 (HHV-8), is associated with Kaposi's tumors seen in immunosuppressed patients including, for example, AIDS patients (Knipe et al. 2007; Mesri et al. 2010). In the case of KSHV, a viral transcript is edited in a manner that affects both a protein coding sequence and a microRNA (Gandy et al. 2007). During lytic infection most KSHV viral genes are transcribed in cascades with temporal regulation, whereas during latency only a few viral genes are expressed and among the most abundant is the K12 kaposin transcript. The K12 transcript encodes three kaposin proteins (A, B, C) and a microRNA (miR-K10) and has oncogenic potential (Damania 2004). A–I editing occurs at genome position117990 in the K12 transcript, and in the kaposin A ORF, changes serine at position 38 to glycine. The nt substitution also changes position 2 at the 5′end of miR-K10, potentially altering targeting. Editing levels are increased nearly 10-fold following treatment with phorbol ester or sodium butyrate to activate lytic virus replication. Transcripts containing an A at 117990 are tumorigenic, while those with a G corresponding to edited RNAs with I are not tumorigenic as measured by focus formation in Rat3 cells and tumor production in nude mice. ADAR1, at least the p110 form expressed using baculovirus and purified from Sf9 insect cells, efficiently edits the K12 transcript (Gandy et al. 2007).

Epstein–Barr virus, EBV or human herpes virus 4 (HHV-4), is a lymphotropic herpesvirus that can infect and transform a range of human B cells and is associated with latent infections and diseases including infectious mononucleosis and Burkitt's lymphoma (Knipe et al. 2007). EBV encodes more than 20 microRNAs and among them is the BART6 miRNA (Pfeffer et al. 2004; Skalsky and Cullen 2010). In the case of EBV, four viral miRNAs including BART6 miRNA primary transcripts are edited in latently EBV-infected cells (Iizasa et al. 2010). Primary miRNA transcripts are processed by the Drosha and Dicer endonucleases that act together with dsRNA binding proteins to generate the mature 20–22-nt miRNAs that function in silencing of gene expression (Filipowicz et al. 2008; Skalsky and Cullen 2010). The BART6 viral miRNA targets the Dicer nuclease at multiple sites in the 3′-UTR of Dicer mRNA. A–I editing of BART6 dramatically reduces loading of miRs onto the RISC silencing complex and inhibits silencing activity. The editing analysis of EBV primary miRNAs in EBV-infected human lymphoblastoid, Daudi Burkitt lymphoma and nasopharyngeal carcinoma cell lines suggest that EBV miR-BART6 RNAs play important roles in the regulation of viral replication and latency, and that A–I editing of BART6 may be an adaptive selection to counteract the targeting of Dicer by miR-BART6 (Iizasa et al. 2010).

Poxviridae and Adenoviridae: Finally, gene products of two viruses with linear dsDNA genomes, vaccinia virus of the *Poxviridae* that multiplies in the cytoplasm and adenovirus 5 of the *Adenoviridae* that multiplies in the nucleus, have been demonstrated to antagonize ADAR1 enzymatic activity. The vaccinia virus E3L protein and the adenovirus VAI RNA inhibit A–I editing activity of ADAR1 (Lei et al. 1998; Liu et al. 2001). However, it is not known whether vaccinia virus or adenovirus viral RNA is edited by an ADAR. Interestingly, the Z-DNA binding domain present in the N-terminal region of ADAR1 p150 was originally described as a poxvirus E3L homology domain (Patterson and Samuel 1995). The Z-DNA binding domain present in the N-terminal region of vaccinia virus E3L plays a role in viral pathogenesis; mutations that decrease Z-DNA binding correlate with decreased viral pathogenicity in the mouse model (Kim et al. 2003). The E3L protein also binds dsRNA and mediates IFN resistance, promotes vaccinia virus growth and impairs virus-mediated apoptosis. Loss of PKR expression in HeLa cells complements the vaccinia virus E3L deletion mutant phenotype by restoration of viral protein synthesis and largely abolishes virus-induced apoptosis (Zhang et al. 2008). ADAR1 suppresses activation of PKR (Nie et al. 2007; Toth et al. 2009) as earlier discussed. Adenovirus VAI RNA, in addition to antagonizing PKR activation (Kitajewski et al. 1986), also antagonizes the activity of ADAR1 (Lei et al. 1998; Taylor et al. 2005).

4.5 Retroviruses

Retroviridae: Retroviruses are enveloped viruses that possess a positive-stranded RNA genome of about 7–10 kbp dependent upon the specific retrovirus. The single-stranded positive-sense RNA genome is converted by the process of reverse transcription into a dsDNA provirus that is subsequently integrated into the host cell's genome. Retroviral RNAs then are produced by the host cell RNA polymerase II from the integrated provirus dsDNA template and processed by the host cell splicing, capping and polyadenylation machineries (Knipe et al. 2007). ADARs and A–I editing have been reported to affect retrovirus-host cell interactions, most recently in studies with human immunodeficiency virus (HIV), a member of the lentivirus genus of the *Retroviridae* family. ADARs are reported to generally display an HIV proviral effect (Phuphuakrat et al. 2008; Doria et al. 2009, 2011; Clerzius et al. 2009). Sequence changes consistent with the catalytic activity of an ADAR, A–I (G) mutations, are described not only for HIV but were first reported for two avian retroviruses (Felder et al. 1994; Hajjar and Linial 1995).

Activation of CD4^{+} T-lymphocytes leads to increased expression of ADAR1 but not of ADAR2 (Phuphuakrat et al. 2008). ADAR1 overexpression was shown to increase HIV production as measured by p24 Gag protein expression in a manner that required ADAR catalytic activity, whereas the silencing of ADAR1 inhibited HIV production (Phuphuakrat et al. 2008). When two regions of HIV RNA with dsRNA character were examined, the TAR and RRE *cis*-regulatory elements, sequences around TAR did not show any repetitive editing, whereas 8 of 30 clones showed A–G mutations in a 3-nt site in the RRE region (Phuphuakrat et al. 2008). An earlier study had shown that the TAR RNA stem-loop structure of HIV was an editing substrate for ADARs in a different assay, microinjected *Xenopus oocytes* (Sharmeen et al. 1991). Extending the observations of Phuphuakrat et al. (2008), ADAR1 expression was also shown to not only increase during HIV replication in lymphocytes, but ADAR1 was found to interact with and inhibit the RNA-dependent protein kinase PKR, and to reverse the PKR-mediated inhibition of HIV LTR expression and viral production (Clerzius et al. 2009). The proviral effect of ADAR1 in these transfection analyses, however, did not require the catalytic domain of the deaminase, but did require the Z-DNA and dsRNA nucleic acid binding domains of ADAR1, suggesting that under this experimental design, either nucleic acid sequestration or protein–protein interaction or both activities of ADAR1 were sufficient to mediate the proviral effects and that deamination of A–I was not necessary (Clerzius et al. 2009).

An independent study found that ADAR1 stimulated HIV replication by both A–I editing-dependent and editing-independent mechanisms (Doria et al. 2009). Overexpression of ADAR1 by transfection was also reported to increase HIV protein production in a manner independent of catalytic activity, but virions produced in the presence of overexpressed catalytically active ADAR1 were released more efficiently and displayed enhanced infectivity in challenge assays.

Sequence analyses revealed editing of HIV RNAs in the 5′-UTR region shared by HIV RNAs as well as the Tat and Rev coding sequences in ADAR1-transfected cells, but in ADAR2-transfected cells only the 5′-UTR sequence changes were observed (Doria et al. 2009, 2011). Analogous to ADAR1, ADAR2 also enhances the release of progeny HIV virions by an editing-dependent mechanism, but unlike ADAR1, ADAR2 did not increase the infectivity of the HIV produced (Doria et al. 2011). The catalytic independent effect of ADAR1 on HIV1 expression seen under transfection conditions (Clerzius et al. 2009; Doria et al. 2009) may indeed relate to the impairment of PKR kinase activation and subsequent phosphorylation of eIF-2α thereby increasing protein production (Nie et al. 2007; Wang and Samuel 2009).

The enhanced release of HIV virions by overexpression of either ADAR1 or ADAR2 is curious (Doria et al. 2009, 2011). Interestingly, ADAR2 deficiency has been shown to impair exocytosis (Yang et al. 2010). Selective knockdown of ADAR2 expression markedly impaired glucose-stimulated insulin secretion in the rat INS-1 cells and primary pancreatic islets and significantly diminished KCl-stimulated protein secretion in rat adrenal pheochromocytoma PC12 cells. Catalytically active ADAR2, but not editing-deficient mutant ADAR2, could rescue the impairment in stimulated secretion from ADAR2 knockdown cells (Yang et al. 2010). An intriguing possibility is that the overexpression of either ADAR1 or ADAR2 enhances exocytotic processes in a manner that facilitates HIV virion assembly and release by budding (Ganser-Pornillos et al. 2008). Potentially also relevant to HIV replication responses, it has been described that inflammation elevates the level of ADAR1 but not ADAR2, and I-containing mRNA, in T-cell lymphocytes activated with TNFα (Yang et al. 2003).

The HIV-1 results taken together are most consistent with a proviral role of ADARs in the interactions of HIV-1 with the host as revealed from studies of cell culture systems including T-lymphocytes (Phuphuakrat et al. 2008; Doria et al. 2009, 2011; Clerzius et al. 2009). The proviral actions of ADARs with the lentivirus HIV-1, both catalytic-dependent and -independent, are in contrast to the antiretroviral actions of members of the APOBEC3 family of DNA mutator editing enzymes. APOBEC3G and some related cytidine deaminases function as host restriction factors and act in an antiviral manner to provide immunity against retroviruses, including HIV-1, as well as to protect the cell from endogenous retroelements (Chiu and Greene 2008; Malim 2009).

5 Conclusions

Virus–host interactions are complex. In some situations, viruses take advantage of ADARs to enhance their replication. That is, the ADAR effect is proviral. This is illustrated by HDV amber/W editing to produce large delta antigen and by measles, vesicular stomatitis and HIV, where ADAR appears to impair activation of antiviral processes including PKR and IRF3 responses. In other situations, the

Table 2 Mechanisms by which ADARs might potentially affect the host response to viral infection

Biological process	Molecular event	Reference[a]
Interferon signaling and action		
PKR	Suppressed PKR activation	Nie et al. (2007), Toth et al. (2009), Wang et al. (2009), Li et al. (2010)
IRF3	Suppressed IRF3 activation	Toth et al. (2009), Vitali and Scadden (2010)
IFN	Suppressed IFN induction	Hartner et al. (2009)
Macromolecular synthesis and degradation		
Pre-mRNA splicing	Altered splicing	Rueter et al. (1999), Raitskin et al. (2001)
Ribosome decoding during translation	Amino acid substitution, termination	Cattaneo and Billeter (1992), Casey (2011)
RNA degradation	Cleavage of I-containing RNA	Scadden (2005), Scadden and O'Connell (2005)
RNA silencing	Altered miR processing or targeting	Iizasa et al. (2010), Wulff and Nishikura (2011)
Adaptive Immunity		
Immune surveillance	Antibody neutralization of viral antigens	Rueda et al. (1994), Martínez and Melero (2002) Zahn et al. (2007)

[a] See text for additional references

action of ADARs appears antiviral. This is illustrated by several myxoviruses with the ADAR-dependent protection against virus-induced CPE, and possibly by HCV and HDV under conditions of ADAR overexpression as would be expected for ADAR1 p150 in IFN-treated cells.

The details of the mechanisms by which the functional effects of ADARs can be both antiviral and proviral, dependent upon the virus-host combination, largely remain to be elucidated. Among the possible mechanisms are effects on the activation of innate immune responses, on macromolecular synthesis and degradation, and on adaptive immune surveillance (Table 2). No doubt the different consequences of ADARs, pro- or anti- viral, in part relate to functionally important differences between viral replication schemes. Functional genomic screens have revealed that many different host cell machineries are utilized by viruses during their replication. Spatio-temporal differences between viruses and how they interact with their hosts during replication, even for two different viruses that may have the capacity to multiply in the same kind of host cell, may lead to different functional outcomes attributed to ADAR actions. It is firmly established that ADARs can act directly, and in a catalytically dependent manner on viral RNA, to alter function and subsequently the biological outcome of an infection. It is also becoming clear that ADARs may act indirectly to impact the infective process. For example, if a consequence of viral infection is the induction of ADAR1, then the induced ADAR p150 protein may act to alter innate immune signaling responses including activation of interferon and dsRNA-mediated cellular responses including PKR and IRF3. Such activations, of PKR and IRF3, would subsequently be expected to have broad effects on the host's transcriptome and proteome.

ADARs were discovered based on their catalytic activity, the ability to deaminate adenosine in duplex RNA structures that affects the subsequent stability of the targeted RNA. Subsequently, examples were provided of highly selective adenosine deamination by ADARs within an open reading frame sequence, which gave rise to protein products with altered function because I is decoded as G instead of A leading to amino acid substitution as illustrated by HDV delta antigen and the glutamate and serotonin receptors. It is now clear, initially from computational approaches and then subsequently biochemical analyses, that a number of non-coding RNAs are also ADAR editing targets, including the expression and function of microRNAs. Finally, results have emerged to suggest that ADARs may function not only by deamination of adensoine in RNA regions of duplex structure, but also in a catalytically independent manner that presumably involves complex formation between ADARs and other proteins as interacting partners or the binding of nucleic acids by ADARs through their Z and dsRBD domains.

The ADAR proteins are fundamentally important determinants that mediate not only gene-selective but also general or non-selective effects in mammalian cells including virus-infected cells. In addition to different mechanisms by which ADARs may act in either a catalytically dependent or independent manner on *qualitatively different targets*, conceivably the *quantitative extent* of an ADAR-mediated action on the host may play a pivotal role in tipping the balance in either a pro- or anti- viral direction. For example, in cultured cells or intact animals, robust action of an ADAR may impair innate antiviral responses by either catalytically destabilizing activator RNAs of PKR, IRF3 or IFN responses or by altering in a potentially catalytic-independent manner protein binding partner or nucleic acid binding interactions. In the whole animal, possibly the immune response to a viral pathogen is modulated in a manner that is antiviral when extensive A–I editing leads to hypermutation and impairment of viral protein production. But possibly, a proviral response emerges when a low level of A–I editing occurs which does not disrupt threshold production of an essential viral protein such as a surface or envelope component, but instead alters the epitope structure in a manner that leads to escape from the host's immune surveillance network. Future studies of the ADARs will likely continue to provide us with many surprises and new insights into biological processes in the context of understanding how they modulate the outcome of virus–host interactions differently for different viruses–host combinations.

Acknowledgments This work was supported by the National Institutes of Health, Research Grants AI-12520 and AI-20611. I would like to thank Dr. Cyril George and Dr. Christian Pfaller for their helpful comments and the present and past members of the Samuel Laboratory together with the many investigators in the ADAR and innate immunity fields whose collective studies made this chapter possible.

References

Amexis G, Rubin S, Chizhikov V, Pelloquin F, Carbone K, Chumakov K (2002) Sequence diversity of Jeryl Lynn strain of mumps virus: quantitative mutant analysis for vaccine quality control. Virology 300:171–179

Appel N, Schaller T, Penin F, Bartenschlager R (2006) From structure to function: new insights into hepatitis C virus RNA replication. J Biol Chem 281:9833–9836

Arnaud N, Dabo S, Maillard P, Budkowska A, Kalliampakou KI, Mavromara P, Garcin D, Hugon J, Gatignol A, Akazawa D, Wakita T, Meurs EF (2010) Hepatitis C virus controls interferon production through PKR activation. PLoS ONE 5:e10575. doi:10.1371

Baczko K, Lampe J, Liebert UG, Brinckmann U, ter Meulen V, Pardowitz I, Budka H, Cosby SL, Isserte S, Rima BK (1993) Clonal expansion of hypermutated measles virus in a SSPE brain. Virology 197:188–195

Barraud P, Allain FH-T (2011) ADAR Proteins: Double-stranded RNA and Z-DNA binding domains. Curr Top Microbiol Immunol 353

Bass BL (2002) RNA editing by adenosine deaminases that act on RNA. Annu Rev Biochem 71:817–846

Bass BL, Weintraub H (1987) A developmentally regulated activity that unwinds RNA duplexes. Cell 48:607–613

Bass BL, Weintraub H (1988) An unwinding activity that covalently modifies its double-stranded RNA substrate. Cell 55:1089–1098

Benjamin TL (2001) Polyoma virus: old findings and new challenges. Virology 289:167–173

Billecocq A, Spiegel M, Vialat P, Kohl A, Weber F, Bouloy M, Haller O (2004) NSs protein of Rift Valley fever virus blocks interferon production by inhibiting host gene transcription. J Virol 78:9798–9806

Boonyaratanakornkit J, Bartlett E, Schomacker H, Surman S, Akira S, Bae YS, Collins P, Murphy B, Schmidt A (2011) The C proteins of human parainfluenza virus type 1 limit double-stranded RNA accumulation that would otherwise trigger activation of MDA5 and protein kinase R. J Virol 85:1495–1506

Borden EC, Sen GC, Uze G, Silverman RH, Ransohoff RM, Foster GR, Stark GR (2007) Interferons at age 50: past, current and future impact on biomedicine. Nat Rev Drug Discov 6:975–990

Brown BA II, Lowenhaupt K, Wilbert CM, Hanlon EB, Rich A (2000) The zalpha domain of the editing enzyme dsRNA adenosine deaminase binds left-handed Z-RNA as well as Z-DNA. Proc Natl Acad Sci USA 97:13532–13536

Carpenter JA, Keegan LP, Wilfert L, O'Connell MA, Jiggins FM (2009) Evidence for ADAR-induced hypermutation of the Drosophila sigma virus (Rhabdoviridae). BMC Genet 10:75. doi:10.1186/1471-2156-10-75

Casey JL (2011) Control of ADAR1 editing of hepatitis delta virus RNAs. Curr Top Microbiol Immunol 353

Cattaneo R, Billeter MA (1992) Mutations and A/I hypermutations in measles virus persistent infections. Curr Top Microbiol Immunol 176:63–74

Cattaneo R, Schmid A, Rebmann G, Baczko K, ter Meulen V, Bellini WJ, Rozenblatt S, Billeter MA (1986) Accumulated measles virus mutations in a case of subacute sclerosing panencephalitis: interrupted matrix protein reading frame and transcription alteration. Virology 154:97–107

Cattaneo R, Schmid A, Eschle D, Baczko L, ter Meulen V, Billeter MA (1988) Biased hypermutation and other genetic changes in defective measles viruses in human brain infections. Cell 55:255–265

Chakrabarti A, Jha BK, Silverman RH (2011) New insights into the role of RNase L in innate immunity. J Interferon Cytokine Res 31:49–57

Chambers P, Rima BK, Duprex WP (2009) Molecular differences between two Jeryl Lynn mumps virus vaccine component strains, JL5 and JL2. J. Gen Virol 90:2973–2981

Chiu YL, Greene WC (2008) The APOBEC3 cytidine deaminases: an innate defensive network opposing exogenous retroviruses and endogenous retroelements. Annu Rev Immunol 26:317–353
Clerzius G, Gélinas JF, Daher A, Bonnet M, Meurs EF, Gatignol A (2009) ADAR1 interacts with PKR during human immunodeficiency virus infection of lymphocytes and contributes to viral replication. J Virol 83:10119–10128
Colby C, Morgan MJ (1971) Interferon induction and action. Annu Rev Microbiol 25:333–360
Damania B (2004) Oncogenic gamma-herpesviruses: comparison of viral proteins involved in tumorigenesis. Nat Rev Microbiol 2:656–668
Desterro JM, Keegan LP, Lafarga M, Berciano MT, O'Connell M, Carmo-Fonseca M (2003) Dynamic association of RNA-editing enzymes with the nucleolus. J Cell Sci 116:1805–1818
Doria M, Neri F, Gallo A, Farace MG, Michienzi A (2009) Editing of HIV-1 RNA by the double-stranded RNA deaminase ADAR1 stimulates viral infection. Nucleic Acids Res 37: 5848–5858
Doria M, Tomaselli S, Neri F, Ciafre SA, Farace MG, Michienzi A, Gallo A (2011) The ADAR2 editing enzyme is a novel Hiv-1 proviral factor. J Gen Virol Feb 2 [Epub ahead of print]
Emmott E, Wise H, Loucaides EM, Matthews DA, Digard P, Hiscox JA (2010) Quantitative proteomics using SILAC coupled to LC-MS/MS reveals changes in the nucleolar proteome in influenza A virus-infected cells. J Proteome Res 9:5335–5345
Felder MP, Laugier D, Yatsula B, Dezélée P, Calothy G, Marx M (1994) Functional and biological properties of an avian variant long terminal repeat containing multiple A to G conversions in the U3 sequence. J Virol 68:4759–4767
Feng S, Li H, Zhao J, Pervushin K, Lowenhaupt K, Schwartz TU, Dröge P (2011) Alternate rRna secondary structures as regulators of translation. Nat Struct Mol Biol 18:169–176
Field AK, Tytell AA, Lampson GP, Hilleman MR (1967) Inducers of interferon and host resistance. II. Multistranded synthetic polynucleotide complexes. Proc Natl Acad Sci USA 58:1004–1010
Fierro-Monti I, Mathews MB (2000) Proteins binding to duplexed RNA: One motif, multiple functions. Trends Biochem Sci 25:241–246
Filipowicz W, Bhattacharyya SN, Sonenberg N (2008) Mechanisms of post-transcriptional regulation by microRNAs: are the answers in sight? Nat Rev Genet 9:102–114
Gandy SZ, Linnstaedt SD, Muralidhar S, Cashman KA, Rosenthal LJ, Casey JL (2007) RNA editing of the human herpesvirus 8 kaposin transcript eliminates its transforming activity and is induced during lytic replication. J Virol 81:13544–13551
Ganser-Pornillos BK, Yeager M, Sundquist WI (2008) The structural biology of HIV assembly. Curr Opin Struct Biol 18:203–217
Garaigorta U, Chisari FV (2009) Hepatitis C virus blocks interferon effector function by inducing protein kinase R phosphorylation. Cell Host Microbe 6:513–522
Garcia-Sastre A (2011) 2 methylate or not 2 methylate: viral evasion of the type I interferon response. Nat Immunol 12:114–115
George CX, Samuel CE (1999a) Human RNA-specific adenosine deaminase ADAR1 transcripts possess alternative exon 1 structures that initiate from different promoters, one constitutively active and the other interferon inducible. Proc Natl Acad Sci USA 96:4621–4626
George CX, Samuel CE (1999b) Characterization of the 5′-flanking region of the human RNA-specific adenosine deaminase *ADAR1* gene and identification of an interferon-inducible *ADAR1* promoter. Gene 229:203–213
George CX, Wagner MV, Samuel CE (2005) Expression of interferon-inducible RNA adenosine deaminase ADAR1 during pathogen infection and mouse embryo development involves tissue-selective promoter utilization and alternative splicing. J Biol Chem 280: 15020–15028
George CX, Das S, Samuel CE (2008) Organization of the mouse RNA-specific adenosine deaminase Adar1 gene 5′-region and demonstration of STAT1-independent, STAT2-dependent transcriptional activation by interferon. Virology 380:338–343
George CX, Gan Z, Liu Y, Samuel CE (2011) Adenosine deaminases acting on RNA (ADARs), RNA editing and interferon action. J Interferon Cytokine Res 31:99–117

Gomatos PJ, Tamm I (1963) The secondary structure of reovirus RNA. Proc Natl Acad Sci USA 49:707–714

Goodman RA, Macbeth MR, Beal PA (2011) ADAR Proteins: Structure and Catalytic Activity. Curr Top Microbiol Immunol 353

Grande-Pérez A, Sierra S, Castro MG, Domingo E, Lowenstein PR (2002) Molecular indetermination in the transition to error catastrophe: systematic elimination of lymphocytic choriomeningitis virus through mutagenesis does not correlate linearly with large increases in mutant spectrum complexity. Proc Natl Acad Sci USA 99:12938–12943

Gu R, Zhang Z, Decerbo JN, Carmichael GG (2009) Gene regulation by sense-antisense overlap of polyadenylation signals. RNA 15:1154–1163

Habjan M, Pichlmair A, Elliott RM, Overby AK, Glatter T, Gstaiger M, Superti-Furga G, Unge H, Weber F (2009) NSs protein of rift valley fever virus induces the specific degradation of the double-stranded RNA-dependent protein kinase. J Virol 83:4365–4375

Hajjar AM, Linial ML (1995) Modification of retroviral RNA by double-stranded RNA adenosine deaminase. J Virol 69:5878–5882

Hall WW, Lamb RA, Choppin PW (1979) Measles and subacute sclerosing panencephalitis virus proteins: lack of antibodies to the M protein in patients with subacute sclerosing panencephalitis. Proc Natl Acad Sci USA 76:2047–2051

Hartner JC, Walkley CR (2011) Roles of ADARs in mouse development. Curr Top Microbiol Immunol 353

Hartner JC, Schmittwolf C, Kispert A, Muller AM, Higuchi M, Seeburg PH (2004) Liver disintegration in the mouse embryo caused by deficiency in the RNA-editing enzyme ADAR1. J Biol Chem 279:4894–4902

Hartner JC, Walkley CR, Lu J, Orkin SH (2009) ADAR1 is essential for the maintenance of hematopoiesis and suppression of interferon signaling. Nat Immunol 10:109–115

Hartwig D, Schoeneich L, Greeve J, Schütte C, Dorn I, Kirchner H, Hennig H (2004) Interferon-alpha stimulation of liver cells enhances hepatitis delta virus RNA editing in early infection. J Hepatol 41:667–672

Hartwig D, Schütte C, Warnecke J, Dorn I, Hennig H, Kirchner H, Schlenke P (2006) The large form of ADAR 1 is responsible for enhanced hepatitis delta virus RNA editing in interferon-alpha-stimulated host cells. J Viral Hepat 13:150–157

Herbert A, Alfken J, Kim YG, Mian IS, Nishikura K, Rich A (1997) A Z-DNA binding domain present in the human editing enzyme, double-stranded RNA adenosine deaminase. Proc Natl Acad Sci USA 94:8421–8426

Higuchi M, Stefan M, Single FN, Hartner J, Rozov A, Burnashev N, Feldmeyer D, Sprengel R, Seeburg PH (2000) Point mutation in an ampa receptor gene rescues lethality in mice deficient in the RNA-editing enzyme ADAR2. Nature 406:78–81

Hundley HA, Bass BL (2010) ADAR editing in double-stranded UTRs and other noncoding RNA sequences. Trends Biochem Sci 35:377–383

Hwang Y, Chen EY, Gu ZJ, Chuang WL, Yu ML, Lai MY, Chao YC, Lee CM, Wang JH, Dai CY, Shian-Jy Bey M, Liao YT, Chen PJ, Chen DS (2006) Genetic predisposition of responsiveness to therapy for chronic hepatitis C. Pharmacogenomics 7:697–709

Iizasa H, Wulff BE, Alla NR, Maragkakis M, Megraw M, Hatzigeorgiou A, Iwakiri D, Takada K, Wiedmer A, Showe L, Lieberman P, Nishikura K (2010) Editing of Epstein-Barr virus-encoded BART6 microRNAs controls their dicer targeting and consequently affects viral latency. J Biol Chem 285:33358–33370

Isaacs A, Lindenmann J (1957) Virus interference. I. The interferon. Proc R Soc Lond B Biol Sci 147:258–267

Joklik WK (1981) Structure and function of the reovirus genome. Microbiol Rev 45:483–501

Kawai T, Akira S (2010) The role of pattern-recognition receptors in innate immunity: update on Toll-like receptors. Nat Immunol 11:373–384

Kawakubo K, Samuel CE (2000) Human RNA-specific adenosine deaminase (*ADAR1*) gene specifies transcripts that initiate from a constitutively active alternative promoter. Gene 258:165–172

Kim U, Wang Y, Sanford T, Zeng Y, Nishikura K (1994) Molecular cloning of c DNA for double-stranded-RNA adenosine-deaminase, a candidate enzyme for nuclear-RNA editing. Proc Natl Acad Sci USA 91:11457–11461

Kim YG, Muralinath M, Brandt T, Pearcy M, Hauns K, Lowenhaupt K, Jacobs BL, Rich A (2003) A role for Z-DNA binding in vaccinia virus pathogenesis. Proc Natl Acad Sci USA 100:6974–6979

Kitajewski J, Schneider RJ, Safer B, Munemitsu SM, Samuel CE, Thimmappaya B, Shenk T (1986) Adenovirus VAI RNA antagonizes the antiviral action of interferon by preventing activation of the interferon-induced eIF-2 alpha kinase. Cell 45:195–200

Knipe D, Howley PM, DE Griffin, Lamb RA, Martin MA, Roizman B, Straus SE (2007) Fields Virology, 5th Ed edn. Lippincott Williams and Wilkins, Philadelphia

Kuhen KL, Samuel CE (1997) Isolation of the interferon-inducible RNA-dependent protein kinase PKR promoter and identification of a novel DNA element within the 5′-flanking region of human and mouse *Pkr* genes. Virology 227:119–130

Kumar M, Carmichael GG (1997) Nuclear antisense RNA induces extensive adenosine modifications and nuclear retention of target transcripts. Proc Natl Acad Sci USA 94:3542–3547

Kumar H, Kawai T, Akira S (2011) Pathogen recognition by the innate immune system. Int Rev Immunol 30:16–34

Lehmann KA, Bass BL (2000) Double-stranded RNA adenosine deaminases ADAR1 and ADAR2 have overlapping specificities. Biochemistry 39:12875–12884

Lei M, Liu Y, Samuel CE (1998) Adenovirus VAI RNA antagonizes the RNA-editing activity of the ADAR adenosine deaminase. Virology 245:188–196

Lemon SM (2010) Induction and evasion of innate antiviral responses by hepatitis C virus. J Biol Chem 285:22741–22747

Li Z, Wolff KC, Samuel CE (2010) RNA adenosine deaminase ADAR1 deficiency leads to increased activation of protein kinase PKR and reduced vesicular stomatitis virus growth following interferon treatment. Virology 396:316–322

Liebert UG, Baczko K, Budka H, ter Meulen V (1986) Restricted expression of measles virus proteins in brains from cases of subacute sclerosing panencephalitis. J Genl Virol 67:2435–2444

Liu Z, Carmichael GG (1993) Polyoma virus early-late switch: regulation of late RNA accumulation by DNA replication. Proc Natl Acad Sci USA 90:8494–8498

Liu Y, Samuel CE (1996) Mechanism of interferon action: Functionally distinct RNA-binding and catalytic domains in the interferon-inducible, double-stranded RNA- specific adenosine deaminase. J Virol 70:1961–1968

Liu Z, Batt DB, Carmichael GG (1994) Targeted nuclear antisense RNA mimics natural antisense-induced degradation of polyoma virus early RNA. Proc Natl Acad Sci USA 91:4258–4262

Liu Y, George CX, Patterson JB, Samuel CE (1997) Functionally distinct double-stranded RNA-binding domains associated with alternative splice site variants of the interferon-inducible double-stranded RNA-specific adenosine deaminase. J Biol Chem 272:419–4428

Liu Y, Wolff KC, Jacobs BL, Samuel CE (2001) Vaccinia virus E3L interferon resistance protein inhibits the interferon-induced adenosine deaminase A-to-I editing activity. Virology 289:378–387

Maas S, Gommans WM (2009) Novel exon of mammalian ADAR2 extends open reading frame. PLoS One 4:e4225

Maas S, Rich A, Nishikura K (2003) A-to-I RNA editing: Recent news and residual mysteries. J Biol Chem 278:1391–1394

Malim MH (2009) APOBEC proteins and intrinsic resistance to HIV-1 infection. Philos Trans R Soc Lond B Biol Sci 364:675–687

Martínez I, Melero JA (2002) A model for the generation of multiple A to G transitions in the human respiratory syncytial virus genome: predicted RNA secondary structures as substrates for adenosine deaminases that act on RNA. J Gen Virol 83(Pt 6):1445–1455

McAllister CS, Toth AM, Zhang P, Devaux P, Cattaneo R, Samuel CE (2010) Mechanisms of protein kinase PKR-mediated amplification of beta interferon induction by C protein-deficient measles virus. J Virol 84:380–386

McCormack SJ, Samuel CE (1995) Mechanism of interferon action: RNA-binding activity of full-length and R-domain forms of the RNA-dependent protein kinase PKR–determination of KD values for VAI and TAR RNAs. Virology 206:511–519

McCormack SJ, Thomis DC, Samuel CE (1992) Mechanism of interferon action: identification of a RNA binding domain within the N-terminal region of the human RNA-dependent P1/eIF-2 alpha protein kinase. Virology 188:47–56

Mesri EA, Cesarman E, Boshoff C (2010) Kaposi's sarcoma and its associated herpesvirus. Nat Rev Cancer 10:707–719

Mittaz L, Scott HS, Rossier C, Seeburg PH, Higuchi M, Antonarakis SE (1997) Cloning of a human RNA editing deaminase (ADARB1) of glutamate receptors that maps to chromosome 21q22.3. Genomics 41:210–217

Moss WJ, Griffin DE (2006) Global measles elimination. Nat Rev Microbiol 4:900–908

Murphy DG, Dimock K, Kang CY (1991) Numerous transitions in human parainfluenza virus 3 RNA recovered from persistently infected cells. Virology 181:760–763

Nakhaei P, Genin P, Civas A, Hiscott J (2009) Rig-I-like receptors: sensing and responding to RNA virus infection. Semin Immunol 21:215–222

Nie Y, Hammond GL, Yang JH (2007) Double-stranded RNA deaminase ADAR1 increases host susceptibility to virus infection. J Virol 81:917–923

Nishikura K (2010) Functions and regulation of RNA editing by ADAR deaminases. Annu Rev Biochem 79:321–349

O'Connell MA, Krause S, Higuchi M, Hsuan JJ, Totty NF, Jenny A, Keller W (1995) Cloning of cDNAs encoding mammalian double-stranded RNA-specific adenosine deaminase. Mol Cell Biol 15:1389–1397

O'Neill LA (2009) DNA makes RNA makes innate immunity. Cell 138:428–430

O'Hara P, Nichol S, Horodyski F, Holland J (1984) Vesicular stomatitis virus defective interfering particles can contain extensive genomic sequence rearrangements and base substitutions. Cell 36:915–924

Oldstone MB (2009) Modeling subacute sclerosing panencephalitis in a transgenic mouse system: uncoding pathogenesis of disease and illuminating components of immune control. Curr Top Microbiol Immunol 330:31–54

Oldstone MB, Lewicki H, Thomas D, Tishon A, Dales S, Patterson J, Manchester M, Homann D, Naniche D, Holz A (1999) Measles virus infection in a transgenic model: virus-induced immunosuppression and central nervous system disease. Cell 98:629–640

Paro S, Li X, O'Connell MA, Keegan LP (2011) Regulation and Functions of ADAR in *Drosophila.* Curr Top Microbiol Immunol 353

Patterson JB, Samuel CE (1995) Expression and regulation by interferon of a double-stranded-RNA-specific adenosine deaminase from human cells: Evidence for two forms of the deaminase. Mol Cell Biol 15:5376–5388

Patterson JB, Thomis DC, Hans SL, Samuel CE (1995) Mechanism of interferon action - double-stranded RNA-specific adenosine-deaminase from human-cells is inducible by alpha-interferon and gamma-interferon. Virology 210:508–511

Patterson JB, Cornu TI, Redwine J, Dales S, Lewicki H, Holz A, Thomas D, Billeter MA, Oldstone MB (2001) Evidence that the hypermutated M protein of a subacute sclerosing panencephalitis measles virus actively contributes to the chronic progressive CNS disease. Virology 291:215–225

Pedersen IM, Cheng G, Wieland S, Volinia S, Croce CM, Chisari FV, David M (2007) Interferon modulation of cellular microRNAs as an antiviral mechanism. Nature 449:919–822

Perez JF, Armstrong GL, Farrington LA, Hutin YJ, Bell BP (2006) The contributions of hepatitis B virus and hepatitis C virus infections to cirrhosis and primary liver cancer worldwide. J Hepatol 45:529–538

Pfeffer S, Zavolan M, Grässer FA, Chien M, Russo JJ, Ju J, John B, Enright AJ, Marks D, Sander C, Tuschl T (2004) Identification of virus-encoded microRNAs. Science 304:734–736
Phuphuakrat A, Kraiwong R, Boonarkart C, Lauhakirti D, Lee TH, Auewarakul P (2008) Double-stranded RNA adenosine deaminases enhance expression of human immunodeficiency virus type 1 proteins. J Virol 82:10864–10872
Pindel A, Sadler A (2011) The role of protein kinase R in the interferon response. J Interferon Cytokine Res 31:59–70
Placido D, Brown BA, Lowenhaupt K, Rich A, Athanasiadis A (2007) A left-handed RNA double helix bound by the Z alpha domain of the RNA-editing enzyme ADAR1. Struct 15:395–404
Poulsen H, Nilsson J, Damgaard CK, Egebjerg J, Kjems J (2001) Crm1 mediates the export of ADAR1 through a nuclear export signal within the Z-DNA binding domain. Mol Cell Biol 21:7862–7871
Pugnale P, Pazienza V, Guilloux K, Negro F (2009) Hepatitis delta virus inhibits alpha interferon signaling. Hepatol 49:398–406
Raitskin O, Cho DS, Sperling J, Nishikura K, Sperline R (2001) RNA editing activity is associated with spliceing factors in InRNP particles: The nuclear pre-mRNA processing machinery. Proc Natl Acad Sci USA 98:6571–6576
Randall RE, Goodbourn S (2008) Interferons and viruses: an interplay between induction, signalling, antiviral responses and virus countermeasures. J Gen Virol 89:1–47
Rebagliati MR, Melton DA (1987) Antisense RNA injections in fertilized frog eggs reveal an RNA duplex unwinding activity. Cell 48:599–605
Rueda P, García-Barreno B, Melero JA (1994) Loss of conserved cysteine residues in the attachment (G) glycoprotein of two human respiratory syncytial virus escape mutants that contain multiple A–G substitutions (hypermutations). Virology 198:653–662
Rueter SM, Dawson RT, Emeson RB (1999) Regulation of alternative splicing by RNA editing. Nature 399:75–80
Sadler AJ, Williams BR (2007) Structure and function of the protein kinase R. Curr Top Microbiol Immunol 316:253–292
Samuel CE (1979) Mechanism of interferon action: Phosphorylation of protein synthesis initiation factor eIF-2 in interferon-treated human cells by a ribosome-associated kinase processing site specificity similar to hemin-regulated rabbit reticulocyte kinase. Proc Natl Acad Sci USA 76:600–604
Samuel CE (2001) Antiviral actions of interferons. Clin Microbiol Rev 14:778–809
Samuel CE (2007) Innate immunity minireview series: making biochemical sense of nucleic acid sensors that trigger antiviral innate immunity. J Biol Chem 282:15313–15314
Samuel CE (2010a) Thematic minireview series: toward a structural basis for understanding influenze virus-host cell interactions. J Biol Chem 285:28399–28401
Samuel CE (2010b) Thematic minireview series: elucidating hepatitis C virus-host interactions at the biochemical level. J Biol Chem 285:22723–22724
Samuel CE (2011) Adenosine deaminases acting on RNA (ADARs) are both antiviral and proviral. Virology 411:180–193
Sansam CL, Wells KS, Emeson RB (2003) Modulation of RNA editing by functional nucleolar sequestration of ADAR2. Proc Natl Acad Sci USA 100:14018–14023
Sarkis PT, Ying S, Xu R, Yu XF (2006) STAT1-independent call type-specific regulation of antiviral APOBEC3G by IFN-alpha. J Immunol 177:4530–4540
Scadden AD (2005) The RISC subunit Tudor-SN binds to hyper-edited double-stranded RNA and promotes its cleavage. Nat Struct Mol Biol 12:489–496
Scadden AD, O'Connell MA (2005) Cleavage of dsRNAs hyper-edited by ADARs occurs at preferred editing sites. Nucleic Acids Res 33:5954–5964
Schindler C, Levy DE, Decker T (2007) JAK-STAT signaling: from interferons to cytokines. J Biol Chem 282:20059–20063
Schmid A, Spielhofer P, Cattaneo R, Baczko K, ter Meulen V, Billeter MA (1992) Subacute sclerosing panencephalitis is typically characterized by alterations in the fusion protein cytoplasmic domain of the persisting measles virus. Virology 188:910–915

Schwartz T, Rould MA, Lowenhaupt K, Herbert A, Rich A (1999) Crystal structure of the Z domain of the human editing enzyme ADAR1 bound to left-handed Z-DNA. Science 284:1841–1845

Sen A, Pruijssers AJ, Dermody TS, Garcia-Sastre A, Greenberg HB (2011) The early interferon response to rotavirus is regulated by PKR and depends on MAVS/IPS-1, RIG-I, MDA-5, and IRF3. J Virol 85:3717–3732

Serra MJ, Smolter PE, Westhof E (2004) Pronounced instability of tandem IU base pairs in RNA. Nucleic Acids Res 32:1824–1828

Sharmeen L, Bass B, Sonenberg N, Weintraub H, Groudine M (1991) Tat-dependent adenosine-to-inosine modification of wild-type transactivation response RNA. Proc Natl Acad Sci USA 88:8096–8100

Shtrichman R, Heithoff DM, Mahan MJ, Samuel CE (2002) Tissue selectivity of interferon-stimulated gene expression in mice infected with dam(+) versus dam(-) *Salmonella enterica* Serovar typhimurium strains. Infect Immun 70:5579–5588

Singh M, Kesterson RA, Jacobs MM, Joers JM, Gore JC, Emeson RB (2007) Hyperphagia-mediated obesity in transgenic mice misexpressing the RNA-editing enzyme ADAR2. J Biol Chem 282:22448–22459

Skalsky RL, Cullen BR (2010) Viruses, microRNAs, and host interactions. Annu Rev Microbiol 64:123–141

Slavov D, Gardiner K (2002) Phylogenetic comparison of the pre-mRNA adenosine deaminase ADAR2 genes and transcripts: Conservation and diversity in editing site sequence and alternative splicing patterns. Gene 299:83–94

Stewart WE (1979) The interferon system. Springer-Verlag. New York

Strehblow A, Hallegger M, Jantsch MF (2002) Nucleocytoplasmic distribution of human RNA-editing enzyme ADAR1 is modulated by double-stranded RNA-binding domains, a leucine-rich export signal, and a putative dimerization domain. Mol Biol Cell 13:3822–3835

Strobel SA, Cech TR, Usman N, Beigelman L (1994) The 2, 6-diaminopurine riboside.5-methylisocytidine wobble base pair: an isoenergetic substitution for the study of G.U pairs in RNA. Biochemistry 33:13824–13835

Suspène R, Renard M, Henry M, Guétard D, Puyraimond-Zemmour D, Billecocq A, Bouloy M, Tangy F, Vartanian JP, Wain-Hobson S (2008) Inversing the natural hydrogen bonding rule to selectively amplify GC-rich ADAR-edited RNAs. Nucleic Acids Res 36(12):e72. doi: 10.1093/nar/gkn295

Suspène R, Petit V, Puyraimond-Zemmour D, Aynaud MM, Henry M, Guétard D, Rusniok C, Wain-Hobson S, Vartanian JP (2011) Double-stranded RNA adenosine deaminase ADAR-1-induced hypermutated genomes among inactivated seasonal influenza and live attenuated measles virus vaccines. J Virol 85:2458–2462

Tanaka H, Samuel CE (1994) Mechanism of interferon action: structure of the mouse PKR gene encoding the interferon-inducible RNA-dependent protein kinase. Proc Natl Acad Sci USA 91:7995–7999

Taylor JM (2003) Replication of human hepatitis delta virus: Recent developments. Trends Microbiol 11:185–190

Taylor DR, Puig M, Darnell MER, Mihalik K, Feinstone SM (2005) New antiviral pathway that mediates hepatitis C virus replicon interferon sensitivity through ADAR1. J Virol 79:6291–6298

tenOever BR, Ng SL, Chua MA, McWhirter SM, García-Sastre A, Maniatis T (2007) Multiple functions of the IKK-related kinase IKKepsilon in interferon-mediated antiviral immunity. Science 315:1274–1278

Toth AM, Zhang P, Das S, George CX, Samuel CE (2006) Interferon action and the double-stranded RNA-dependent enzymes ADAR1 adenosine deaminase and PKR protein kinase. Prog Nucleic Acid Res Mol Biol 81:369–434

Toth AM, Li Z, Cattaneo R, Samuel CE (2009) RNA-specific adenosine deaminase ADAR1 suppresses measles virus-induced apoptosis and activation of protein kinase PKR. J Biol Chem 284:29350–29356

Tytell AA, Lampson GP, Field AK, Hilleman MR (1967) Inducers of interferon and host resistance. 3. Double-stranded RNA from reovirus type 3 virions (reo 3-RNA). Proc Natl Acad Sci USA 58:1719–1722

Uematsu S, Akira S (2007) Toll-like receptors and Type I interferons. J Biol Chem 282: 15319–15323

Vitali P, Scadden AD (2010) Double-stranded RNAs containing multiple IU pairs are sufficient to suppress interferon induction and apoptosis. Nat Struct Mol Biol 17:1043–1050

Wagner RW, Smith JE, Cooperman BS, Nishikura K (1989) A double-stranded RNA unwinding activity introduces structural alterations by means of adenosine to inosine conversions in mammalian cells and Xenopus eggs. Proc Natl Acad Sci USA 86:2647–2651

Wang Q, Miyakoda M, Yang W, Khillan J, Stachura DL, Weiss MJ, Nishikura K (2004) Stress-induced apoptosis associated with null mutation of *Adar1* RNA editing deaminase gene. J Biol Chem 279:4952–4961

Wang Y, Samuel CE (2009) Adenosine deaminase ADAR1 increases gene expression at the translational level by decreasing protein kinase PKR-dependent eIF-2alpha phosphorylation. J Mol Biol 393:777–787

Wang Y, Zeng Y, Murray JM, Nishikura K (1995) Genomic organization and chromosomal location of the human dsRNA adenosine deaminase gene: The enzyme for glutamate-activated ion channel RNA editing. J Mol Biol 254:184–195

Ward SV, George CX, Welch JJ, Liou LY, Hahm B, Lewicki H, de la Torre JC, Samuel CE, Oldstone MB (2011) RNA editing enzyme adenosine deaminase is a restriction factor for controlling measles virus replication that also is required for embryogenesis. Proc Natl Acad Sci USA 108:331–336

Weier HUG, George CX, Greulich KM, Samuel CE (1995) The interferon-inducible, double-stranded RNA-specific adenosine deaminase gene (DSRAD) maps to human chromosome 1q21.1–21.2. Genomics 30:372–375

Weier HUG, George CX, Lersch RA, Breitweser S, Cheng JF, Samuel CE (2000) Assignment of the RNA-specific adenosine deaminase gene (*Adar*) to mouse chromosome 3f2 by in situ hybridization. Cytogenet. Cell Genet 89:214–215

Welzel TM, Morgan TR, Bonkovsky HL, Naishadham D, Pfeiffer RM, Wright EC, Hutchinson AA, Crenshaw AT, Bashirova A, Carrington M, Dotrang M, Sterling RK, Lindsay KL, Fontana RJ, Lee WM, Di Bisceglie AM, Ghany MG, Gretch DR, Chanock SJ, Chung RT, O'Brien TR HALT-C, Group Trial (2009) Variants in interferon-alpha pathway genes and response to pegylated interferon-Alpha2a plus ribavirin for treatment of chronic hepatitis C virus infection in the hepatitis C antiviral long-term treatment against cirrhosis trial. Hepatology 49(6):1847–1858

Wong TC, Ayata M, Ueda S, Hirano A (1991) Role of biased hypermutation in evolution of subacute sclerosing panencephalitis virus from progenitor acute measles virus. J Virol 65:2191–2199

Wong TC, Ayata M, Hirano A, Yoshikawa Y, Tsuruoka H, Yamanouchi K (1994) Generalized and localized biased hypermutation affecting the matrix gene of a measles virus strain that causes subacute sclerosing panencephalitis. J Virol 63:5464–5468

Wong SK, Sato S, Lazinski DW (2003) Elevated activity of the large form of ADAR1 in vivo: very efficient RNA editing occurs in the cytoplasm. RNA 9:586–598

Wulff B-E, Nishikura K (2011) Modulation of Micro RNA Expression and Function by ADARs. Curr Top Microbiol Immunol 353

XuFeng R, Boyer MJ, Shen H, Li Y, Yu H, Gao Y, Yang Q, Wang Q, Cheng T (2009) ADAR1 is required for hematopoietic progenitor cell survival via RNA editing. Proc Natl Acad Sci USA 106:17763–17768

Yang JH, Nie Y, Zhao Q, Su Y, Pypaert M, Su H, Rabinovici R (2003) Intracellular localization of differentially regulated RNA-specific adenosine deaminase isoforms in inflammation. J Biol Chem 278:45833–45842

Yang L, Zhao L, Gan Z, He Z, Xu J, Gao X, Wang X, Han W, Chen L, Xu T, Li W, Liu Y (2010) Deficiency in RNA editing enzyme ADAR2 impairs regulated exocytosis. FASEB J. 24:3720–3732

Yoneyama M, Fujita T (2007) Function of RIG-I-like receptors in antiviral innate immunity. J Biol Chem 282:15315–15318

Yoneyama M, Fujita T (2010) Recognition of viral nucleic acids in innate immunity. Rev Med Virol 20:4–22

You S, Murray CL, Luna JM, Rice CM (2011) End game: Getting the most out of microRNAs. Proc Natl Acad Sci USA 108:3101–3102

Zahn RC, Schelp I, Utermöhlen O, von Laer D (2007) A-to-G hypermutation in the genome of lymphocytic choriomeningitis virus. J Virol 81:457–464

Zhang P, Jacobs BL, Samuel CE (2008) Loss of protein kinase PKR expression in human HeLa cells complements the vaccinia virus E3L deletion mutant phenotype by restoration of viral protein synthesis. J Virol 82:840–848

Role of ADARs in Mouse Development

Carl R. Walkley, Brian Liddicoat and Jochen C. Hartner

Abstract RNA editing by deamination of adenosine to inosine (A-to-I editing) is a physiologically important posttranscriptional mechanism that can regulate expression of genes by modifying their transcripts. A-to-I editing is mediated by adenosine deaminases acting on RNA (ADAR) that can catalytically exchange adenosines to inosines, with varying efficiency, depending on the structure of the RNA substrates. Significant progress in understanding the biological function of mammalian ADARs has been made in the past decade by the creation and analysis of gene-targeted mice with disrupted or modified ADAR alleles. These studies have revealed important roles of ADARs in neuronal and hematopoietic tissue during embryonic and postnatal stages of mouse development.

Contents

C. R. Walkley · B. Liddicoat
St. Vincent's Institute of Medical Research and Department of Medicine,
St. Vincent's Hospital, University of Melbourne, Melbourne, VIC 3065, Australia

J. C. Hartner (✉)
TaconicArtemis GmbH, Neurather Ring 1, 51063 Koeln, Germany
e-mail: jochen.hartner@taconicartemis.com

Current Topics in Microbiology and Immunology (2012) 353: 197–220
DOI: 10.1007/82_2011_150

Published Online: 3 July 2011

1 Introduction

It is becoming increasingly appreciated that RNA plays an active role in the regulation of gene expression. Posttranscriptional gene regulation by RNA editing is an evolutionary conserved mechanism involving insertion, deletion, or modification of nucleotides in RNA transcripts. In higher eukaryotes, the most widespread modification is the deamination of adenosine to inosine (A-to-I editing) which is catalyzed by a family of enzymes collectively termed ADAR (adenosine deaminase acting on RNA). ADARs were originally identified through their unwinding activity on double-stranded RNA (dsRNA) substrates (Bass and Weintraub 1987, 1988; Rebagliati and Melton 1987; Wagner et al. 1989). The RNA editing function of ADARs was discovered by serendipity in a nuclear transcript expressed in the brain (Sommer et al. 1991), where A-to-I editing was observed in a protein-coding transcript by comparison of genomic and cDNA sequence. This analysis revealed a guanosine in the cDNA transcript at a position that specified an adenosine at the corresponding position in genomic DNA. This RNA modification mechanism is now known to regulate gene expression at multiple levels and influence the diversity and expression of the proteome both directly and indirectly. A-to-I editing can modulate gene expression in multiple ways, depending on the nature of the RNA substrate and the position of the targeted adenosine within the RNA (Fig. 1).

A-to-I editing in mammals is mediated by a small family of candidate enzymes, termed ADAR1-3 (adenosine deaminase acting on RNA-1-3). Each ADAR has regions for binding double-stranded RNA and a carboxy terminal catalytic domain distantly related to bacterial cytidine deaminase (Melcher et al. 1996b; Mittaz et al. 1997). The catalytic domain is highly conserved between ADARs. ADAR1 and ADAR2 are widely expressed, whereas expression of ADAR3 appears to be

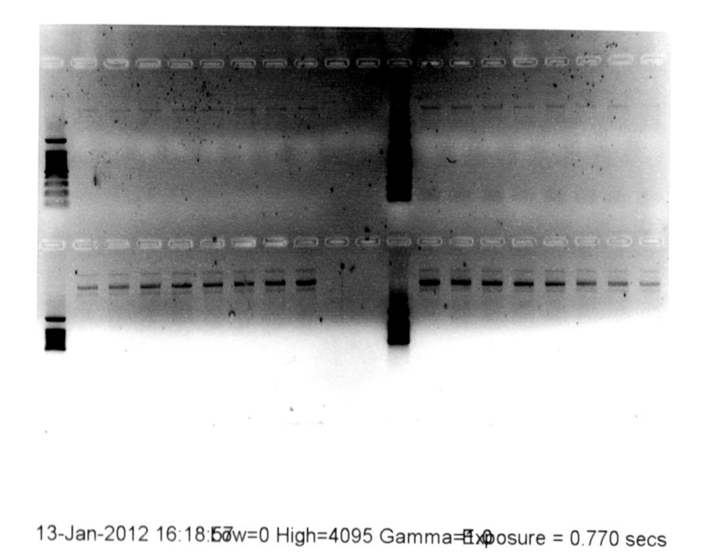
13-Jan-2012 16:18:57 Low=0 High=4095 Gamma=1.0 Exposure = 0.770 secs

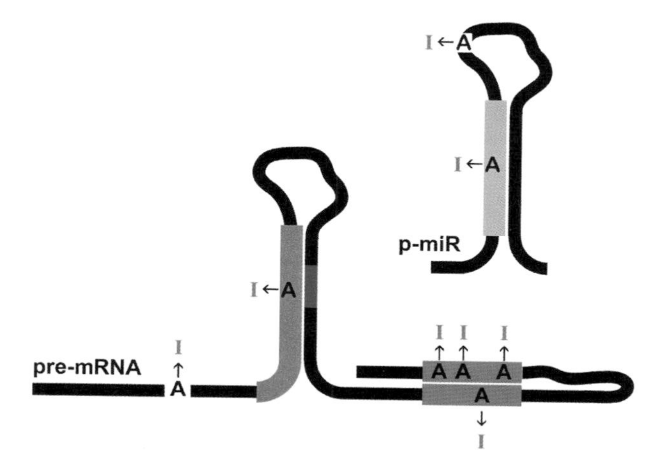

Fig. 1 A-to-I RNA editing. ADARs bind to dsRNA through their dsRBD and deaminate a specific adenosine to inosine via its deaminase domain. A-to-I editing can occur in exons (*red*) when duplex with an intron (*black*) containing an ECS (*blue*). Alu repeat regions (*green*) located within introns and UTRs form inverted intramolecular RNA duplexes in several pre-mRNA transcripts that are subject to widespread A-to-I editing by ADAR1 and ADAR2. Editing of micro RNA precursors (p-miR) can lead to edited mature miRNA (*orange*) resulting in altered target specificity, which may have significant biological consequences. Alternatively, A-to-I editing in p-miR regions flanking the mature miRNA coding region can alter processing of p-miR transcripts by Drosha and Dicer

restricted to the brain (Chen et al. 2000; Melcher et al. 1996a). Site-selective A-to-I editing of primary transcripts in the central nervous system is an established in vivo function of ADAR1 (Hartner et al. 2004) and ADAR2 (Higuchi et al. 1993, 2000). No function has been delineated for ADAR3 and it lacks detectable editing activity on synthetic dsRNA or known ADAR substrates.

1.1 Editing of Protein-Coding Sequence

The classical example of site-selective A-to-I editing mediated recoding of a protein-coding transcript concerns a glutamine codon (CAG) in the gene of the glutamate receptor subunit GluA2 (GluR-B, GluR2) (Higuchi et al. 1993; Lomeli et al. 1994; Sommer et al. 1991). At the corresponding position in the transcript, an arginine codon (CGG) is found. Editing by ADAR2 at this position, termed Q/R site, strictly depends on the presence of an editing site complementary sequence (ECS) located in a downstream intron that is required for intramolecular duplex formation. A-to-I editing at Q/R site occurs with ≈100% efficiency, whereas the extent of A-to-I editing for other positions identified in diverse mammalian transcripts ranges from <10% to ≈40%. The requirement of intronic sequence for site-selective editing dictates that editing has to occur in the nucleus before the transcript is spliced and exported to the nucleus for translation in the cytoplasm.

A-to-I editing can also create new protein variants resulting from the creation or elimination of intronic splice sites. ADAR2 was found to edit its own pre-mRNA resulting in the loss of functional ADAR2 protein expression due to premature translation termination in an alternate reading frame (Rueter et al. 1999). The application of high-throughput sequencing technology has increased the known repertoire of substrates for A-to-I editing which results in changes in the encoded proteins (Levanon and Eisenberg 2006; Levanon et al. 2004; Li et al. 2009; Sakurai et al. 2010). When coupled with information gained from genetically-engineered models, such as murine models or shRNA approaches, it will be possible to more clearly define the extent and cellular consequences of A-to-I editing on the diversity of the proteome.

1.2 Editing of MicroRNA

MicroRNAs (miRNA) are a newly recognized class of genomically encoded small RNAs originating from larger dsRNA precursors. miRNAs regulate gene expression by binding to complementary target mRNAs priming them for degradation or unproductive translation (Bartel 2004). miRNA precursors have been demonstrated to be targets for A-to-I editing (Heale et al. 2009; Luciano et al. 2004; Yang et al. 2006). The consequences of ADAR activity on miRNA can be diverse. Firstly if the targeted adenosine is contained within the seed sequence of the miRNA then the edited mature miRNA can have altered target profiles compared to the unedited transcript. This has been demonstrated for miR-376 in human and several others to date. Secondly, the edited miRNA can have altered processing through reducing in the cleavage by Drosha and Dicer during miRNA biogenesis. Lastly, the editing of RNA can alter the efficiency of miRNA binding to the target RNA. This could have two possible outcomes, establishing a consensus binding site on the target to facilitate miRNA binding or disrupting the site and reducing miRNA regulation of the target RNA (Blow et al. 2006; Borchert et al. 2009; Das and Carmichael 2007; Kawahara et al. 2008; Kawahara et al. 2007a, b; Scadden and Smith 2001).

1.3 Editing of Repeats

More recently, transcriptome analyses have identified widespread A-to-I editing in Alu repeats. Alu repeats are short interspersed repetitive elements and the editing of these has been suggested to have broader implications in gene regulation. The numbers and frequencies of Alu repeats have expanded rapidly during primate evolution and the biological significance of these regions is not clearly defined

(Athanasiadis et al. 2004; Eisenberg et al. 2005; Kim et al. 2004; Mattick and Mehler 2008; Paz-Yaacov et al. 2010). Use of bioinformatics and high-throughput sequencing technologies has revealed a high frequency of A-to-I editing in Alu repeats and that this is largely restricted to primates (Levanon et al. 2004, 2005; Mattick and Mehler 2008; Moller-Krull et al. 2008; Paz-Yaacov et al. 2010). These findings highlight potentially divergent effects of ADAR mediated RNA editing in primates and other mammals.

1.4 Physiological Significance of A-to-I Editing

The physiological significance of A-to-I editing is best exemplified by the editing of the GluA2 transcript. While A-to-I editing in the brain is generally believed to fine-tune neuro-physiological processes (Seeburg et al. 1998), ADAR2-mediated editing of a single adenosine within a codon that specifies a functionally critical position in the channel lining segment of GluA2 protein becomes a matter of life and death. Homozygous disruption of the ADAR2 gene in mice leads to the development of seizures and neuro-degeneration in the hippocampus followed by the early postnatal death. Remarkably, the genomic replacement of the edited adenosine in GluA2 results in a complete rescue of the ADAR2 phenotype, not anticipated from the widespread expression of ADAR2. The physiological role of ADAR1 appears to contrast with that of ADAR2, where a single edit can account for the majority of the knockout phenotype. As described in more detail, ADAR1 deficiency results in embryonic death, associated with defects in hematopoiesis and hepatocyte survival. The development and analysis of genetically-modified ADAR alleles in mice has led to a more detailed understanding of the cellular and molecular effects of loss of ADAR function. Genetically-engineered murine ADAR alleles generated to ablate the function of the encoded proteins are summarized in Fig. 2.

2 ADAR1: Required for Embryonic and Postnatal Development

2.1 Unique Features of ADAR1

ADAR1 is widely expressed and the development and characterization of mutant mouse models has demonstrated that it is essential for the development of hematopoietic and hepatocyte lineages (Hartner et al. 2004; Wang et al. 2004). As noted there are two distinct isoforms of ADAR1 that are generated by alternative promoters; a full-length interferon-inducible 150 kDa protein (ADAR1p150) and a constitutive N-terminal truncated 110 kDa form (ADAR1p110) (Kim et al. 1994a, b; Kim and Nishikura 1993). ADAR1p110 localizes to the nucleus, whereas p150 is

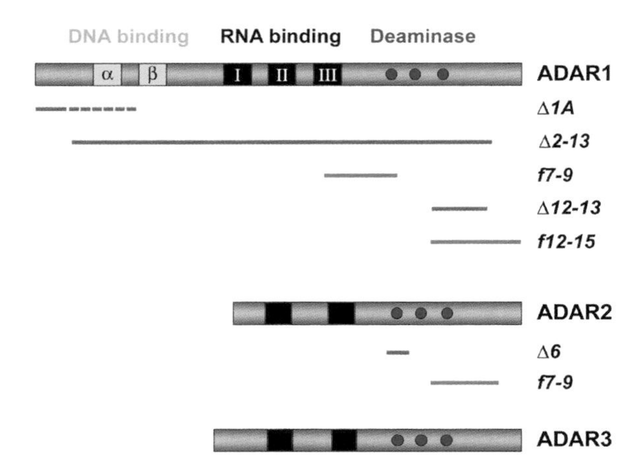

Fig. 2 Summary of ADAR gene-targeted alleles in mice. Domain structures of ADAR1–3 protein and summary of loss-of-function ADAR alleles that have been generated in mice. The extent of exon deletion is indicated by *red* (constitutive alleles) and *green* (conditional alleles after Cre-recombinase mediated excision of the loxP-flanked gene segment) horizontal lines beneath the ADAR1 and ADAR2 domain structures. Five knockout alleles of ADAR1 encompassing a range of functions of the enzyme have been described. In the *Adar1 Δ1A* allele, the interferon-inducible promoter and exon1A containing the translation initiation codon of the full-length ADAR1 p150 protein have been deleted so that translation is initiated exclusively from a downstream ATG resulting in expression of the amino-terminally truncated ADAR1 p110 isoform that lacks the Zα domain. The constitutive *Adar1* ***Δ2–13*** allele lacks exons 2–13 encoding the DNA and dsRNA binding domains and most of the deaminase domain. Exons flanked by loxP sites in the conditional *Adar1 f7–9* allele encode part of the third dsRNA binding and deaminase domain, or part of the deaminase domain in the conditional *f12–15* allele. Exons deleted in the constitutive *Adar1* Δ12–13 allele code for part of deaminase domain. Two null alleles of ADAR2 have been described, one with a constitutive deletion of exon 6 (*Adar2 Δ6*) encoding an essential part of the deaminase domain, the other (*Adar2 f7–9*) with exons 7–9, also encoding part of the deaminase domain, flanked by loxP elements for Cre-recombinase mediated disruption

found predominantly in the cytoplasm (Patterson and Samuel 1995). The biological significance of ADAR1 cellular localization is as yet unknown.

Several structural features distinguish ADAR1 from the two more closely-related ADAR2 and ADAR3: ADAR1 contains a third dsRNA binding domain (dsRBD), which may explain RNA editing site-selectivity between ADAR1 and ADAR2 as seen in the pre-mRNA of the 5-HT_{2C} receptor in neurons (Hartner et al. 2004). A nuclear localization signal (NLS) was discovered to be located within this third dsRBD (Eckmann et al. 2001). ADAR1 also contains a nucleic acid binding motif at the amino terminus that can bind either DNA or RNA in the Z-conformation (Herbert et al. 1995). This Z-DNA binding domain comprises of two subdomains Z_a and Z_b. ADAR1 p150 contains both, whereas only the Z_b subdomain is present in the p110 isoform. Recent studies suggest this domain targets ADAR1 to transcriptionally active regions, perhaps allowing ADAR1 to edit transcripts as they are being synthesized (Eckmann and Jantsch 1999).

The generation of mutant alleles of ADAR1 has lead to important observations regarding its cellular role and participation in various biological processes. In contrast to ADAR2, to date no known target or targets have been defined which can account for the ADAR1 null phenotype. The ADAR1 mutant alleles have been important in defining cell types in which ADAR1 function is critical, a necessary first step in the characterization of the biological function of ADAR1. The current focus of ADAR1 studies has been on an understanding of its role in RNA editing, largely driven by the elegant paradigm of ADAR2. Few studies have attempted to explicate a non-catalytic role of ADAR1. It is important to assess all possible aspects of ADAR1 to dissect the physiological importance of its function. The range of ADAR1 mutant alleles that have been generated will begin to allow assessment of the possible functions of ADAR1 in vivo.

2.2 Haploinsufficiency for ADAR1?

The first-described null allele of ADAR1 led to death of heterozygous chimeric embryos (Wang et al. 2000). The targeting approach partially disrupted exon 11 and completely disrupted exons 12 and 13 by the insertion of a PGK-neomycin cassette in reverse orientation (Wang et al. 2000). Analysis of chimeric embryos revealed a scarcity of enucleated erythrocytes in the blood of embryos with a high contribution of cells with one targeted Adar allele, indicating a requirement for ADAR1 in fetal-liver erythropoiesis. While defective hematopoiesis described in chimeric embryos was also detected in three mouse lines carrying differently configured ADAR1 null alleles, the claim of haploinsufficiency for ADAR1 could not be reproduced and appears to have resulted from faulty expression of the manipulated $Adar1^{\Delta 12-13}$ allele, potentially leading to insufficient ADAR1 protein levels in highly-chimeric embryos.

2.3 ADAR1 Deficiency Causes Embryonic Death

It is now apparent that there is no haploinsufficieny for ADAR1, confirmed in three independent targeted alleles, one disrupting sequence encoding the carboxyterminal portion of ADAR1 protein, one the carboxyterminal part of the third dsRBD along with part of the catalytic domain, and one disrupting most of the gene, including the DBMs, dsRBDs and most of the catalytic domain (Hartner et al. 2004, 2009; Wang et al. 2004). Following on from the initially reported ADAR1 mutant allele, several additional alleles were generated. Three germ-line deficient alleles have been generated: exons 12 through 15 encoding the carboxyterminal portion of ADAR1 protein were deleted by germ-line deletion of loxP elements (Wang et al. 2004); exons 7 through 9 coding for the carboxyterminal part of dsRBDIII and part of the catalytic domain were deleted by Cre mediated recombination in the germ line (Hartner et al. 2004) and a replacement of exons

2–13 with a neomycin cassette deleting most of the ADAR1 protein-coding sequence were established (Hartner et al. 2004).

Heterozygous null animals of all three null lines were viable and ostensibly normal. By contrast, homozygosity for the differently configured ADAR1 null alleles consistently caused embryonic death between E11.5 and E12.5. Beginning at early E11 and strikingly apparent by E12, ADAR1-deficient embryos displayed a significant reduction in fetal liver size and, shortly before death, were slightly retarded in development when compared with controls. Closer inspection of ADAR1-deficient embryos revealed from E11.5 onward a significantly reduced cell density in the fetal liver as a consequence of massive cell death (Hartner et al. 2004; Wang et al. 2004). Increased apoptosis in ADAR1-deficient embryos has also been detected in tissues other than fetal liver and appeared to be most pronounced in vertebrae and heart that express high levels of ADAR1 protein in wild-type embryos. However, it cannot be ruled out that widespread apoptosis is secondary to disintegration of the fetal liver structure, which is likely to interrupt the blood circulation, ultimately resulting in anoxia and cell death throughout the embryo. Consistent with the null allele findings, chimeric studies with ADAR1-deficient embryonic stem (ES) cells showed a failure in contribution to both the hepatic and hematopoietic lineages in adult chimeric mice (Hartner et al. 2004). These complementary studies indicated a cell-autonomous requirement for ADAR1 in both hepatocytes and hematopoietic lineages, but they were not able to clearly discriminate a role in formation of the lineage or maintenance.

Cells cultured from ADAR1-deficient embryos exhibited increased apoptosis when a stress response was induced by serum deprivation (Wang et al. 2004). Expression of the interferon-inducible ADAR1 p150 increased upon serum deprivation, perhaps implicating a role of p150 for cell survival upon induction of cellular stress. As ADAR1 is a dsRNA binding and modifying protein, the cell death induced by ADAR1 deficiency has led to speculation that absence of ADAR1 may trigger cell activation of death pathways induced by dsRNA. However, cell death induced by ADAR1 deficiency does not appear to be mediated by pathways involving the dsRNA-activated serine/threonine kinase PKR, as PKR deficiency did not rescue the death of ADAR1-deficient embryos. RNase L represents another pathway induced by dsRNA during viral infection or cellular stress responses. This pathway is unlikely to be affected, as judged from the presence of stable general rRNA and mRNA in ADAR1-deficient embryos and cultured cells (Wang et al. 2004). The analysis of ADAR1 knockout mice has demonstrated that ADAR1 has a distinct biological function that cannot be compensated for by ADAR2 or ADAR3 and that it plays an essential role outside of the central nervous system.

2.4 ADAR1 in Hematopoiesis

ADAR1-deficient embryos died at day E11-12 of embryonic development, when hematopoiesis shifts from the extraembryonic yolk sac to the fetal liver to produce all blood cell lineages that are found in the adult mouse. ADAR1-deficient fetal liver from early E11 onward displayed a pronounced scarcity of hematopoietic cells. Of note, primitive erythropoiesis in the yolk sac that provides for the first blood cells of the developing embryo does not require ADAR1. Thus, differential dependence on ADAR1 becomes an additional distinctive feature of primitive and definitive erythropoiesis (Hartner et al. 2004). More detailed analysis of fetal-liver hematopoiesis in $Adar1^{-/-}$ embryos revealed a severely impaired colony-forming potential of multi-potential progenitors both in vitro and in vivo. The impaired colony-forming potential of cultured progenitors isolated from hematopoietic tissues of ADAR1-deficient embryos suggested a role of ADAR1 in proliferation and/or survival of these progenitors. Immunophenotyping and quantification of HSCs in early-E11 $Adar1^{-/-}$ fetal livers revealed that ADAR1 is dispensable for the emergence of phenotypic HSCs and their migration from production sites to the fetal liver (Hartner et al. 2004, 2009; Wang et al. 2004).

The generation of mouse lines for inducible ADAR1 deficiency provided new insight to the function of ADAR1 in embryonic and adult hematopoiesis in vivo. In absence of ADAR1, the development of multiple blood cell lineages was impaired, consistent with a requirement for ADAR1 in an immature hematopoietic progenitor or hematopoietic stem cells (HSC) that give rise to all blood cell lineages.

2.5 Hematopoietic Specific Deletion Reveals a Cell-Autonomous Role of ADAR1

To study the function of ADAR1-deficient HSCs in vivo, a series of experiments was performed employing hematopoietic reconstitution, in combination with flow cytometry that permits tracking of transplanted cells and their descendants (Fig. 3). Transplantation of irradiated wild-type recipients with fetal liver or bone-marrow cells from mice carrying one $Adar1^{\Delta 2-13}$ knockout and one conditional $Adar1^{f7-9}$ allele in combination with an inducible Cre recombinase transgene demonstrated that ADAR1 is essential for the in vivo maintenance of the hematopoietic stem cell and immature progenitor cell compartments in fetal liver and adult bone marrow (Hartner et al. 2004, 2009). ADAR1 appears to be dispensible for the maintenance of HSCs but becomes essential as they progress to the progenitor stage of hematopoietic differentiation. Immature progenitors undergo massive cell death, resulting in the block of multi-lineage hematopoiesis.

2.6 **Mx1-*Cre* ADAR1** *and h***Scl-*Cre*^{ERT2} ADAR1**

Two distinct strategies have been used to generate in vivo somatic inactivation of ADAR1 in hematopoietic cells. Firstly, the *Mx1*-Cre transgene was used which elicits efficient gene deletion throughout the hematopoietic system (and many

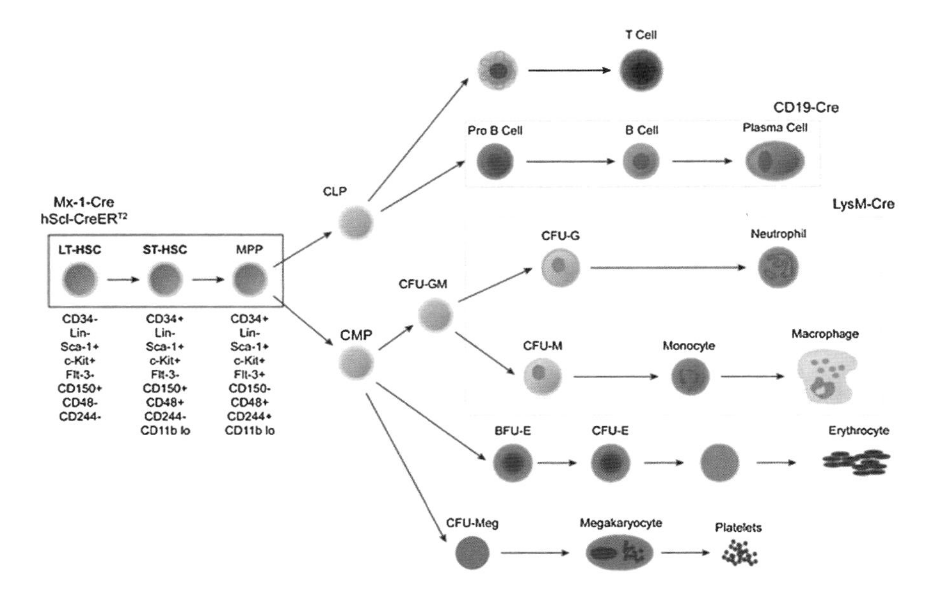

Fig. 3 Conditional deletion of ADAR1 in hematopoietic lineages. Schematic representation of differentiation of hematopoietic lineages from LT-HSCs to mature blood cells. Surface markers used to distinguish progenitor cells by flow cytometry are listed below each progenitor cell type. Cells enclosed within rectangles depict lineages where ADAR1 is disrupted by conditional Cre-recombinase expression. Mx1-Cre and hScl-CreER^{T2} delete ADAR1 within HSCs and immature progenitor cells so that all cell derived from these will lack functional ADAR1 alleles. CD19-Cre is expressed from the pro-B cell stage of B-lymphoid differentiation onward, and LysM-Cre deletes ADAR1 predominantly from myeloid lineage. *Abbreviations* *LT-HSC* long-term hematopoietic stem cell, *ST-HSC* short-term hematopoietic stem cell, *MPP* multipotent progenitor, *CLP* common lymphoid progenitor, *CMP* common myeloid progenitor, *CFU-GM* colony-forming unit granulocyte monocyte, *CFU-G* colony-forming unit granulocyte, *CFU-M* colony-forming unit monocyte, *BFU-E* blast-forming unit erythroid, *CFU-E* colony-forming unit erythroid, *CFU-Meg* colony-forming unit megakaryocyte

other cell types in the mouse) after administration of the interferon inducer polyinosinicpolycytidylic acid (poly(I:C) (Kuhn et al. 1995)). *Mx1*-Cre efficiently excised conditional (floxed) ADAR1 alleles throughout the hematopoietic system (Hartner et al. 2009). Untreated A*dar1*$^{fl/-}$; *Mx1*-Cre mice were born at Mendelian frequency but, unexpectedly, died shortly after birth. No definitive pathology was found to explain the death of the neonatal animals but it was possibly a consequence of low-level Cre induction by endogenous interferon leading to ADAR1 deletion (Hartner et al. 2009). However, *Adar1*$^{fl/-}$; *Mx1*-Cre embryos appeared grossly normal at E14.5, despite having slightly lower total fetal-liver cellularity. To circumvent the unanticipated early death of untreated *Adar1*$^{fl/-}$; *Mx1*-Cre neonates, transplantation studies with fetal liver-derived HSCs were performed. This strategy has successfully been employed to assess hematopoiesis in otherwise lethal alleles. Transplantation of *Adar1*$^{fl/-}$; *Mx1*-Cre allowed for fetal-liver cells to engraft a congenic host, where hematopoietic cells can be identified by differential expression of the congenic CD45.1/CD45.2 antigens present on all nucleated leukocytes. This approach also permits an assessment of the cell-autonomous nature of the requirement for ADAR1 as only cells derived from *Adar1*$^{fl/-}$; *Mx1*-Cre HSCs will be able to inducibly delete ADAR1.

At 5 weeks after transplant, the recipients were treated with the poly(I:C) to induce Cre expression from the interferon-inducible *Mx1* transgene. Strikingly, within the two-week period of treatment with poly(I:C) there was a dramatic reduction in the contribution of *Adar1*$^{fl/-}$; *Mx1*-Cre to hematopoiesis. The contribution of cells with induced ADAR1 deficiency to peripheral blood cells, including myeloid, B-lymphoid and T-lymphoid lineages, progressively decreased and was negligible by 23 weeks after induction of ADAR1 deficiency. These data demonstrate a cell-autonomous requirement for ADAR1 in fetal liver-derived hematopoiesis (Hartner et al. 2009).

A second strategy using a HSC-specific tamoxifen inducible transgene was also employed. The h*Scl*-CreER^{T2} trangene contains an enhancer of the Scl/Tal1 gene to drive Cre-ER^{T2} expression in the hematopoietic stem cell population (Gothert et al. 2005). This strategy allowed somatic inactivation of ADAR1 within the context of adult bone-marrow HSCs (Hartner et al. 2009). When crossed to *Adar1*$^{fl/-}$ mice, no neonatal death, as observed in *Adar1*$^{fl/-}$; *Mx1*-Cre mice, ocurred. Cre activity was induced by the administration of tamoxifen either by injection or orally. As was observed with the *Adar1*$^{fl/-}$; *Mx1*-Cre, the *Adar1*$^{fl/-}$; h*Scl*-CreER^{T2} model demonstrated a significant depletion of short-term HSCs and multi-potent progenitors in ADAR1-deficient bone marrow. This second approach also yielded important information regarding the biological consequences of ADAR1 depletion at the cellular level. Cell cycle status analysis of HSCs and immature progenitors revealed that ADAR1-deficient HSCs undergo an increased rate of apoptosis and the majority of the population is induced to enter the cell cycle in contrast to control bone marrow. Moreover, ADAR1-deficient bone marrow had a much larger fraction of actively cycling phenoytypic short-term HSCs, but not multi-potent progenitors and was depleted of HSCs, suggesting exhaustion of HSCs due to continuous activation of the stem cell compartment. Collectively it could be determined that ADAR1 is an essential regulator of HSC maintenance but is dispensable for the emergence of long-term repopulating (LT)-HSCs and, perhaps, their self-renewal. ADAR1 becomes essential for the survival of HSCs as they progress to the multi-potent progenitor stage of blood cell differentiation. These findings suggest a cell-autonomous requirement for ADAR1 in the maintenance of HSCs in both fetal liver and adult bone marrow, consistent with the observation that ADAR1-deficient embryonic stem cells did not contribute to hematopoietic tissues in adult chimeric mice (Hartner et al. 2009).

Similar results were obtained by transplanting an *Adar1*$^{fl/fl}$ HSC-enriched population after MSCV Cre-mediated deletion of a conditional ADAR1 knockout allele (XuFeng et al. 2009). Of note, it appears that ADAR1 deficiency does not affect HSC homing which refers to the capacity of HSCs to migrate from the blood stream to the bone-marrow microenvironment. The proliferative capacity of ADAR1-deficient HSCs in in vitro culture systems appeared normal as well, whereas rapid death of these cells was observed as they progressed to the multipotent progenitor stage of blood cell differentiation. These data collectively demonstrated that ADAR1 was specifically required at the transition of long-term repopulating HSCs to short-term repopulating cells.

In contrast to the absolute requirement for ADAR1 in immature hematopoietic populations, conditional ablation of ADAR1 in mature blood lineages has to date revealed no apparent phenotype. Deletion of ADAR1 in B cells using CD19-Cre was not reported to result in an abnormal phenotype although the immune responses in these mice have not been tested (Yang et al. 2006). CD19 is expressed by pro-B cells and yet the lack of ADAR1 does not apparently impair the maturation process of B cell development. Disruption of ADAR1 in mature myeloid cells by using Lysosome M-Cre (LysM) did not significantly impact on macrophages and granulocytes (Hartner et al. 2009). The role of ADAR1 in erythropoiesis is yet to be tested, but recently developed erythroid-specific Cre recombinase alleles would allow this. Of note, HSCs and primitive progenitors display higher expression of interferon-inducible full-length ADAR1p150 than of the constitutively expressed shorter ADAR1p110 in HSCs, which may indicate a possible requirement for ADAR1p150 in interferon-induced pathways in hematopoiesis (Yang et al. 2003). Indeed, it may be this differential expression of the two isoforms that determines a cell's sensitivity to ADAR1 depletion. Based on current in vivo and cell culture data, ADAR1 deficiency ultimately results in apoptotic death of cells that depend on ADAR1 function. Future studies will have to address how apoptotic pathways are activated upon deletion of ADAR1.

2.7 The Role of ADAR1 in Hematopoiesis: Interferon Pathway Regulation

Based on the identification of a specific requirement for ADAR1 in the regulation of hematopoiesis, genome-wide transcriptome combined with gene set-enrichment analysis has shed new light on the molecular mechanism underlying the severe phenotype caused by ADAR1 deficiency (Fig. 4) (Hartner et al. 2009). Comparison of the gene expression profiles obtained from ADAR1 knockout and control fetal-liver HSCs revealed a strong association between ADAR1 deficiency and the gene expression 'signatures' of interferon-treated or virally-infected cells. ADAR1-deficient HSCs showed global upregulation of transcripts inducible by type I interferon (IFN-α and IFN-β), type II interferon (IFN-γ), or both. Genes that were significantly induced (up to 300-fold) by ADAR1 deficiency encompassed a spectrum of interferon associated pathways and transcripts. These include the genes of the transcription factors STAT1 and STAT2 and IRF1, IRF7 and IRF9; the GTPases Mx1 and Mx2; the RNA-activated protein kinase PKR (EIF2AK2); the 2′,5′-oligoadenylate synthetases OAS1, OAS2 and OAS3; the ubiquitin-like modifiers Isg15 and Isg20; the interferon-induced proteins with tetratricopeptide

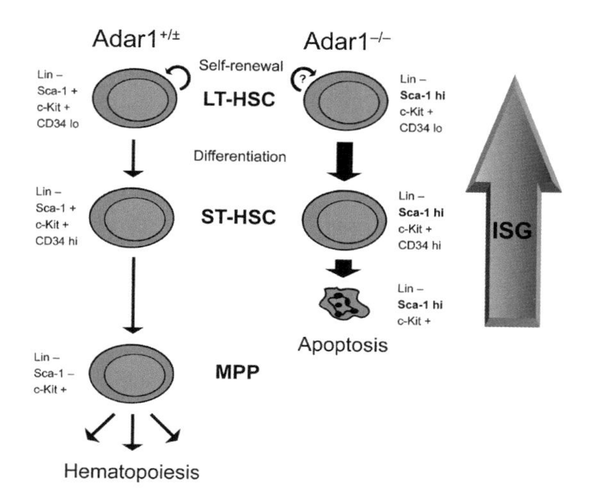

Fig. 4 Suggested model for the role of ADAR1 in HSCs. *Vertical arrows* mark the transition from long-term (LT-) HSC via short-term (ST-) HSC to multi-potent progenitor (MPP) as defined by lineage marker (Lin), cKit (K), Sca-1 (S), and CD34 surface expression. *Semi-circled arrows* indicate self-renewal. Both ADAR1-deficient ($Adar1^{-/-}$) HSCs and MPPs exhibit a global upregulation of transcripts expressed from interferon-stimulated genes (ISGs), including Sca-1, as compared with wild-type and $Adar1^{+/-}$ ($Adar1^{+/\pm}$) cells. Differentiation of lineage-negative, cKit-positive, Sca-1-positive, CD34-negative/low (LKS^{+} $CD34^{lo}$) LT-HSCs via ST-HSCs (LKS^{+} $CD34^{hi}$) and MPPs (LKS^{-} $CD34^{hi}$) is accompanied by an increase in CD34 and a decrease in Sca-1 surface expression. Perhaps as a consequence of the interferon pathway activation, ADAR1-deficient LKS^{+} HSCs do not downregulate Sca-1 and undergo rapid apoptosis as they progress to the LKS^{-} progenitor stage of hematopoiesis

repeats Ifit1-Ifit3; and ADAR1. This intersection of ADAR biology and interferon pathways has previously been demonstrated in the context of viral infection (George et al. 2009, 2011; Samuel 2011) but not in the context of normal biology in non-infected settings. Among those transcripts that were induced by ADAR1 deficiency, only those known to be inducible by interferon were significantly changed in *Adar1*$^{-/-}$ HSCs relative to wild-type and *Adar1*$^{+/-}$ control HSCs. Likewise, among transcripts encoding dsRNA-binding proteins, only those expressed from interferon-inducible promoters, such as those encoding PKR (EIF2AK2), IFIH1 (RIG-1) and TLR3, were upregulated in ADAR1-deficient HSCs. Of note, among the deregulated genes the dsRNA-binding protein NF90 has been shown to interact with ADAR1 (Nie et al. 2005). Intriguingly there is a significant overlap between the signature of ADAR1 deficiency and gene expression changes associated with over-expression of a variant of NF90 protein that suggests a biologically relevant link of NF90 and ADAR1. It is noteworthy that ADAR1 deficiency in macrophages and neutrophils did not affect the transcript abundance of Stat1, one of the genes most substantially upregulated in *Adar1*$^{-/-}$ HSCs. This observation is consistent with the finding that ADAR1 is dispensable in the myeloid lineage as assessed by conditional ADAR1 gene disruption directed by LysM-Cre (Hartner et al. 2009). These findings hint at cell-specific differences in how ADAR1 participates in regulation of interferon signaling that require further investigation.

The global induction of interferon in ADAR1-deficient embryos from E11.5 coincided with increasing levels of apoptosis suggesting that death signal pathways may be induced by interferon. Collectively, the analyses of animals deficient in ADAR1 have demonstrated an essential cell-intrinsic requirement for ADAR1 in the suppression of interferon-inducible pathways. Supporting this observation, a recent study found that inosine-uridine (IU)-dsRNA can suppress interferon stimulated genes (ISGs) and apoptosis (Vitali and Scadden 2010). IU mismatches are formed in dsRNA as a result of ADAR editing. Activation of interferon regulatory factor 3 (IRF3), a gene essential for the induction of ISGs and apoptosis, was found to be inhibited by IU-dsRNA (Vitali and Scadden 2010).

2.8 The Role of ADAR1 in Hepatocytes

In addition to the hematopoietic phenotype in the fetal liver, ADAR1-deficient embryos also display abnormalities in hepatoblasts. Early hepatogenesis appeared intact in ADAR1 knockout embryos, as judged from morphological examination and expression analysis of hepatic marker genes (Hartner et al. 2004). The breakdown of the fetal-liver architecture beginning at E11.5 suggested that, in addition to the hematopoietic cells, hepatoblasts were also undergoing cell death. Consistent with a cell-autonomous requirement for ADAR1 in hepatic cells, ADAR1-deficient ES cells did not contribute to liver in adult chimeric mice. Moreover, disruption of the ADAR1 gene in mice with transgenic Cre expression directed by the Albumin promoter led to disorganized liver architecture and increased cell death in the liver of adult ADAR1 mutant mice, along with decreased glucose and increased liver enzyme levels (Wang et al. 2004). Perhaps as a consequence of these impairments, ADAR1 Albumin-Cre mutant mice displayed a significantly reduced size when compared with controls. More detailed study of the hepatic lineage will be required and ADAR1 targets will need to be identified.

2.9 Transcript Editing is a Biological Function of ADAR1

The generation of mutant ADAR1 alleles has allowed for the definitive demonstration of in vivo A-to-I editing capacity for ADAR1. Analysis of A-to-I editing within candidate transcripts expressed in neurons cultured from *Adar1*$^{-/-}$ embryonic brain (Hartner et al. 2004) and teratomas derived from ADAR1-deficient ES cells (Wang et al. 2004) unequivocally established that site-selective A-to-I editing is a biological function of ADAR1. These studies revealed a remarkable selectivity of A-to-I editing for closely-spaced adenosines in serotonin

2C receptor pre-mRNA. Neurons cultured from embryos with dual deficiency for ADAR1 and ADAR2 allowed unambigious assignment of edited adenosines to either ADAR1, ADAR2, or as substrates for both enzymes (Hartner et al. 2004).

2.10 ADAR1p150 Specific Functions?

It is not clear to what extent the two ADAR1 p110 and p150 isoforms differ in their biological properties. Prominent differences include the interferon regulation of ADAR1p150 and the cellular localization of the isoforms with ADAR1p110 being predominantly nuclear and ADAR1p150 being found in both the nucleus and cytoplasm. Homozygosity for a targeted ADAR1 allele that lacks the interferon-inducible promoter providing for ADAR1p150 transcripts has been reported to recapitulate the embryonic death caused by total ADAR1 deficiency encompassing both ADAR1 isoforms. This finding indicates that the interferon-inducible ADAR1 may be the essential isoform during embryonic development (Ward et al. 2011). ADAR1p150-deficient cells demonstrated a specific role for this isoform in controlling measles virus replication (Ward et al. 2011). Indeed, a critical role of the interferon-inducible ADAR1 p150 is consistent with its more abundant expression in HSCs and immature hematopoietic progenitors and the strong activation of the interferon pathway by total ADAR1 deficiency. However, further analysis of this mutant will be required to exclude a potential effect of the retained selection marker in the ADAR1 p150 knockout allele on expression of ADAR1 p110 in the cells that depend on ADAR1 or mutations in the p110 protein-coding region that may have been introduced during the gene-targeting maneuver (Wang et al. 2000; Ward et al. 2011).

3 ADAR2 (Adarb1, RED1): Important for Brain Function

The generation of ADAR2-deficient animals and their subsequent analysis has defined the in vivo paradigm for A-to-I editing activity. Through elegant genetic models it was demonstrated that a single editing event in the glutamate receptor subunit GluA2 could account for the pronounced phenotype induced by ADAR2 deficiency (Higuchi et al. 2000).

3.1 ADAR2 Deficiency Causes Epileptic Seizures and Early Postnatal Death

The ADAR2 gene was disrupted by replacing the majority of exon 6 (originally annotated as exon 4) with a PGK-neomycin cassette. This exon encodes a

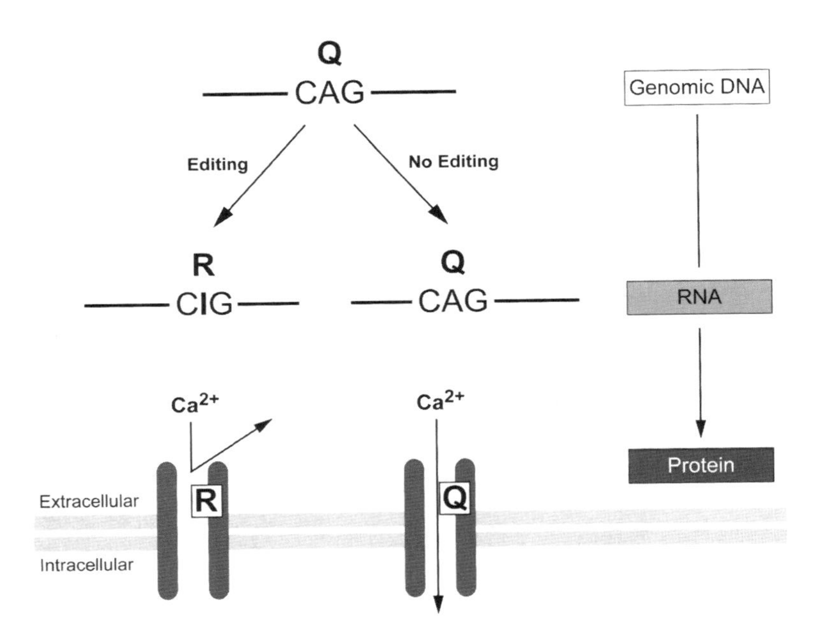

Fig. 5 Consequences of ADAR2 editing on GluA2 function. Editing of the Q/R site in the transcript for the glutamate receptor subunit GluA2 leads to a change in protein-coding sequence. ADAR2 deficiency results in a loss of editing at this site and physiologically relevant changes in the function of GluA2, which results in seizures and death. The Q/R site is a critical channel determinant that, among other parameters, controls influx of divalent cations, including Ca^{2+}. The Q/R-site unedited version of GluA2 (Q) renders glutamate receptors with GluA2 participation permeable to Ca^{2+}, whereas channels that contain the edited (R) GluA2 subunit are impermeable to Ca^{2+}. Uncontrolled Ca^{2+} influx through channels with Q/R-site unedited GluA2 is likely to contribute to the development of seizures in ADAR2-deficient animals

functionally essential adenosine deaminase motif of ADAR2 (Higuchi et al. 2000). *Adar2*$^{+/-}$ intercrosses produced *Adar2*$^{-/-}$ mice at the expected Mendelian frequency indicating that loss of ADAR2, unlike ADAR1 deficiency, is compatible with normal embryonic development. While *Adar2*$^{+/-}$ mice were outwardly normal, *Adar2*$^{-/-}$ mice developed epileptic seizures 2 weeks after birth and died by approximately postnatal day 21. *Adar2*$^{-/-}$ animals bear a close phenotypic resemblance to GluA2$^{+/\Delta ECS}$ mice carrying a genetically-engineered GluA2 allele that lacks an intronic editing site complementary sequence (ECS) required for editing at the Q/R site (Brusa et al. 1995). In addition to severely reduced editing at the GluA2 Q/R site, ADAR2-deficient brain displayed substantially reduced editing at most of 25 positions in diverse neuronal transcripts (Higuchi et al. 2000).

Despite the reduction or loss of A-to-I editing in a number of substrates, the dramatic phenotype of *ADAR2*$^{-/-}$ mice reverted to normal when the edited CGG codon replaced the unedited CAG version in both GluA2 alleles (*Gria2*$^{R/R}$), defining the Q/R site as the physiologically most critical substrate of ADAR2 (Fig. 5). Another interesting observation in *ADAR2*$^{-/-}$ brain was that 10% of

GluA2 pre-mRNA was edited as compared with 40% of spliced transcripts, suggesting preferential splicing of Q/R-site-edited transcripts. Moreover, editing at the Q/R site was shown to modulate GluA2 protein processing within the endoplasmic reticulum (Greger et al. 2002). This demonstrates that all editing events are not equivalent in their cellular consequences and that a single editing event can account for the most pronounced aspects of the phenotype in ADAR2-deficient animals. The physiological significance of the changes in the editing status of the other brain-derived transcripts in ADAR2-deficient mice remains largely to be explored. A comprehensive physiological examination of 320 parameters in $Adar2^{-/-}/Gria2^{R/R}$ compound mutant mice has revealed a spectrum of subtle changes associated with diverse tissue types in these animals (Horsch et al. 2011). Many of these changes are not outside of the normal physiological ranges for wild-type mice of the same genetic background. Nevertheless these analyses have revealed additional phenotypes associated with ADAR2 deficiency, including significant changes in behavior, hearing, serum markers of allergic responses (IgE levels) and changes in gene expression profiles. The changes in hearing were associated with increased levels of the signal transduction regulator Rgs3. It had been reported that ADAR2 may play a role in metabolic control in pancreatic β-cells, but the detailed physiological analysis of ADAR2-deficient animals recently reported would suggest that ADAR2 does not have an essential role in normal pancreatic function (Gan et al. 2006). While these reported physiological changes do not impinge on normal functioning, an assessment of ADAR2 knockout mice under non-physiological or pathological conditions, where alterations in editing of additional targets may become of organismal consequence, is warranted (Horsch et al. 2011).

3.2 Auto-Editing by ADAR2

ADAR2 was found to edit its own transcript at an intronic position creating an AI dinucleotide that serves as a splice acceptor. Utilization of this splice acceptor leads to inclusion of an additional 47 nucleotides in the ADAR2 protein-coding sequence and premature termination of translation (Rueter et al. 1999). The hypothesis that ADAR2 can modulate its protein expression by editing its own transcript was tested in gene-targeted mice that lack a stretch of intronic ADAR2 genomic sequence involved in the formation of a duplex structure that ADAR2 protein requires to edit the intronic position within the ADAR2 primary transcript (Feng et al. 2006). However, heterozygosity or homozygosity for the $Adar2^{\Delta ECS}$ allele did not produce a discernible phenotype, despite an increase in ADAR2 protein and A-to-I editing of known ADAR target RNAs in the brain of mutant mice. Compensatory effects during embryonic development that may mask potential consequences of increased ADAR2 expression in absence of autoregulation could be circumvented by inducible disruption of the conditional $Adar2^{\Delta ECS}$ allele (Feng et al. 2006).

3.3 ADAR2 and Amyotrophic Lateral Sclerosis

Amyotrophic lateral sclerosis (ALS) is the most common form of adult onset motor neuron disease in humans. The vast majority of cases is sporadic and the underlying genetic mechanisms associated with its initiation are not well defined. An intriguing observation was made in spinal neurons from *post-mortem* analysis of ALS patients where it was found that a substantial proportion of GluA2 mRNA is unedited at the Q/R site-edited by ADAR2 (Kawahara et al. 2006; Kwak and Kawahara 2005; Takuma et al. 1999). This observation appears to be specific for ALS as all GluA2 transcripts from controls were edited (Kawahara et al. 2003; Kwak and Kawahara 2005; Takuma et al. 1999) and raises the possibility that ADAR2 may play an important role in the pathogenesis of human ALS.

To test this hypothesis in a murine model, a conditional ADAR2 allele was generated with exons 7–9 flanked by loxP elements (Hideyama et al. 2010). These exons encode part of the deaminase domain of ADAR2. Deletion of ADAR2 in the brain was achieved by using VAChT-Cre.Fast mice that express Cre recombinase in cholinergic neurons. VAChT-Cre.Fast-mediated gene disruption results in a tissue mosaic where deletion occurs with approximately 50% efficiency in a given cell population. Using this model, Cre activity would progressively occur up to 5 weeks of age and be selective for a subset of motor neurons in the spinal column and central nervous system. Mutant mice were viable and presented with a phenotype of being hypokinetic, along with abnormal posture. *Adar2$^{fl/fl}$*; VAChT-Cre.Fast animals did not display signs of paralysis and had a normal survival up until around 18 months of age when they declined more rapidly than control animals. Loss of ADAR2 led to the death of the large neurons in the anterior horn of the brain. This observation was similar to that in the human ALS patients. The apoptotic phenotype could be prevented by the introduction of the Q/R site-edited form of GluA2. These experiments formally demonstrate that ADAR2 deficiency leads to death of motor neurons induced by failure of ADAR2 mediated editing at the GluA2 Q/R site (Hideyama et al. 2010; Kawahara et al. 2004).

3.4 Transgenic Expression of ADAR2

Ectopic expression of rat ADAR2 protein from a randomly integrated transgene in mice revealed a pronounced phenotype that was strikingly different from the neurological syndrome associated with ADAR2 deficiency or the lack of an overt phenotype in mice that cannot control ADAR2 expression by editing the ADAR2 transcript (Singh et al. 2007). Transgenic mice expressing either wild-type or editing-deficient ADAR2 isoforms displayed adult onset obesity characterized by hyperglycemia hyperleptinemia and increased adiposity. The drastic weight gain of these animals appeared to result predominantly from hyperphagia rather than a metabolic derangement (Singh et al. 2007). Interestingly, manifestation of the obese phenotype was independent of the deaminase activity of ADAR2, indicating that non-physiologic RNA binding rather than adenosine deamination is involved. The striking phenotype of transgenic ADAR2 mice appears to contrast with the normal phenotype of ADAR2 mutant mice with increased ADAR2 expression and A-to-I editing due to failure of ADAR2 autoregulation (Feng et al. 2006). Indeed, the obesity observed in ADAR2 transgenic mice may result from CMV-driven misexpression of the transgene outside cells normally expressing ADAR2. The possibility of a gain-of-function effect of rat ADAR2 ectopically expressed in mouse may also need to be considered, not only in view of recent data demonstrating that RNA binding by ADARs can be dependent on the primary sequence of RNA (Stefl et al. 2010).

4 ADAR3 (Adarb2, RED2): A Non-functional Family Member?

In contrast to ADAR1 and ADAR2 that are expressed in many tissues, expression of ADAR3 is restricted to the brain. ADAR3 harbors a 13–15 amino-acid, arginine-rich sequence motif referred to as R-domain that mediates binding of single-stranded RNA (Chen et al. 2000; Melcher et al. 1996a) and also acts as a nuclear localization signal.

The biological role of ADAR3 remains enigmatic. While structurally an RNA editing enzyme that is well conserved in vertebrate evolution and shares 50% protein sequence identity with ADAR2, no catalytic activity has to date been documented for ADAR3. Failure of ADAR3 to edit synthetic and endogenous dsRNA in vitro (Melcher et al. 1996a) has led to speculation that ADAR3 may negatively regulate ADAR1 and/or ADAR2 editing of transcripts in the brain (Chen et al. 2000).

Deficiency for ADAR3 in mice did not interfere with apparently normal embryonic and postnatal development (C. Faul, M. Higuchi; P.H. Seeburg, unpublished), even though functional compensation for ADAR3 deficiency by ADAR1 and/or ADAR2 has not been ruled out. Investigating the function of ADAR3 under non-physiological or pathological conditions may reveal a function for this evolutionarily conserved ADAR.

5 Outlook

Research in the past decade involving functional knockout studies in gene-targeted mice has significantly advanced our understanding of the biological functions of ADARs and revealed the importance of these proteins during embryonic and postnatal development. Future research will have to address the molecular mechanisms underlying the requirement for ADAR1 in hematopoiesis and hepatic tissue and ultimately identify the critical substrates.

The generation and analysis of lineage-restricted deficiency of ADAR1 has highlighted the highly specific requirement for A-to-I editing within the hematopoietic system at the transition from hematopoietic stem cells to progenitors. The use of genetically modified mice to identify specific cell subsets that are dependent on ADAR1 will facilitate the enrichment of these cell types and combining with new transcriptome-wide approaches. Transcriptome sequencing and biochemical approaches that permit enrichment of inosine-containing transcripts in tissues and cell types that require ADAR function should be helpful to elucidate the full cellular repertoire of adenosine deamination (Wulff et al. 2010). The convergence of these technologies will allow the identification of genes regulated by ADAR1 and their functional validation.

The role of ADARs in non-physiological settings, of both individual and combined ADAR deficiency, has not been significantly explored. Studies in humans have yielded interesting associations of human traits with modifications of ADAR family members. Recent studies from human cancer suggest roles for ADAR proteins in both solid and hematological tumors (Ma et al. 2011; Maas et al. 2001). Observations of altered editing activity in tumors, both increased and decreased, highlight a potentially fascinating role for these enzymes in the diversification of the tumor proteome and regulation of tumor transcription through effects on miRNA pathways (Shah et al. 2009). Ongoing improvements in the modeling of human cancer in the mouse and siRNA/shRNA technologies lend themselves to a functional assessment of the contribution of ADARs to cancer pathogenesis. Genetic association studies of human centenarians have identified variations in ADAR2 that are associated with longevity (Sebastiani et al. 2009). The functional effects of these polymorphisms have not been clearly defined, although it would appear from the murine models that these cannot be loss of function alleles. Mutant ADAR1 alleles have been identified in the dominantly inherited human skin pigmentation disorder *dyschromatosis symmetrica hereditaria*. These human alleles appear to disrupt the function of ADAR1 leading to a loss of function or dominant negative effect and present as a range of point mutations (Hou et al. 2007; Liu et al. 2006a, b). The modeling of these disease-associated alleles in murine models may yield important information regarding the cellular roles and targets of ADARs in the pathology of human disease.

It will be important to attempt to define functions for ADARs beyond RNA editing. The uncoupling of the deaminase function of ADAR1 and ADAR2 from a perhaps more general role in RNA metabolism mediated by their dsRNA and DNA binding domains is yet to be fully explored. Recent data suggest an involvement of ADAR1's Z-alpha domain in the regulation of protein translation efficiency at the ribosome (Feng et al. 2011), which warrants further assessment in vivo. Given the diversity of features and functions exhibited by the members of the ADAR family, future studies may well uncover a secret life of these intriguing proteins.

Acknowledgments Work in C.R.W's laboratory is supported by grants from the National Health and Medical Research Council (NHMRC) Australia; C.R.W. is the Philip Desbrow Seniro Research Fellow of the Leukaemia Foundation.

References

Athanasiadis A, Rich A, Maas S (2004) Widespread A-to-I RNA editing of Alu-containing mRNAs in the human transcriptome. PLoS Biol 2:e391

Bartel DP (2004) MicroRNAs: genomics, biogenesis, mechanism, and function. Cell 116:281–297

Bass BL, Weintraub H (1987) A developmentally regulated activity that unwinds RNA duplexes. Cell 48:607–613

Bass BL, Weintraub H (1988) An unwinding activity that covalently modifies its double-stranded RNA substrate. Cell 55:1089–1098

Blow MJ, Grocock RJ, van Dongen S, Enright AJ, Dicks E, Futreal PA, Wooster R, Stratton MR (2006) RNA editing of human microRNAs. Genome Biol 7:R27

Borchert GM, Gilmore BL, Spengler RM, Xing Y, Lanier W, Bhattacharya D, Davidson BL (2009) Adenosine deamination in human transcripts generates novel microRNA binding sites. Hum Mol Genet 18:4801–4807

Brusa R, Zimmermann F, Koh DS, Feldmeyer D, Gass P, Seeburg PH, Sprengel R (1995) Early-onset epilepsy and postnatal lethality associated with an editing-deficient GluR-B allele in mice. Science 270:1677–1680

Chen CX, Cho DS, Wang Q, Lai F, Carter KC, Nishikura K (2000) A third member of the RNA-specific adenosine deaminase gene family, ADAR3, contains both single- and double-stranded RNA binding domains. RNA 6:755–767

Das AK, Carmichael GG (2007) ADAR editing wobbles the microRNA world. ACS Chem Biol 2:217–220

Eckmann CR, Jantsch MF (1999) The RNA-editing enzyme ADAR1 is localized to the nascent ribonucleoprotein matrix on Xenopus lampbrush chromosomes but specifically associates with an atypical loop. J Cell Biol 144:603–615

Eckmann CR, Neunteufl A, Pfaffstetter L, Jantsch MF (2001) The human but not the Xenopus RNA-editing enzyme ADAR1 has an atypical nuclear localization signal and displays the characteristics of a shuttling protein. Mol Biol Cell 12:1911–1924

Eisenberg E, Nemzer S, Kinar Y, Sorek R, Rechavi G, Levanon EY (2005) Is abundant A-to-I RNA editing primate-specific? Trends Genet 21:77–81

Feng Y, Sansam CL, Singh M, Emeson RB (2006) Altered RNA editing in mice lacking ADAR2 autoregulation. Mol Cell Biol 26:480–488

Feng S, Li H, Zhao J, Pervushin K, Lowenhaupt K, Schwartz TU, Droge P (2011) Alternate rRNA secondary structures as regulators of translation. Nat Struct Mol Biol 18:169–176

Gan Z, Zhao L, Yang L, Huang P, Zhao F, Li W, Liu Y (2006) RNA editing by ADAR2 is metabolically regulated in pancreatic islets and beta-cells. J Biol Chem 281:33386–33394

George CX, Li Z, Okonski KM, Toth AM, Wang Y, Samuel CE (2009) Tipping the balance: antagonism of PKR kinase and ADAR1 deaminase functions by virus gene products. J Interferon Cytokine Res 29:477–487

George CX, Gan Z, Liu Y, Samuel CE (2011) Adenosine deaminases acting on RNA, RNA editing, and interferon action. J Interferon Cytokine Res 31:99–117

Gothert JR, Gustin SE, Hall MA, Green AR, Gottgens B, Izon DJ, Begley CG (2005) In vivo fate-tracing studies using the Scl stem cell enhancer: embryonic hematopoietic stem cells significantly contribute to adult hematopoiesis. Blood 105:2724–2732

Greger IH, Khatri L, Ziff EB (2002) RNA editing at arg607 controls AMPA receptor exit from the endoplasmic reticulum. Neuron 34(5):759–772

Hartner JC, Schmittwolf C, Kispert A, Muller AM, Higuchi M, Seeburg PH (2004) Liver disintegration in the mouse embryo caused by deficiency in the RNA-editing enzyme ADAR1. J Biol Chem 279:4894–4902

Hartner JC, Walkley CR, Lu J, Orkin SH (2009) ADAR1 is essential for the maintenance of hematopoiesis and suppression of interferon signaling. Nat Immunol 10:109–115

Heale BS, Keegan LP, McGurk L, Michlewski G, Brindle J, Stanton CM, Caceres JF, O'Connell MA (2009) Editing independent effects of ADARs on the miRNA/siRNA pathways. EMBO J 28:3145–3156

Herbert A, Lowenhaupt K, Spitzner J, Rich A (1995) Double-stranded RNA adenosine deaminase binds Z-DNA in vitro. Nucleic Acids Symp Ser 13(33):16–19

Hideyama T, Yamashita T, Suzuki T, Tsuji S, Higuchi M, Seeburg PH, Takahashi R, Misawa H, Kwak S (2010) Induced loss of ADAR2 engenders slow death of motor neurons from Q/R site-unedited GluR2. J Neurosci 30:11917–11925

Higuchi M, Single FN, Kohler M, Sommer B, Sprengel R, Seeburg PH (1993) RNA editing of AMPA receptor subunit GluR-B: a base-paired intron-exon structure determines position and efficiency. Cell 75:1361–1370

Higuchi M, Maas S, Single FN, Hartner J, Rozov A, Burnashev N, Feldmeyer D, Sprengel R, Seeburg PH (2000) Point mutation in an AMPA receptor gene rescues lethality in mice deficient in the RNA-editing enzyme ADAR2. Nature 406:78–81

Horsch M, Seeburg PH, Adler T, Aguilar-Pimentel JA, Becker L, Calzada-Wack J, Garrett L, Gotz A, Hans W, Higuchi M, Holter SM, Naton B, Prehn C, Puk O, Racz I, Rathkolb B, Rozman J, Schrewe A, Adamski J, Busch DH, Esposito I, Graw J, Ivandic B, Klingenspor M, Klopstock T, Mempel M, Ollert M, Schulz H, Wolf E, Wurst W, Zimmer A, Gailus-Durner V, Fuchs H, Hrabe de Angelis M, Beckers J (2011) Requirement of the RNA editing enzyme ADAR2 for normal physiology in mice. J Biol Chem 286(21):18614–18622

Hou Y, Chen J, Gao M, Zhou F, Du W, Shen Y, Yang S, Zhang XJ (2007) Five novel mutations of RNA-specific adenosine deaminase gene with dyschromatosis symmetrica hereditaria. Acta Derm Venereol 87:18–21

Kawahara Y, Kwak S, Sun H, Ito K, Hashida H, Aizawa H, Jeong SY, Kanazawa I (2003) Human spinal motoneurons express low relative abundance of GluR2 mRNA: an implication for excitotoxicity in ALS. J Neurochem 85:680–689

Kawahara Y, Ito K, Sun H, Aizawa H, Kanazawa I, Kwak S (2004) Glutamate receptors: RNA editing and death of motor neurons. Nature 427:801

Kawahara Y, Sun H, Ito K, Hideyama T, Aoki M, Sobue G, Tsuji S, Kwak S (2006) Underediting of GluR2 mRNA, a neuronal death inducing molecular change in sporadic ALS, does not occur in motor neurons in ALS1 or SBMA. Neurosci Res 54:11–14

Kawahara Y, Zinshteyn B, Chendrimada TP, Shiekhattar R, Nishikura K (2007a) RNA editing of the microRNA-151 precursor blocks cleavage by the Dicer-TRBP complex. EMBO Rep 8:763–769

Kawahara Y, Zinshteyn B, Sethupathy P, Iizasa H, Hatzigeorgiou AG, Nishikura K (2007b) Redirection of silencing targets by adenosine-to-inosine editing of miRNAs. Science 315:1137–1140

Kawahara Y, Megraw M, Kreider E, Iizasa H, Valente L, Hatzigeorgiou AG, Nishikura K (2008) Frequency and fate of microRNA editing in human brain. Nucleic Acids Res 36:5270–5280

Kim U, Nishikura K (1993) Double-stranded RNA adenosine deaminase as a potential mammalian RNA editing factor. Semin Cell Biol 4:285–293

Kim U, Garner TL, Sanford T, Speicher D, Murray JM, Nishikura K (1994a) Purification and characterization of double-stranded RNA adenosine deaminase from bovine nuclear extracts. J Biol Chem 269:13480–13489

Kim U, Wang Y, Sanford T, Zeng Y, Nishikura K (1994b) Molecular cloning of cDNA for double-stranded RNA adenosine deaminase, a candidate enzyme for nuclear RNA editing. Proc Natl Acad Sci USA 91:11457–11461

Kim DD, Kim TT, Walsh T, Kobayashi Y, Matise TC, Buyske S, Gabriel A (2004) Widespread RNA editing of embedded alu elements in the human transcriptome. Genome Res 14:1719–1725

Kuhn R, Schwenk F, Aguet M, Rajewsky K (1995) Inducible gene targeting in mice. Science 269:1427–1429

Kwak S, Kawahara Y (2005) Deficient RNA editing of GluR2 and neuronal death in amyotropic lateral sclerosis. J Mol Med 83:110–120

Levanon EY, Eisenberg E (2006) Algorithmic approaches for identification of RNA editing sites. Brief Funct Genomic Proteomic 5:43–45

Levanon EY, Eisenberg E, Yelin R, Nemzer S, Hallegger M, Shemesh R, Fligelman ZY, Shoshan A, Pollock SR, Sztybel D, Olshansky M, Rechavi G, Jantsch MF (2004) Systematic identification of abundant A-to-I editing sites in the human transcriptome. Nat Biotechnol 22:1001–1005

Levanon EY, Hallegger M, Kinar Y, Shemesh R, Djinovic-Carugo K, Rechavi G, Jantsch MF, Eisenberg E (2005) Evolutionarily conserved human targets of adenosine to inosine RNA editing. Nucleic Acids Res 33:1162–1168

Li JB, Levanon EY, Yoon JK, Aach J, Xie B, Leproust E, Zhang K, Gao Y, Church GM (2009) Genome-wide identification of human RNA editing sites by parallel DNA capturing and sequencing. Science 324:1210–1213

Liu Q, Jiang L, Liu WL, Kang XJ, Ao Y, Sun M, Luo Y, Song Y, Lo WH, Zhang X (2006a) Two novel mutations and evidence for haploinsufficiency of the ADAR gene in dyschromatosis symmetrica hereditaria. Br J Dermatol 154:636–642

Liu Y, Xiao SX, Peng ZH, Lei XB, Wang JM, Li Y, Li XL (2006b) Two frameshift mutations of the double-stranded RNA-specific adenosine deaminase gene in Chinese pedigrees with dyschromatosis symmetrica hereditaria. Br J Dermatol 155:473–476

Lomeli H, Mosbacher J, Melcher T, Hoger T, Geiger JR, Kuner T, Monyer H, Higuchi M, Bach A, Seeburg PH (1994) Control of kinetic properties of AMPA receptor channels by nuclear RNA editing. Science 266:1709–1713

Luciano DJ, Mirsky H, Vendetti NJ, Maas S (2004) RNA editing of a miRNA precursor. RNA 10:1174–1177

Ma CH, Chong JH, Guo Y, Zeng HM, Liu SY, Xu LL, Wei J, Lin YM, Zhu XF, Zheng GG (2011) Abnormal expression of ADAR1 isoforms in Chinese pediatric acute leukemias. Biochem Biophys Res Commun 406:245–251

Maas S, Patt S, Schrey M, Rich A (2001) Underediting of glutamate receptor GluR-B mRNA in malignant gliomas. Proc Natl Acad Sci USA 98:14687–14692

Mattick JS, Mehler MF (2008) RNA editing, DNA recoding and the evolution of human cognition. Trends Neurosci 31:227–233

Melcher T, Maas S, Herb A, Sprengel R, Higuchi M, Seeburg PH (1996a) RED2, a brain-specific member of the RNA-specific adenosine deaminase family. J Biol Chem 271:31795–31798

Melcher T, Maas S, Herb A, Sprengel R, Seeburg PH, Higuchi M (1996b) A mammalian RNA editing enzyme. Nature 379:460–464

Mittaz L, Scott HS, Rossier C, Seeburg PH, Higuchi M, Antonarakis SE (1997) Cloning of a human RNA editing deaminase (ADARB1) of glutamate receptors that maps to chromosome 21q22.3. Genomics 41:210–217

Moller-Krull M, Zemann A, Roos C, Brosius J, Schmitz J (2008) Beyond DNA: RNA editing and steps toward Alu exonization in primates. J Mol Biol 382:601–609

Nie Y, Ding L, Kao PN, Braun R, Yang JH (2005) ADAR1 interacts with NF90 through double-stranded RNA and regulates NF90-mediated gene expression independently of RNA editing. Mol Cell Biol 25:6956–6963

Patterson JB, Samuel CE (1995) Expression and regulation by interferon of a double-stranded-RNA-specific adenosine deaminase from human cells: evidence for two forms of the deaminase. Mol Cell Biol 15:5376–5388

Paz-Yaacov N, Levanon EY, Nevo E, Kinar Y, Harmelin A, Jacob-Hirsch J, Amariglio N, Eisenberg E, Rechavi G (2010) Adenosine-to-inosine RNA editing shapes transcriptome diversity in primates. Proc Natl Acad Sci USA 107:12174–12179

Rebagliati MR, Melton DA (1987) Antisense RNA injections in fertilized frog eggs reveal an RNA duplex unwinding activity. Cell 48:599–605

Rueter SM, Dawson TR, Emeson RB (1999) Regulation of alternative splicing by RNA editing. Nature 399:75–80

Sakurai M, Yano T, Kawabata H, Ueda H, Suzuki T (2010) Inosine cyanoethylation identifies A-to-I RNA editing sites in the human transcriptome. Nat Chem Biol 6:733–740

Samuel CE (2011) Adenosine deaminases acting on RNA (ADARs) are both antiviral and proviral. Virology 411(2):180–193
Scadden AD, Smith CW (2001) RNAi is antagonized by A–>I hyper-editing. EMBO Rep 2:1107–1111
Sebastiani P, Montano M, Puca A, Solovieff N, Kojima T, Wang MC, Melista E, Meltzer M, Fischer SE, Andersen S, Hartley SH, Sedgewick A, Arai Y, Bergman A, Barzilai N, Terry DF, Riva A, Anselmi CV, Malovini A, Kitamoto A, Sawabe M, Arai T, Gondo Y, Steinberg MH, Hirose N, Atzmon G, Ruvkun G, Baldwin CT, Perls TT (2009) RNA editing genes associated with extreme old age in humans and with lifespan in C. elegans. PLoS One 4:e8210
Seeburg PH, Higuchi M, Sprengel R (1998) RNA editing of brain glutamate receptor channels: mechanism and physiology. Brain Res Brain Res Rev 26:217–229
Shah SP, Morin RD, Khattra J, Prentice L, Pugh T, Burleigh A, Delaney A, Gelmon K, Guliany R, Senz J, Steidl C, Holt RA, Jones S, Sun M, Leung G, Moore R, Severson T, Taylor GA, Teschendorff AE, Tse K, Turashvili G, Varhol R, Warren RL, Watson P, Zhao Y, Caldas C, Huntsman D, Hirst M, Marra MA, Aparicio S (2009) Mutational evolution in a lobular breast tumour profiled at single nucleotide resolution. Nature 461:809–813
Singh M, Kesterson RA, Jacobs MM, Joers JM, Gore JC, Emeson RB (2007) Hyperphagia-mediated obesity in transgenic mice misexpressing the RNA-editing enzyme ADAR2. J Biol Chem 282:22448–22459
Sommer B, Kohler M, Sprengel R, Seeburg PH (1991) RNA editing in brain controls a determinant of ion flow in glutamate-gated channels. Cell 67:11–19
Stefl R, Oberstrass FC, Hood JL, Jourdan M, Zimmermann M, Skrisovska L, Maris C, Peng L, Hofr C, Emeson RB, Allain FH (2010) The solution structure of the ADAR2 dsRBM-RNA complex reveals a sequence-specific readout of the minor groove. Cell 143:225–237
Takuma H, Kwak S, Yoshizawa T, Kanazawa I (1999) Reduction of GluR2 RNA editing, a molecular change that increases calcium influx through AMPA receptors, selective in the spinal ventral gray of patients with amyotrophic lateral sclerosis. Ann Neurol 46:806–815
Vitali P, Scadden AD (2010) Double-stranded RNAs containing multiple IU pairs are sufficient to suppress interferon induction and apoptosis. Nat Struct Mol Biol 17:1043–1050
Wagner RW, Smith JE, Cooperman BS, Nishikura K (1989) A double-stranded RNA unwinding activity introduces structural alterations by means of adenosine to inosine conversions in mammalian cells and Xenopus eggs. Proc Natl Acad Sci USA 86:2647–2651
Wang Q, Khillan J, Gadue P, Nishikura K (2000) Requirement of the RNA editing deaminase ADAR1 gene for embryonic erythropoiesis. Science 290:1765–1768
Wang Q, Miyakoda M, Yang W, Khillan J, Stachura DL, Weiss MJ, Nishikura K (2004) Stress-induced apoptosis associated with null mutation of ADAR1 RNA editing deaminase gene. J Biol Chem 279:4952–4961
Ward SV, George CX, Welch MJ, Liou LY, Hahm B, Lewicki H, de la Torre JC, Samuel CE, Oldstone MB (2011) RNA editing enzyme adenosine deaminase is a restriction factor for controlling measles virus replication that also is required for embryogenesis. Proc Natl Acad Sci USA 108:331–336
Wulff BE, Sakurai M, Nishikura K (2010) Elucidating the inosinome: global approaches to adenosine-to-inosine RNA editing. Nat Rev Genet 12:81–85
XuFeng R, Boyer MJ, Shen H, Li Y, Yu H, Gao Y, Yang Q, Wang Q, Cheng T (2009) ADAR1 is required for hematopoietic progenitor cell survival via RNA editing. Proc Natl Acad Sci USA 106:17763–17768
Yang JH, Luo X, Nie Y, Su Y, Zhao Q, Kabir K, Zhang D, Rabinovici R (2003) Widespread inosine-containing mRNA in lymphocytes regulated by ADAR1 in response to inflammation. Immunology 109:15–23
Yang W, Chendrimada TP, Wang Q, Higuchi M, Seeburg PH, Shiekhattar R, Nishikura K (2006) Modulation of microRNA processing and expression through RNA editing by ADAR deaminases. Nat Struct Mol Biol 13:13–21

Regulation and Functions of ADAR in *Drosophila*

Simona Paro, Xianghua Li, Mary A. O'Connell and Liam P. Keegan

Abstract *Drosophila melanogaster* has a single *Adar* gene encoding a protein related to mammalian ADAR2 that edits transcripts encoding glutamate receptor subunits. We describe the structure of the *Drosophila Adar* locus and use ModENCODE information to supplement published data on *Adar* gene transcription, and splicing. We discuss the roles of ADAR in *Drosophila* in terms of the two main types of RNA molecules edited and roles of ADARs as RNA-binding proteins. Site-specific RNA editing events in transcripts encoding ion channel subunits were initially found serendipitously and subsequent directed searches for editing sites and transcriptome sequencing have now led to 972 edited sites being identified in 597 transcripts. Four percent of *D. melanogaster* transcripts are site-specifically edited and these encode a wide range of largely membrane-associated proteins expressed particularly in CNS. Electrophysiological studies on the effects of specific RNA editing events on ion channel subunits do not suggest that loss of RNA editing events in ion channels consistently produce a particular outcome such as making *Adar* mutant neurons more excitable. This possibility would have been consistent with neurodegeneration seen in *Adar* mutant fly brains. A further set of ADAR targets are dsRNA intermediates in siRNA generation, derived from transposons and from structured RNA loci. Transcripts with convergent overlapping 3′ ends are also edited and the first discovered instance of RNA editing in *Drosophila*, in the *Rnp4F* transcript, is an example. There is no evidence yet to show that *Adar* antagonizes RNA interference in *Drosophila*. Evidence has been obtained that catalytically inactive ADAR proteins exert effects on microRNA generation and RNA interference. Whether all effects of inactive ADARs are due to RNA-binding or to even further roles of these proteins remains to be determined.

S. Paro · X. Li · M. A. O'Connell · L. P. Keegan (✉)
MRC Human Genetics Unit, Institute of Genetics and Molecular Medicine,
Western General Hospital, Crewe Road, Edinburgh EH4 2XU, UK
e-mail: Liam.Keegan@hgu.mrc.ac.uk

Current Topics in Microbiology and Immunology (2012) 353: 221–236
DOI: 10.1007/82_2011_152

Published Online: 15 July 2011

Keywords RNA editing · ADAR · Drosophila · RNA interference · Genetics · Development

Contents

1 Introduction

Drosophila has a single *Adar* gene encoding a protein closely related to vertebrate ADAR2. This makes *Drosophila* an excellent model to study conserved roles of ADAR2–type proteins in site-specific editing of CNS transcripts. This role of ADARs appears to have developed strongly in the evolution of *Drosophila* with many edited transcripts identified. Other roles of ADARs in non-specific RNA editing related to microRNA processing and RNA interference or as RNA-binding proteins are likely to be conserved also.

2 *Drosophila Adar* Gene Transcription, Splicing and RNA Editing

2.1 Adar *Gene Transcription*

The single *Adar* gene in *Drosophila melanogaster (D. melanogaster)* lies at cytogenetic position 2B6-7, near the tip of the X chromosome (Palladino et al. 2000a). Expression is highest in the CNS but also widespread outside the CNS at lower levels. Expression of *Adar* increases at metamorphosis. It was proposed that two different promoters, 4A and 4B, control the transcription of the *Adar* gene (Fig. 1). The constitutive 4A promoter is active all through fly development and transcription increases at the pupal stage. The 4B promoter was proposed to be approximately 1 kb downstream, within a large intron of transcripts from the 4A

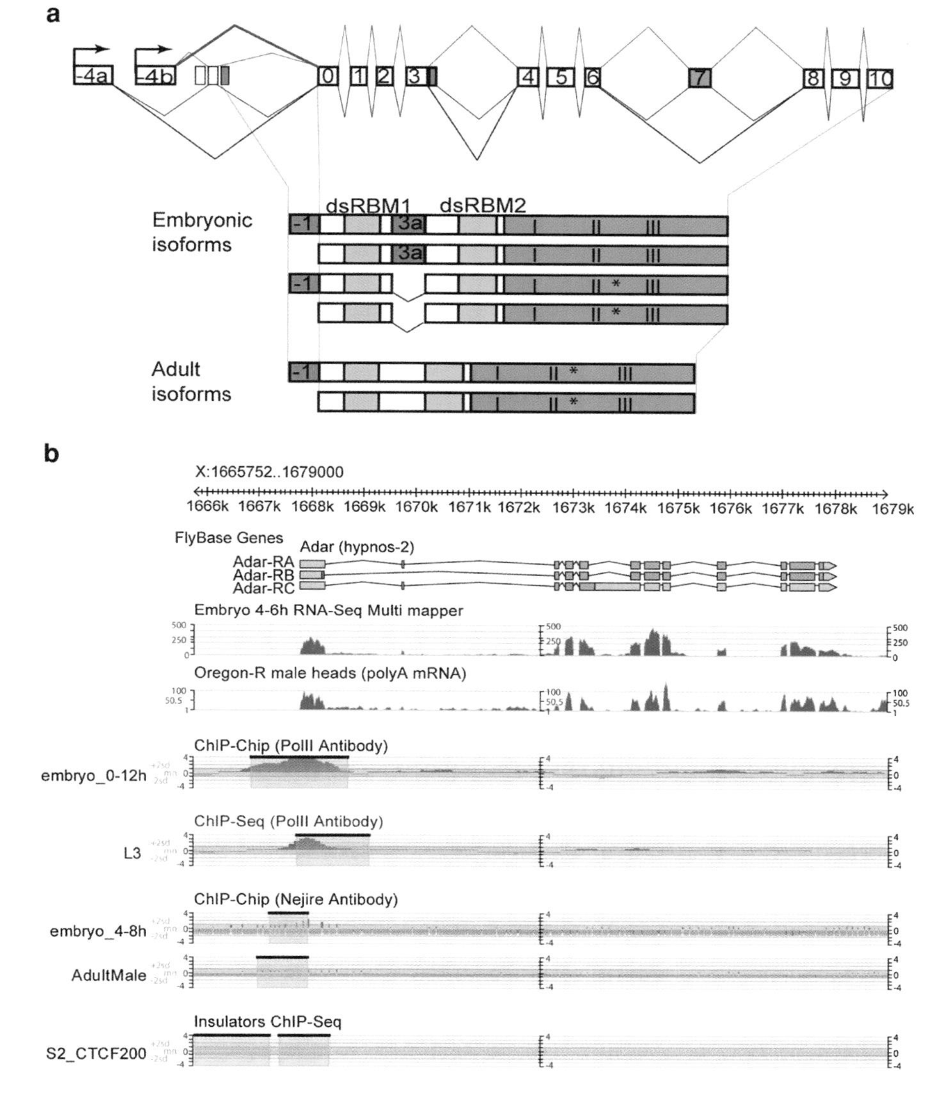

Fig. 1 **a**. *Adar* gene structure, embryonic splicing pattern (*below the gene*) and adult splicing pattern (*above the gene*), and ADAR protein isoforms expressed in embryos and adults. **b**. Selected *Adar* gene region tracks from ModENCODE browser showing embryonic and adult transcription, and binding patterns of RNA polymerase II, the enhancer-locating transcriptional coactivator P300/CBP and the insulator protein CTCF

promoter, based on finding cDNAs with an alternative 5′ exon derived from this region and 5′ RACE analysis (Palladino et al. 2000a).

For *Drosophila* genes and chromosomes a great deal of new information has been provided by the Model Organisms component of the Encyclopedia of DNA Elements project (ModENCODE), which covers the entire fly genome (Roy et al. 2010). Developmental transcription data from the *Drosophila* ModENCODE project does not show a dramatic increase in transcripts corresponding to the proposed first exon—4B of the adult-specific transcript in adult flies, (see Fig. 1 and *Adar* data at FlyBase at http://flybase.org and GBrowse links to data for *Adar* on their mirror site for ModENCODE at http://modencode.oicr.on.ca/fgb2/gbrowse/fly/?name=Adar). Some exons may be underrepresented in RNA-Seq data for various reasons.

Other data from the ModENCODE project shows that the *Adar* locus lies in an open chromatin region, actively transcribed, with expected enrichments of histone H3K4Me1, H3K4Me3 and H3K27Ac modifications at the constitutive promoter as well as RNA Polymerase II accumulation at the promoter in both embryos and adults, strong CTCF with some extension of Polymerase II more 3′ in the adult data (Fig. 1). Upstream of the constitutive promoter there is a very strong prediction of a chromatin insulator based on CTCF protein binding in embryos and adults. Insulators may establish chromatin loops and form boundaries between regions of gene regulation. Other insulator predictions are about 180 kb downstream and 110 kb upstream of the *Adar* promoter. The promoter region also binds Origin Recognition Complex (ORC) proteins in embryo and at metamorphosis and this and other evidence suggests that the promoter region contains an origin of replication active at these times.

A possible enhancer immediately upstream of the constitutive promoter is suggested by binding of the *Drosophila* homolog of the transcriptional coactivator P300/CBP, which is encoded by the *Nejire* gene in *Drosophila* (Akimaru et al. 1997). This protein has been extremely valuable in locating enhancers in human and vertebrate genomes (Visel et al. 2009). CBP is CREB-binding protein, a transcriptional coactivator that binds to the DNA-binding cAMP response element binding protein CREB as well as to many other transcription activators bound at enhancers (Vo and Goodman 2001). The CBP coactivator has histone acetyltransferase activity at H3K27 sites and other sites on histones. Most of the transcription regulators, particularly neural transcription regulators, that are likely to regulate *Adar* specifically have not been mapped yet and the *Adar* transcriptional control sequences have not been defined. The cAMP response protein CREB is a possible regulator of *Adar*, based on mammalian data (Gan et al. 2006; Peng et al. 2006) and this could provide a link between *Adar* expression and neuronal activity.

2.2 *Embryonic and Adult* Adar *Splice forms and ADAR Protein Isoforms*

The *Adar* transcripts have long 5′ UTRs with alternatively spliced exons. Based on the estimated relative abundances of different splice forms these transcripts are expected to generate predominantly two different protein isoforms starting specifically at the alternative exons −1 or +1; the inclusion of alternative exon −1 results in a protein being expressed with an additional 12 amino acids at the amino terminus. Two other starting methionines, in the more rarely included exon −2 and exon 0, produce two different protein isoforms that share high homology at the amino terminus (MKFDS and MKFEC) (Palladino et al. 2000b).

A constitutive splicing pattern is seen clearly in embryos that persists in the background also in adults but an adult-specific splicing pattern in a subset of transcripts is superimposed on this. Transcripts are spliced to include or exclude alternative exon 3a with exclusion of this exon occurring in the adult-specific splicing pattern. The ADAR 3/4 isoform predominates after metamorphosis (Palladino et al. 2000a). Exon 3a has a rare nonconsensus splice donor site (GCAAG vs. GTAAG) and it may be that a specific splicing enhancer contributes to the inclusion of exon 3a (Marcucci et al. 2009). Interestingly, the inclusion of exon 3a introduces an additional 38 amino acids, modifying the distance between the two double strand RNA binding motifs (dsRBM1 and dsRBM2), to a spacing that resembles that of vertebrate ADAR1 rather than ADAR2. There is a very strong correlation between the presence of adult exon 4b in the 5′ UTR and the adult splicing pattern deleting exon 3a. The adult splicing pattern also correlates strongly with RNA editing at exon 7 in the *Adar* transcript.

Also, in embryos particularly, transcripts accumulate in which exon 7 is spliced out. This may serve to restrain ADAR activity in embryos as truncated ADAR proteins are predicted (Ma et al. 2002). Most of exon 7, though not the splice junctions, are predicted to form a large dsRNA structure involved in editing here (Keegan et al. 2005). This structure may affect the splicing of exon 7.

2.3 Adar *Mutant Phenotypes and Outstanding Questions in* Adar *Regulation*

The $Adar^{5G1}$ deletion removes the entire *Adar* gene. Under ideal conditions, $Adar^{5G1}$ mutants develop into morphologically normal adults and they perform functions necessary to sustain life (eating, respiration and metabolism) (Palladino et al. 2000b). However they display severe neuro-behavioural deficits such as slow uncoordinated locomotion, tremors and alteration of normal posture; furthermore they obsessively and frequently clean their wings and they are able to jump and fly but only when repeatedly provoked. The earlier characterized $Adar^{1F4}$ deletion mutant is intriguing; it deletes only the promoters and not the coding sequence and has some residual transcript expressed at a low level. It is phenotypically indistinguishable from $Adar^{5G1}$ but it edits the *Adar* transcript only and not any other target transcript that has been examined.

The main outstanding questions about *Adar* gene expression relate to how expression is controlled. Is transcription regulated by CREB or by neuronal factors needed for ubiquitous neural expression? Is *Adar* expression or self-editing regulated by neuronal activity?

3 The *Drosophila* ADAR Protein Isoforms

Drosophila ADAR contains two double strand RNA binding domains within the amino terminal half of the protein: dsRBM1 (53-133aa) and dsRBM2 (196-273aa). dADAR protein with the alternative exon 3a inserted between the two dsRBMs rescues *Adar* mutant phenotypes less efficiently than the adult-typical ADAR 3/4 isoform (Keegan et al. 2005).

Binding to RNA is necessary for formation of vertebrate ADAR homo- or hetero-dimers and for editing activity. Sequences within the first 46 amino acids and the first dsRBM are required for dimerization of dADAR (Gallo et al. 2003).

However, based on domain exchange experiments between mammalian ADAR1 and ADAR2, the main determinant of ADAR specificity lie in the deaminase domain at the carboxyl terminus. The dADAR deaminase domain contains three zinc-binding motifs (at positions 372, 430 and 493) that are essential to coordinate zinc near the active site glutamate at position 374.

The self-editing event that takes place in the catalytic domain of the protein changes a serine residue (S) close to the zinc-chelating motif II to a glycine (G). In adult flies, ADAR edits its own mRNA with 40% efficiency to encode an ADAR 3/4 G edited isoform that is eightfold less active by in vitro measurements and that rescues *Adar* mutant phenotypes less efficiently than the unedited isoform (Keegan et al. 2005). It is not known what the physiological role of the self-editing event is. Understanding this will require further study of factors regulating the activity of ADAR itself.

4 Roles of *Drosophila* ADAR

There are three general categories of effects that we can distinguish for ADARs: site-specific RNA editing in transcripts, non-specific RNA editing in long dsRNA precursors in RNA interference pathways and potential RNA editing-independent roles, probably as RNA-binding proteins.

4.1 Site-Specific RNA Editing in Drosophila Transcripts and Consequences

4.1.1 Serendipitously Discovered Editing Sites Led to Searches for Further Sites

Site-specific RNA editing events were first detected serendipitously in *Drosophila* transcripts encoding ion channel subunits such as *cacophony (cac)* encoding the large, pore-forming subunit of the voltage-gated CNS calcium channel (Smith et al. 1996) and *paralytic (para)* encoding the large, pore-forming subunit of the voltage-gated sodium channel (Hanrahan et al. 2000). Other individually identified edited transcripts included *DrosGluCl* encoding a glutamate-gated chloride channel subunit gene (Semenov and Pak 1999), the *Adar* transcript itself (Palladino et al. 2000a) and the *Dalpha5* transcript encoding the pore-forming subunit of a nicotinic acetylcholine receptor (Grauso et al. 2002).

Although no definite signature sequence motif was found for an ADAR editing site, editing site complementary sequences (ECSs) usually located in an adjacent intron form imperfect duplex RNA by base-pairing with the exon that contains the adenosine to be edited. This is as expected from studies of vertebrate glutamate receptor transcript editing (Higuchi et al. 1993). Based on the hypothesis that cis-elements required for editing site/ECS duplex formation will be conserved where RNA editing of particular sites is conserved between species. Hoopengardner et al. (2003) identified 16 new edited targets in *Drosophila* by comparing genome sequences of *D. melanogaster* and *D. pseudoobscura* to identify highly conserved exons. They examined 914 genes annotated as ion channels ($n = 135$), G protein-coupled receptors ($n = 178$), proteins involved in synaptic transmission ($n = 102$), and transcription factors ($n = 499$). All the edited transcripts they discovered by this method encode proteins functioning in rapid electrical and chemical neurotransmission, among which were seven voltage-gated ion channels (VGIC), five components of the synaptic release machinery, and four ligand-gated ion channels (LGIC). The number of edited sites differed from one to seven in each transcript. Nevertheless, due to the limited size of the screen pool and the possibility that there are some rapidly evolving ADAR editing events, this approach was not able to detect all the ADAR targets. It was found that in *Drosophila* some ECS elements are not a single sequence unit as in the vertebrate glutamate receptor transcripts but consist of fragments that are not arranged sequentially in the genome but come together in the transcript to pair with the edited region and stack along it (Reenan 2005).

Another systematic approach to identify ADAR targets was carried out using sequence data from the *Drosophila* Gene Collection project which set out to provide a sequence of one individual adult head cDNA with a complete protein-coding sequence for each gene in the genome (DGC; http://www.fruitfly.org/DGC). Stapleton et al. (2006) compared the cDNA clone sequences with genomic DNA and further experimentally verified 27 new targets of ADAR, expanding the categories of edited transcripts to seven. They identified three more classes of ADAR target transcripts: encoding vesicular trafficking proteins, ion homeostasis proteins and cytoskeletal components. However, it remained likely that not all edited transcripts were yet detected, partly because sites edited less than 100% might not be detected in individual cDNA sequences.

4.1.2 Four Percent of all *Drosophila* Transcripts have Site-Specific RNA Editing

The list of known site-specifically edited transcripts in *Drosophila* has recently been very dramatically increased by the publication of the ModENCODE study of the developmental transcriptome based on extensive RNA Seq analyses of

RNA from 72 samples and 30 distinct developmental stages. By analyzing the poly(A)+ RNA Seq data, Graveley et al. (2011) identified 972 edited positions within transcripts of 597 genes, which is around 4% of the *Drosophila* genes.

Graveley and colleagues observed several important common features of the edited sites in their sequencing data. Firstly, consistent with the earlier studies (Hoopengardner et al. 2003; Jepson and Reenan 2007), exons containing editing sites are more highly conserved than unedited exons. Secondly the frequency of editing increases throughout development; editing often begins in late pupal stages and many of the newly discovered sites are edited only in adult flies. Thirdly, editing levels do not correlate with the expression levels of the genes. Lastly, the majority of the edited sites (630) alter amino acid coding, 201 sites are silent, and 141 are within untranslated regions.

In addition, Graveley and colleagues identified by computational analysis three length classes of a potential editing-associated sequence motif having the edited A near the 3′ end. Although motifs A and B are more common, Motif C, the shortest one, is observed to be most strongly associated with the editing sites and over-represented in early developmental editing events (Graveley et al. 2011). The other two motifs are longer than Motif C but rather similar and tend to have a G immediately 3′ of the edited A and further Gs running 5′ at −2, −5, −8 and −11 from the edited position i.e. G residues at every third base. Oddly, these conserved motifs are mostly 5′ of the edited A, whereas the ADAR dsRBDs bind mainly 3′ of the edited A (Stefl et al. 2010). It is not clear that these motifs will necessarily contribute to dsRNA duplex stretches as editing site/ECS duplexes tend to be short in *Drosophila* compared to those seen in vertebrate transcripts. Possibly the motifs reflect further interactions of substrate RNAs with ADARs or with other proteins.

Functional categories highly represented among the edited transcripts based on the classification of molecular functions of encoded proteins include transporter activity ($n = 66$), enzyme regulator activity ($n = 31$, mainly GTPase regulator activity), binding activity, catalytic activity and structural molecule activity ($n = 5$, all are genes encoding structural constituents of muscle). The most widely studied edited transcripts encode proteins with transporter activities. However, binding activity is the biggest category of molecular function among the edited transcripts, consisting of protein binding ($n = 132$), nucleotide binding ($n = 76$), lipid binding ($n = 14$) and ion binding ($n = 22$) classes. Edited transcripts included in catalytic activity categories include 31 genes with kinase activity and 17 genes with phosphatase activity. Analyzed from the cellular component aspect, most edited transcripts reside in membrane structures including ion channel complexes, plasma membranes, membrane bounded vesicles and mitochondrial membranes. Also, there are edited transcripts encoding components involved in cell projections, synapses, and cytoskeleton. (AmiGO analyses, and statistical analyses were carried out using the FlyMine website http://www.flymine.org)

4.1.3 Effects of Individual RNA Editing Events on Ion Channel Subunits and Other Proteins

Intriguing suggestions for the overall function of site-specific RNA editing have been made that now need to be re-examined with larger numbers of sites. One proposal is that editing events tend to change less conserved residues in highly conserved functional regions of proteins (Reenan 2005; Yang et al. 2008). A somewhat related suggestion is that editing events tend to alter evolutionarily conserved amino acid sequences in such a way as to introduce an evolutionarily novel residue at a conserved position in the genomic sequence (Tian et al. 2008). RNA editing is then evolutionarily restorative—as though a new, unedited, functional protein isoform is provided from the unedited transcripts while the isoform with the evolutionary consensus residue is provided by RNA editing.

It is not always obvious how significant the functional consequences of editing events in individual proteins are. Nevertheless, to our knowledge, where effects of RNA editing changes on protein functions have been sought they have been found. This suggests that editing events have been selected for effects on protein function even though the effects are sometimes subtle. Several extensively studied editing events include the one in ADAR itself which undergoes self-editing to reduce enzymatic activity, possibly as a fine-tuning mechanism for RNA editing regulation (Keegan et al. 2005).

If there are hints of patterns in the evolutionary selection of editing sites in protein domains then are there also conserved patterns in the effects of RNA editing on protein or neuron function? There are so many editing events in *Drosophila* transcripts that for most the biophysical or physiological consequences are, at best, merely predicted depending on the domains where the edited sites reside. However, recent studies of effects of RNA editing on several *Drosophila* ion channels do now allow these questions to be considered. The GABA receptor, for instance, is generally inhibitory with regard to neuronal excitability. Loss of RNA editing at sites in Rdl leads to increased responsiveness to GABA so that a lower concentration of GABA is sufficient for a half-maximal channel opening response i.e. loss of *Rdl* RNA editing is expected to make neurons less excitable (Jones et al. 2009). Does loss of RNA editing have parallel effects on other channels?

A very detailed study of the biophysical consequences of editing was conducted on *Shab*. *Shab* belongs to the voltage-gated potassium channel family, one of which contains the only specific adenosine position known to be edited by ADARs in chordates, mollusks and arthropods. The RNA structure that directs editing in that case is not conserved between chordates and arthropods so this may be an example of convergent evolution (Bhalla et al. 2004). The original discovery of editing sites in *Shab* by comparing their cDNA with genomic DNA revealed five highly edited sites but the ModENCODE data detects eight edited sites in *Shab* including two silent sites.

Four of the sites are fully edited so, using a two-microelectrode voltage clamp in *Xenopus* oocytes, Ryan et al. (2008) compared the effects of single unediting at each of the edited sites to the genomic construct with no editing and to the fully edited version. The original five sites were the I583V site in the S4 voltage sensor, the T643A site in the pore helix, Y660C in the extracellular turret and T671A and I681V in the S6 segment. One functional consequence of RNA editing in *Shab* is to change the voltage dependence so that the edited channel is less prone to open, which would enhance the excitability of a neuron containing the edited channels. From this the predicted effect of loss of RNA editing is decreasing neuronal excitability.

The effects of loss of editing on the kinetics of channel gating seem to predict an opposite effect on neuronal excitability however. Loss of RNA editing in *Shab* slows both activation and deactivation. The authors suggest that slower activation resulting from loss of editing would tend to make neurons more excitable. Therefore it is unclear whether loss of *Shab* editing would tend to make a neuron more or less excitable overall.

Fully understanding the functional consequences of A-to-I conversion in each transcript is still challenging, especially for the transcripts that have multiple editing sites. The editing events are not only temporally but also spatially tightly regulated to give combinations of isoforms with different sites edited at different levels. For instance, a predominantly expressed edited isoform (68%) of Shaker in male wing tissue is found to have very low (1%) expression in the male head, and the most abundant isoform (27%) in the male head is not detected in the male wing tissue (Ingleby et al. 2009). Homologous recombination in *Drosophila* may be useful to distinguish roles of edited and unedited forms of *Adar* and other edited transcripts (Jepson et al. 2011).

4.2 RNA Editing and RNA Interference

4.2.1 Types of RNA Interference and Production of Different Small RNAs in *Drosophila*

RNA interference is a process of silencing gene expression at the transcriptional or posttranscriptional levels. Small RNAs (21–29 nucleotides) are involved in this process of gene silencing and several classes have been well described in *Drosophila* (Czech and Hannon 2011). These include short interfering RNAs (siRNAs), micro RNAs (miRNAs), repeat associated RNAs (rasiRNAs) and piwi-interacting RNAs (piRNAs).

Small interfering RNAs (siRNAs) are generated by the activity of Dicer2 enzyme which binds longer dsRNA precursors and releases RNA duplexes 21 nucleotides long on each strand with 2 base 3′ overhangs on each end. One strand is discarded and a single stranded siRNA remains in a RNA induced silencing complex (RISC) containing Argonaute2 (AGO2) protein, which then cleaves target RNA molecules complementary to the siRNA strand. This process can act on exogenously supplied dsRNA but when it acts on internally generated dsRNA the products are referred to as endogenous siRNAs (esiRNAs) or repeat associated siRNAs (rasiRNAs).

The processing of pre-miRNAs is similar but these are first cleaved from an endogenously expressed transcript by Drosha enzyme and transported to the cytoplasm for cleavage by Dicer1 enzyme and maturation into mature miRNAs in a miRISC complex containing AGO1 protein. Mature miRNAs inhibit the translation of the complementary mRNA (most often binding to the 3′UTR).

piRNAs are generated particularly from transposon-associated RNAs by Dicer-independent processes and in *Drosophila* these are processed into complexes containing Piwi, Aubergine (Aub) or AGO3 proteins. One amplification process for piRNAs inolves Aub and AGO3 proteins binding opposite strands of triggering RNAs and engaging in ping-pong cleavage reactions that load further RNA copies into the silencing complexes. In mammals this class of RNAs are expressed only in the germline but in *Drosophila* piRNAs are found in germline and also to a lesser extent in somatic tissues (Li et al. 2009; Malone et al. 2009).

4.2.2 RNA Editing in esiRNAs Derived from Transposons, Structured RNA Loci and Convergently Transcribed Genes

The Siomis and their colleagues in Japan have shown that, among *Drosophila* endogenous siRNAs (esiRNAs) recovered from RISC complexes immunoprecipitated with an antibody to AGO2, 18% of all the 21 mer sequences showed A–G changes reflecting probable RNA editing of dsRNA precursors (Kawamura et al. 2008). This corresponds to an adenosine to inosine conversion once every 130 base pairs in precursor dsRNA and this level is similar to estimates of editing rates in mammalian microRNAs (Kawahara et al. 2008): editing of microRNAs has not been studied in *Drosophila*.

ADAR interactions with RNA interference pathways are expected since ADARs and Dicers both act on dsRNA and potentially compete for this substrate (Yang et al. 2005). In addition to a potential for competitive binding of the proteins it has been demonstrated that hyper editing of dsRNA in vitro inhibits cleavage by Dicer (Scadden and Smith 2001). Another experiment had shown that the Tudor-SN component of RISC binds and promotes degradation of hyper edited dsRNAs (Scadden 2005). While some proportion of dsRNA precursors that get edited still go on to contribute to RNA interference pathways with potential to alter the targeting of RISC complexes some portion of edited dsRNAs may be degraded.

The full range of sequences able to contribute to esiRNAs all seem to be equally editable (Kawamura et al. 2008). esiRNAs are derived primarily from transposons and from structured transcripts with potential to form long dsRNA. A very intriguing category of esiRNAs that are more abundant in *Drosophila* than in mice is derived from convergent transcripts with overlapping 3′ ends (Czech et al. 2008; Petschek et al. 1996). The *D. melanogaster* genome has 998 convergently transcribed gene pairs with annotated overlapping transcripts and different but partly overlapping subsets of these produce esiRNAs in ovaries and in Schneider S2 cell cultures. Probably not all convergent overlapping transcript pairs are expressed in the same cells or pairing of UTRs may not be efficient because the numbers of esiRNAs produced are not as high as from structured RNA loci.

The very first A-to-I edited transcripts identified in *Drosophila* were discovered serendipitously as A to G discrepancies between cloned cDNAs and the corresponding genomic sequences of *RNA-binding protein 4F (Rnp4F)*, at 4F5 on the X chromosome (Petschek et al. 1996). *Rnp4F* encodes a protein homologous to human P110/Sart3 protein and to the U4/U6 snRNP recycling factor. Editing in this transcript arises because of convergent transcription of *Rnp4F* and another gene *Something about silencing 10 (Sas10)*, which encodes a nuclear, positively charged, protein. The *S. cerevisiae* ortholog of *Sas10* inhibits chromosomal silencing at the mating-type loci when overexpressed. *Sas10* shares a conserved domain with RNA-binding components of the exosome and U3.

The *Rnp4F* transcript is expressed early in embryogenesis but later in embryogenesis a longer *Sas10* transcript is produced that overlaps with the 3′ end of *Rnp 4F*, leading to a drop in *Rnp4F* transcript levels (Peters et al. 2003). The *Rnp4F* transcript now has adenosines converted to guanosine when cDNA and genomic DNA sequences are compared. In embryonic and larval stages *Sas10* is the much more strongly expressed of the two transcripts but in adults the level of both transcripts is low. The overlapping transcripts appear to trigger RNA interference. modENCODE data now shows that small RNAs are expressed that correspond to the region of transcript overlap in adult tissues, particularly in mutants of *Ago 2* or *r2d2*. Small RNAs from this region are present in AGO 1 complexes immunoprecipitated from adult cells such as ovarian somatic cell (OSC) cultures and from ovaries and these are particulary prominent in AGO 1 complexes immunoprecipitated from cells in which Ago 2 or *r2d2* are mutant or knocked down. Whether loss of RNA editing would help or hinder silencing at *Rnp4F* has not been determined.

The presence of RNA editing events in siRNAs is part of the evidence that these small RNAs are generated from dsRNA. There have not been any reports of editing events in piRNAs. In the case of piRNAs there may be no very extensive dsRNA involved in their formation or any dsRNA that is formed during their biogenesis may be bound within protein complexes and inaccessible to ADARs.

4.2.3 Consequences of RNA Editing for RNA Interference Phenomena

There is no clear published evidence that loss of *Adar* in *Drosophila* influences the potency of RNA interference effects. In the simplest case of pure antagonism loss of *Adar* function should make RNA interference more active, as occurs in *C. elegans* (Knight and Bass 2002). In *Drosophila*, mutations in genes encoding RNAi components did not rescue locomotion defects of *Adar* mutant flies but RNA interference has not been shown to have any relevance for locomotion defects in flies so this is not surprising (Jepson and Reenan 2009). The finding does not rule out the possibility of antagonistic effects of *Adar* on aspects of RNA interference. Such an antagonism has been shown in the case of *white* hairpin-directed RNA interference in *Drosophila* when the cytoplasmically localized human ADAR1 p150 protein is overexpressed but neither *Drosophila Adar* mutations nor overexpression of the nuclear localized *Drosophila* ADAR or human ADAR2 have an effect in this assay. This is presumably because the *white* hairpin is cytoplasmic in this case and ADAR p150 is the only protein with a matched localization.

4.3 RNA Editing-Independent Roles of ADARs

ADARs edit microRNA precursors and thereby redirect RISC complexes containing edited microRNAs to new targets (Kawahara et al. 2007). A follow up study found however that the effect of ADAR binding to inhibit the processing of microRNAs from their precursors is stronger than the effect of retargeting (Heale et al. 2009a). This antagonism is independent of adenosine deamination activity. Stable, catalytically inactive ADAR proteins can be generated by mutating a glutamate residue in the deaminase active site to alanine. Such a catalytically inactive human ADAR1 protein was shown to inhibit processing of micro RNA precursors in vitro and in cultured human cells. This mutant ADAR1 was also shown to retain a substantial portion of the antagonistic effect against RNA interference in *Drosophila* that is exhibited by the wildtype ADAR1 protein. This data joins a range of other evidence that ADARs have important roles independent of deamination that probably arise mainly from their roles as RNA-binding proteins.

A different catalytically inactive ADAR1 mutation in two Japanese families was proposed to have more severe effects than other ADAR1 loss of function mutations because of a dominant negative effect on residual active ADAR1 in those patients (Heale et al. 2009b; Kondo et al. 2008). In *Drosophila* the inactive dADAR protein has been shown to be insufficient to rescue locomotion defects, consistent with the need to edit CNS transcripts. Further study of the relationship between ADAR and RNA interference requires a naturally-occurring, ideally nuclear-based, RNA interference phenomenon to act as a reporter. Whether there are aspects of ADAR function other than antagonizing RNA interference that the inactive protein can provide remains to be determined.

5 Conclusion

In conclusion, study of ADAR RNA editing in *Drosophila* developed from serendipitous findings to systematic discovery of the amazing 596 edited transcripts with 972 sites edited to date. Significant progress has been made in studying the effects of editing on some transcripts, most of which encode ion channels or other membrane proteins. However, there is still quite a long way to go to completely understand the physiological effects of editing on affected proteins in *Drosophila*. Whether editing events are evolutionarily selected to some common purposes and how the editing profiles are fine-tuned remain to be determined. In addition, the effect of ADAR on small RNAs is still not clear. Sequencing of endogenous siRNAs detected vastly over-represented adenosine-to-guanosine mismatches reflecting ADAR editing of dsRNA precursors. Further investigations are needed to test whether cross regulation between A-to-I editing and other post transcriptional modification mechanisms like RNA interference exist.

References

Akimaru H, Hou DX, Ishii S (1997) *Drosophila* CBP is required for dorsal-dependent twist gene expression. Nat Genet 17:211–214

Bhalla T, Rosenthal JJ, Holmgren M, Reenan R (2004) Control of human potassium channel inactivation by editing of a small mRNA hairpin. Nat Struct Mol Biol 11:950–956

Czech B, Hannon GJ (2011) Small RNA sorting: matchmaking for Argonautes. Nat Rev Genet 12:19–31

Czech B, Malone CD, Zhou R, Stark A, Schlingeheyde C, Dus M, Perrimon N, Kellis M, Wohlschlegel JA, Sachidanandam R et al (2008) An endogenous small interfering RNA pathway in *Drosophila*. Nature 453:798–802

Gallo A, Keegan LP, Ring GM, O'Connell MA (2003) An ADAR that edits transcripts encoding ion channel subunits functions as a dimer. Embo J 22:3421–3430

Gan Z, Zhao L, Yang L, Huang P, Zhao F, Li W, Liu Y (2006) RNA editing by ADAR2 is metabolically regulated in pancreatic islets and beta-cells. J Biol Chem 281:33386–33394

Grauso M, Reenan RA, Culetto E, Sattelle DB (2002) Novel putative nicotinic acetylcholine receptor subunit genes, Dalpha5, Dalpha6 and Dalpha7, in *D. melanogaster* identify a new and highly conserved target of adenosine deaminase acting on RNA-Mediated A-to-I Pre-mRNA Editing. Genetics 160:1519–1533

Graveley BR, Brooks AN, Carlson JW, Duff MO, Landolin JM, Yang L, Artieri CG, van Baren MJ, Boley N, Booth BW et al (2011) The developmental transcriptome of *Drosophila melanogaster*. Nature 471:473–479

Hanrahan CJ, Palladino MJ, Ganetzky B, Reenan RA (2000) RNA editing of the *Drosophila* para Na(+) channel transcript. Evolutionary conservation and developmental regulation. Genetics 155:1149–1160

Heale BS, Keegan LP, McGurk L, Michlewski G, Brindle J, Stanton CM, Caceres JF, O'Connell MA (2009a) Editing independent effects of ADARs on the miRNA/siRNA pathways. Embo J 28:3145–3156

Heale BS, Keegan LP, O'Connell MA (2009b) ADARs have effects beyond RNA editing. Cell Cycle 8:4011–4012

Higuchi M, Single FN, Kohler M, Sommer B, Sprengel R, Seeburg PH (1993) RNA editing of AMPA receptor subunit GluR-B: a base-paired intron-exon structure determines position and efficiency. Cell 75:1361–1370

Hoopengardner B, Bhalla T, Staber C, Reenan R (2003) Nervous system targets of RNA editing identified by comparative genomics. Science 301:832–836

Ingleby L, Maloney R, Jepson J, Horn R, Reenan R (2009) Regulated RNA editing and functional epistasis in Shaker potassium channels. J Gen Physiol 133:17–27

Jepson JE, Reenan RA (2007) Genetic approaches to studying adenosine-to-inosine RNA editing. Methods Enzymol 424:265–287

Jepson JE, Reenan RA (2009) Adenosine-to-inosine genetic recoding is required in the adult stage nervous system for coordinated behavior in *Drosophila*. J Biol Chem 284:31391–31400

Jepson JE, Savva YA, Yokose C, Sugden AU, Sahin A, Reenan RA (2011) Engineered alterations in RNA editing modulate complex behavior in *Drosophila*: regulatory diversity of adenosine deaminase acting on RNA (ADAR) Targets. J Biol Chem 286:8325–8337

Jones AK, Buckingham SD, Papadaki M, Yokota M, Sattelle BM, Matsuda K, Sattelle DB (2009) Splice-variant- and stage-specific RNA editing of the *Drosophila* GABA receptor modulates agonist potency. J Neurosci 29:4287–4292

Kawahara Y, Zinshteyn B, Sethupathy P, Iizasa H, Hatzigeorgiou AG, Nishikura K (2007) Redirection of silencing targets by adenosine-to-inosine editing of miRNAs. Science 315: 1137–1140

Kawahara Y, Megraw M, Kreider E, Iizasa H, Valente L, Hatzigeorgiou AG, Nishikura K (2008) Frequency and fate of microRNA editing in human brain. Nucleic Acids Res 36:5270–5280

Kawamura Y, Saito K, Kin T, Ono Y, Asai K, Sunohara T, Okada TN, Siomi MC, Siomi H (2008) *Drosophila* endogenous small RNAs bind to Argonaute 2 in somatic cells. Nature 453:793–797

Keegan LP, Brindle J, Gallo A, Leroy A, Reenan RA, O'Connell MA (2005) Tuning of RNA editing by ADAR is required in *Drosophila*. Embo J 24:2183–2193

Knight SW, Bass BL (2002) The role of RNA editing by ADARs in RNAi. Mol Cell 10:809–817

Kondo T, Suzuki T, Ito S, Kono M, Negoro T, Tomita Y (2008) Dyschromatosis symmetrica hereditaria associated with neurological disorders. J Dermatol 35:662–666

Li C, Vagin VV, Lee S, Xu J, Ma S, Xi H, Seitz H, Horwich MD, Syrzycka M, Honda BM et al (2009) Collapse of germline piRNAs in the absence of Argonaute3 reveals somatic piRNAs in flies. Cell 137:509–521

Ma E, Tucker MC, Chen Q, Haddad GG (2002) Developmental expression and enzymatic activity of pre-mRNA deaminase in *Drosophila melanogaster*. Brain Res Mol Brain Res 102:100–104

Malone CD, Brennecke J, Dus M, Stark A, McCombie WR, Sachidanandam R, Hannon GJ (2009) Specialized piRNA pathways act in germline and somatic tissues of the *Drosophila* ovary. Cell 137:522–535

Marcucci R, Romano M, Feiguin F, O'Connell MA, Baralle FE (2009) Dissecting the splicing mechanism of the *Drosophila* editing enzyme; dADAR. Nucleic Acids Res 37:1663–1671

Palladino MJ, Keegan LP, O'Connell MA, Reenan RA (2000a) dADAR, a *Drosophila* double-stranded RNA-specific adenosine deaminase is highly developmentally regulated and is itself a target for RNA editing. RNA 6:1004–1018

Palladino MJ, Keegan LP, O'Connell MA, Reenan RA (2000b) A-to-I pre-mRNA editing in *Drosophila* is primarily involved in adult nervous system function and integrity. Cell 102:437–449

Peng PL, Zhong X, Tu W, Soundarapandian MM, Molner P, Zhu D, Lau L, Liu S, Liu F, Lu Y (2006) ADAR2-dependent RNA editing of AMPA receptor subunit GluR2 determines vulnerability of neurons in forebrain ischemia. Neuron 49:719–733

Peters NT, Rohrbach JA, Zalewski BA, Byrkett CM, Vaughn JC (2003) RNA editing and regulation of *Drosophila 4f-rnp* expression by sas-10 antisense readthrough mRNA transcripts. RNA 9:698–710

Petschek JP, Mermer MJ, Scheckelhoff MR, Simone AA, Vaughn JC (1996) RNA editing in *Drosophila 4f-rnp* gene nuclear transcripts by multiple A-to-G conversions. J Mol Biol 259: 885–890

Reenan RA (2005) Molecular determinants and guided evolution of species-specific RNA editing. Nature 434:409–413

Roy S, Ernst J, Kharchenko PV, Kheradpour P, Negre N, Eaton ML, Landolin JM, Bristow CA, Ma L, Lin MF et al (2010) Identification of functional elements and regulatory circuits by *Drosophila* modENCODE. Science 330:1787–1797

Ryan MY, Maloney R, Reenan R, Horn R (2008) Characterization of five RNA editing sites in *Shab* potassium channels. Channels (Austin) 2:202–209

Scadden AD (2005) The RISC subunit Tudor-SN binds to hyper-edited double-stranded RNA and promotes its cleavage. Nat Struct Mol Biol 12:489–496

Scadden AD, Smith CW (2001) RNAi is antagonized by A $\rightarrow$ I hyper-editing. EMBO Rep 2:1107–1111

Semenov EP, Pak WL (1999) Diversification of *Drosophila* chloride channel gene by multiple posttranscriptional mRNA modifications. J Neurochem 72:66–72

Smith LA, Wang XJ, Peixoto AA, Neumann EK, Hall LM, Hall JC (1996) A *Drosophila* calcium channel $\alpha 1$ subunit gene maps to a genetic locus associated with behavioural and visual defects. J Neurosci 16:7868–7879

Stapleton M, Carlson JW, Celniker SE (2006) RNA editing in *Drosophila melanogaster*: new targets and functional consequences. RNA 12:1922–1932

Stefl R, Oberstrass FC, Hood JL, Jourdan M, Zimmermann M, Skrisovska L, Maris C, Peng L, Hofr C, Emeson RB, Allain FH (2010) The solution structure of the ADAR2 dsRBM-RNA complex reveals a sequence-specific readout of the minor groove. Cell 143:225–237

Tian N, Wu X, Zhang Y, Jin Y (2008) A-to-I editing sites are a genomically encoded G: implications for the evolutionary significance and identification of novel editing sites. RNA 14:211–216

Visel A, Blow MJ, Li Z, Zhang T, Akiyama JA, Holt A, Plajzer-Frick I, Shoukry M, Wright C, Chen F et al (2009) ChIP-seq accurately predicts tissue-specific activity of enhancers. Nature 457:854–858

Vo N, Goodman RH (2001) CREB-binding protein and p300 in transcriptional regulation. J Biol Chem 276:13505–13508

Yang W, Wang Q, Howell KL, Lee JT, Cho DS, Murray JM, Nishikura K (2005) ADAR1 RNA deaminase limits short interfering RNA efficacy in mammalian cells. J Biol Chem 280: 3946–3953

Yang Y, Lv J, Gui B, Yin H, Wu X, Zhang Y, Jin Y (2008) A-to-I RNA editing alters less-conserved residues of highly conserved coding regions: implications for dual functions in evolution. RNA 14:1516–1525

Index

Current Topics in Microbiology and Immunology (2012) 353: 237–238
DOI: 10.1007/978-3-642-22801-8